Challenges and Future Trends of Magnetic Sensors

Challenges and Future Trends of Magnetic Sensors

Guest Editor

Galina V. Kurlyandskaya

MDPI

Basel • Beijing • Wuhan • Barcelona • Belgrade • Novi Sad • Cluj • Manchester

Guest Editor
Galina V. Kurlyandskaya
Department of Magnetism
and Magnetic Nanomaterials
Ural Federal University
Ekaterinburg
Russia

Editorial Office
MDPI AG
Grosspeteranlage 5
4052 Basel, Switzerland

This is a reprint of the Special Issue, published open access by the journal *Sensors* (ISSN 1424-8220), freely accessible at: https://www.mdpi.com/journal/sensors/special_issues/2D8GBQWZPF.

For citation purposes, cite each article independently as indicated on the article page online and as indicated below:

Lastname, A.A.; Lastname, B.B. Article Title. *Journal Name* **Year**, *Volume Number*, Page Range.

ISBN 978-3-7258-3927-8 (Hbk)
ISBN 978-3-7258-3928-5 (PDF)
https://doi.org/10.3390/books978-3-7258-3928-5

Contents

Preface

Various magnetic materials were investigated in the research presented herein, including magnetic steels, thin permalloy films, and FeNi-based multilayered structures. Studies also explored the relationship between alternating current permeability and magnetic domain structure in amorphous $Co_{85}Nb_{12}Zr_3$ films shaped as magnetic stripes, thin-film spin valve-type sensitive elements, Co-based amorphous ribbons, the magnetoimpedance tomography approach, as well as the Anomalous Nernst and longitudinal spin-Seebeck effects, and other phenomena observed in glass-coated magnetic microwires. In addition, they studied a two-point magnetic gradient localization method for remote, single-magnetic dipole targets based on SQUID magnetometers and the printed electronics direction as well as the hot research topic related to the biomedical applications of magnetic materials, in particular to the problem of the detection of magnetic nanoparticles in a liquid medium and the quantification of their concentration using the magnetoimpedance effect. Reprint counts with thirteen contributions: one communication, eleven regular articles, and one editorial article. They represent an international multidisciplinary community from Brazil, China, Greece, Japan, Portugal, Russia, Spain, Turkey, and the United Kingdom. It can be useful for PhD students and studies working in the field of magnetic nanomaterials, sensor devices, biomedical applications, and environmental applications. I would like to thank all authors, reviewers, and editorial assistants—an international team who made/evaluated/processed submissions and, therefore, supported the sensor materials research line and this particular Special Issue reprint.

Galina V. Kurlyandskaya
Guest Editor

sensors

MDPI

Editorial

Special Issue "Challenges and Future Trends of Magnetic Sensors"

Galina V. Kurlyandskaya

Institute of Natural Sciences and Mathematics, Ural Federal University, Ekaterinburg 620002, Russia; galinakurlyandskaya@urfu.ru

In recent years, "sensor materials" have become a hot research topic in the field of applied research and industrial applications. Related works have appeared in many journals, and *Sensors* (MDPI) joined this list, responding to multiple requests. The list of particular topical advisory panel keywords truly reflects the multidisciplinary nature of this direction addressed to physicists, chemists, biologists, physicians, engineers for the large variety of problems that require solving, and even economists, psychologists, and specialists in educational problems.

Many traditional materials and composites are not suitable for the increasingly complex requirements of the fast-growing number of magnetic sensors and microsystems with magnetic components designed for automatization, navigation, industrial processes' control, environmental control, biosensing and biomedical applications, drug delivery, and many others. The need for magnetic devices continues to challenge the materials science community to develop novel magnetic and composite materials that are suitable for such purposes. The principal requirements for a new generation of sensors are well known: high sensitivity, small size, low power consumption, stability, quick response, resistance to aggressive media, low price, and operated by non-skilled personnel. The increase in the number of nanomaterials available for research and applications requires that the methods of their characterization be even more precise than before. This Special Issue, entitled "Challenges and Future Trends of Magnetic Sensors", intends to summarize the existing and new concepts related to material science, modelling, and technological achievements in the field of magnetic sensors in order to better understand the foreseeable future of these devices.

Around a year ago, we invited researchers to be an important part of this Special Issue, offering the following list of main keywords: magnetic effects; magnetic materials for sensor applications; magnetic field sensors; magnetic biosensors; hybrid sensors; modelling for magnetic sensor applications; complex structures; and composites for magnetic sensor applications. After thorough international reviewing, twelve contributions [Contributions 1–12] (one communication and eleven articles) were published. They represent international multidisciplinary teams from Brazil, China, Greece, Japan, Portugal, Russia, Spain, Turkey, and the United Kingdom.

Interestingly, different magnetic materials were the subject of the research published herein. Magnetic steels cover a large range of present-day applications, with transportation, energy, metallic buildings and bridges, and oil- and gas-related structures being vital for society in many manifestations; this topic was discussed by Hristoforou [Contribution 1]. Thin permalloy films and FeNi-based multilayered structures can be called two of the "eternal" subjects of magnetism, which is still receiving special attention [1,2]. In this Special Issue, a quantitative comparison of the model prediction and the measurement data for the microwave permeability of permalloy films was carefully

Received: 18 February 2025
Accepted: 20 February 2025
Published: 21 February 2025

Citation: Kurlyandskaya, G.V. Special Issue "Challenges and Future Trends of Magnetic Sensors". *Sensors* **2025**, *25*, 1307. https://doi.org/10.3390/s25051307

analysed by Buznikov et al. [Contribution 2], indicating that changes in the magnetic structure of permalloy films appearing with an increase in film thickness are important. The observed decrease in the static permeability and frequency of the ferromagnetic resonance is related to the appearance of perpendicular magnetic anisotropy and the formation of stripe magnetic domains in the films.

The connection between the alternating current permeability and the structure of the magnetic domain was also analysed by Nakai [Contribution 9] for the case of amorphous $Co_{85}Nb_{12}Zr_3$ films in the shape of magnetic stripes. The obtained results are promising for applications in the areas of tunable inductors, electromagnetic shielding, or magnetic sensing for the detection of the stray fields generated by magnetic nanoparticles.

Another example of the thin film sensitive element is given by Milyaev et al. [Contribution 10]. Spin valves, i.e., the structures consisting of two ferromagnetic layers, the free layer and the pinned layer, which are separated by a nonmagnetic spacer, are well-known materials in sensor applications [3,4]. It was shown [Contribution 10] that the fabricated rhombus-shaped CoFeNi/Ru/CoFeNi spin valves were synthetic antiferromagnets. The fabricated sensor elements in which each side of the rhombus was the shoulder of a Wheatstone bridge enabled the device to operate as a full Wheatstone bridge. The sensor had a high sensitivity to the external field changes, and significant magnetic hysteresis is suitable for switching devices.

Co-based amorphous ribbons were the subject of interest of the studies of Bukreev et al. [Contribution 11] and Correa et al. [Contribution 12]. In the first case, the authors compared the results of a computer simulation and experimental study of the magnetoimpedance effect (MI) in amorphous $Co_{68.5}Fe_{4.0}Si_{15.0}B_{12.5}$ and $Co_{68.6}Fe_{3.9}Mo_{3.0}Si_{12.0}B_{12.5}$ ribbons, showing that the maximum MI value exceeds 200%. Such a high value may be of interest for the development of magnetic field sensors. In addition, it was shown that the practically significant characteristics of the MI response strongly depend on the driving current frequency due to the inhomogeneous distribution of magnetic properties over the cross-section. This distribution was studied using the magnetoimpedance tomography approach, also applied by Buznikov et al. [Contribution 7].

Correa et al. [Contribution 12] described thermoelectric phenomena, such as the Anomalous Nernst and longitudinal spin-Seebeck effects in Co-rich rapidly quenched ribbons and ribbon/Pt heterostructures, showing that Pt cover layer deposition leads to an enhancement of the thermoelectric response. The advantages of designing thermopiles consisting of Co-rich ribbon/Pt heterostructures in a parallel association were also shown. The longitudinal spin-Seebeck effect has been the subject of research for many years. For example, Uchida et al. [5] studied a longitudinal spin-Seebeck effect (SSE) in the case of ferrimagnetic insulator $Y_3Fe_5O_{12}$, where the spin current was injected along the direction of a temperature gradient from a ferromagnet into a covering paramagnetic Pt layer. More recently, Ravi el al. [6] described this closely related research direction as flexible spin-caloritronic materials with thermoelectric conversion by nanostructure engineering.

Contributions 7–8 are devoted to the study of glass-coated magnetic microwires, which can be described as versatile magnetic composites with a large variety of ways to tune the response of a magnetic sensitive element adapting a microdevice to the particular needs of the application [7,8].

Zhang et al. [Contribution 4] described a two-point magnetic gradient localization method for remote, single-magnetic dipole targets based on SQUID magnetometers. A linear localization model based on the spatial position relationship between a magnetic moment vector and relative position vectors—which allow for the realization of the high-precision localization of a magnetic target and the calculation of its magnetic moment vector—was used. The simulations data and experimental results demonstrated very high localization performance for remote magnetic targets.

Although printed electronics have been used for decades, improvements in the corresponding fabrication techniques have recently gained a lot of attention when it comes to printing smart magnetic materials for sensors and actuators [9,10]. In the present Special Issue, the printed electronics direction was represented by Ahmed et al. [Contribution 6], who described the advantages of the inkjet printing of magnetostrictive materials for structural health monitoring in the case of carbon fibre-reinforced polymer composites. For example, it was shown how the change in the design and number of layers affects the value of inductance under a particular applied strain.

Barrera et al. [Contribution 5] directed our focus to the hot research topic related to the biomedical applications of magnetic materials, in particular to the problem of the detection of magnetic nanoparticles in a liquid medium and the quantification of their concentration using the magnetoimpedance effect (MI). MI sensors have attracted much attention due to their high sensitivity to the stray magnetic field generated by magnetic nanoparticles, their simple fabrication process, and their relatively low cost [5,11,12]. In Contribution 5, the authors described an MI sensor with an amorphous $Fe_{73.5}Nb_3Cu_1Si_{13.5}B_9$ wire-sensitive element integrated into the millifluidic chip to detect the presence of magnetic nanoparticles of the stabilized magnetic suspension by the quantification of stray fields generated by single-domain superparamagnetic iron oxide or magnetically blocked Co-ferrite nanoparticles. In fact, directly or indirectly, not only Contribution 5 but also Contributions 2, 3, 7, 8 and 11 touch on issues related to the dynamic properties of magnetic materials and MI. In 2024, it was the 30-year anniversary of the discovery of MI, i.e., the change in the complex impedance Z under the application of a constant magnetic field [13–15]. Z depends on the frequency f and amplitude I_{ac} of the alternating current flowing through the ferromagnetic conductor. The effect is understood in the frame of the classic electrodynamics interpretation through the connection of a change in the skin depth and magnetic permeability of a soft ferromagnet being very close to the L.D. Landau approach [16].

The year 2024 also marks 30 years since the re-discovery of the magnetoimpedance effect. Variation in the complex impedance Z, under the application of a constant external magnetic field H, is called magnetoimpedance (MI). Complex impedance Z depends on the frequency and amplitude of the alternating driving current flowing through the ferromagnetic conductor. The phenomenon of the dependence of the alternating current (ac) resistance of cold drawn iron–nickel wires on the value of the applied constant magnetic field was reported by the authors of [17,18]. They proposed a classic electrodynamics interpretation of the observed effect through the connection of a change in the skin depth and magnetic permeability of a soft ferromagnet being very close to the L.D. Landau approach [16]. However, the existence of some poorly controlled technological parameters caused a large variation in the dynamic magnetic properties, even in the cold-drawn wires of the same batch. Therefore, for about six decades, the resistance of variations in the ferromagnetic conductor under the application of a magnetic field has not attracted the attention of many researchers. Later, new soft magnetic materials with stable properties appeared, and fabrication techniques were significantly improved. In 1991, Makhotkin et al. [13] published results related to a low-magnetic-field sensor operating with an FeCoSiB amorphous ribbon-sensitive element describing the functionality principle as a change in the electrical impedance of the ribbon under the application of the magnetic field. In 1994, a set of publications related to MI appeared, and the terms giant magnetoimpedance and magnetoimpedance become well established [14,15]. Thirty years later, we can see that MI sensors have extraordinary sensitivity with respect to the applied magnetic field [11,19], are they are useful for operations control, current and position sensing, intelligent system monitoring, gradient field detection, non-destructive and automobile control, as well as for bio-magnetic measurements at room temperature and other biomedical applications [Contribution 5].

The year 2024 also marks the experimental confirmation of the idea of "altermagnetism", an emerging magnetic phase characterized by robust time-reversal symmetry breaking. In this type of magnetic, it was noticed that materials with mixed properties were different from other types of materials of the magnetic phenomenon. Similarly to antiferromagnets, electrons in altermagnets spin in alternating directions and do not producing magnetization. However, the energy bands also have alternating spins from neighbouring bands. Altermagnetism is now considered the third elementary type of magnetic phases, in addition to the conventional ferromagnets and antiferromagnets [20]. According to experts opinions, altermagnets have the capacity to revolutionize spintronics, data storage, and magnetic sensing devices, providing enhanced efficiency and durability in comparison with traditional ferro- and antiferromagnets [20].

We expect this Special Issue and its corresponding book to be useful for graduate and PHD students, researchers working in the field of magnetic materials and nanocomposites and electronic engineering, and even personnel connected with biomedical applications.

Acknowledgments: As the Academic Editor of this Special Issue, I would like to thank all authors and reviewers who made/evaluated submissions and therefore supported the sensor materials research line and this particular Special Issue.

Conflicts of Interest: The author declares no conflicts of interest.

List of Contributions

1. Hristoforou, E.V. Permeability Sensors for Magnetic Steel Structural Health Monitoring. *Sensors* **2025**, *25*, 606. https://doi.org/10.3390/s25030606.
2. Melnikov, G.Y.; Komogortsev, S.V.; Svalov, A.V.; Gorchakovskiy, A.A.; Vazhenina I.G. Kurlyandskaya, G.V. Effects of Magnetostatic Interactions in FeNi-Based Multilayered Magnetoimpedance Elements. *Sensors* **2024**, *24*, 6308. https://doi.org/10.3390/s24196308.
3. Buznikov, N.A.; Lagarkov, A.N.; Maklakov, S.A.; Maklakov, S.S.; Osipov, A.V.; Rozanov, K.N. Zezyulina, P.A. Analysis of Relationship between Microwave Magnetic Properties and Magnetic Structure of Permalloy Films. *Sensors* **2024**, *24*, 6165. https://doi.org/10.3390/s24196165.
4. Zhang, Y.; Liu, G.; Wang, C.; Qiu, L.; Wang, H.; Liu, W. A Remote Two-Point Magnetic Localization Method Based on SQUID Magnetometers and Magnetic Gradient Tensor Invariants. *Sensors* **2024**, *24*, 5917. https://doi.org/10.3390/s24185917.
5. Barrera, G.; Celegato, F.; Vassallo, M.; Martella, D.; Coïsson, M.; Olivetti, E.S.; Martino, L. Sözeri, H.; Manzin, A.; Tiberto, P. Microfluidic Detection of SPIONs and Co-Ferrite Ferrofluid Using Amorphous Wire Magneto-Impedance Sensor. *Sensors* **2024**, *24*, 4902. https://doi.org/10.3390/s24154902.
6. Ahmed, N.; Smith, P.J.; Morley, N.A. Inkjet Printing Magnetostrictive Materials for Structural Health Monitoring of Carbon Fibre-Reinforced Polymer Composite. *Sensors* **2024**, *24*, 4657 https://doi.org/10.3390/s24144657.
7. Buznikov, N.A., Kurlyandskaya G.V. Theoretical Study of Microwires with an Inhomogeneous Magnetic Structure Using Magnetoimpedance Tomography. *Sensors* **2024**, *24*, 3669. https://doi.org/10.3390/s24113669.
8. Chizhik, A.; Corte-Leon, P.; Zhukova, V.; Blanco, J.M.; Zhukov, A. Monitoring the Velocity of Domain Wall Motion in Magnetic Microwires. *Sensors* **2024**, *24*, 1326. https://doi.org/10.3390/s24041326.
9. Nakai, T. Relationship of Magnetic Domain and Permeability for Clustered Soft Magnetic Narrow Strips with In-Plane Inclined Magnetization Easy Axis on Distributed Magnetic Field. *Sensors* **2024**, *24*, 706. https://doi.org/10.3390/s24020706.
10. Milyaev, M.; Naumova, L.; Germizina, A.; Chernyshova, T.; Pavlova, A.; Krinitsina, T.; Proglyado, V.; Ustinov, V. A Spin Valve-Based Rhombus-Shaped Micro-Object Implementing a Full Wheatstone Bridge. *Sensors* **2024**, *24*, 625. https://doi.org/10.3390/s24020625.

11. Bukreev, D.A.; Derevyanko, M.S.; Semirov, A.V. Magnetoimpedance Effect in Cobalt-Based Amorphous Ribbons with an Inhomogeneous Magnetic Structure. *Sensors* **2023**, *23*, 8283. https://doi.org/10.3390/s23198283.
12. Correa, M.A.; Svalov, A.V.; Ferreira, A.; Gamino, M.; da Silva, E.F.; Bohn, F.; Vaz, F.; de Oliveira, D.F.; Kurlyandskaya, G.V. Longitudinal Spin Seebeck Effect Thermopiles Based on Flexible Co-Rich Amorphous Ribbons/Pt Thin-Film Heterostructures. *Sensors* **2023**, *23*, 7781. https://doi.org/10.3390/s23187781.

References

1. Pan, L.; Xie, H.; Cheng, X.; Zhao, C.; Feng, H.; Cao, D.; Wang, J.; Liu, Q. Tuning the ferromagnetic resonance frequency of soft magnetic film by patterned permalloy micro-stripes with stripe-domain. *J. Magn. Magn. Mater.* **2018**, *457*, 46–51. [CrossRef]
2. Izotov, A.V.; Belyaev, B.A.; Boev, N.M.; Burmitskikh, A.V.; Leksikov, A.A.; Skomorokhov, G.V.; Solovev, P.N. Tailoring the microwave properties of thin Permalloy films using a periodically grooved substrate. *Phys. B* **2022**, *629*, 413654. [CrossRef]
3. Freitas, P.P.; Ferreira, R.; Cardoso, S. Spintronic sensors. *Proc. IEEE* **2016**, *104*, 1894–1918. [CrossRef]
4. Mabarroh, N.; Alfansuri, T.; Wibowo, N.A.; Istiqomah, N.I.; Tumbelaka, R.M.; Suharyadi, E. Detection of green-synthesized magnetite nanoparticles using spin-valve GMR-based sensor and their potential as magnetic labels. *J. Magn. Magn. Mater.* **2022**, *560*, 169645. [CrossRef]
5. Uchida, K.-I.; Adachi, H.; Ota, T.; Nakayama, H.; Maekawa, S.; Saitoh, E. Observation of longitudinal spin-Seebeck effect in magnetic insulators. *Appl. Phys. Lett.* **2010**, *97*, 172505. [CrossRef]
6. Gautam, R.; Hirai, T.; Alasli, A.; Nagano, H.; Ohkubo, T.; Uchida, K.-I.; Sepehri-Amin, H. Creation of flexible spin-caloritronic material with giant transverse thermoelectric conversion by nanostructure engineering. *Nat. Commun.* **2024**, *15*, 2184. [CrossRef] [PubMed]
7. Baranov, S.A. Magnetic models of cast microwires. *Surf. Eng. Appl. Electrochem.* **2011**, *47*, 316–330. [CrossRef]
8. Lofland, S.E.; Garcia-Miquel, H.; Vázquez, M.; Bragat, S.M. Microwave magnetoabsortion in glass-coated amorphous micro wires with radii close to skin depth. *J. Appl. Phys.* **2002**, *92*, 2058–2063. [CrossRef]
9. Bastola, A.K.; Hossain, M. The shape—Morphing performance of magnetoactive soft materials. *Mater. Des.* **2021**, *211*, 110172. [CrossRef]
10. Ahmed, N.; Deffley, R.; Kundys, B.; Morley, N. 3D printing of magnetostrictive property in 17/4 ph stainless steel. *J. Magn. Magn. Mater.* **2023**, *585*, 171115. [CrossRef]
11. Mohri, K.; Uchiyama, T.; Panina, L.V.; Yamamoto, M.; Bushida, K. Recent advances of amorphous wire CMOS IC magnetoim pedance sensors: Innovative high-performance micromagnetic sensor chip. *J. Sens.* **2015**, *2015*, 718069. [CrossRef]
12. Bukreev, D.A.; Derevyanko, M.S.; Moiseev, A.A.; Svalov, A.V.; Semirov, A.V. The study of the distribution of electrical and magnetic properties over the conductor cross-section using magnetoimpedance tomography: Modeling and experiment. *Sensors* **2022**, *22*, 9512. [CrossRef] [PubMed]
13. Makhotkin, V.E.; Shurukhin, B.P.; Lopatin, V.A.; Marchukov, P.Y.; Levin, Y.K. Magnetic field sensors based on amorphous ribbons. *Sens. Actuators A* **1991**, *21*, 759–762. [CrossRef]
14. Beach, R.S.; Berkowitz, A.E. Giant magnetic field dependent impedance of amorphous FeCoSiB wire. *Appl. Phys. Lett.* **1994**, *64*, 3652–3654. [CrossRef]
15. Panina, L.V.; Mohri, K. Magneto-impedance effect in amorphous wires. *Appl. Phys. Lett.* **1994**, *65*, 1189–1191. [CrossRef]
16. Landau, L.D.; Lifshitz, E.M. *Course of Theoretical Physics, Vol. 8: Electrodynamics of Continuous Media*, 2nd ed.; Butterworth-Heinemann: Oxford, UK, 1984.
17. Harrison, E.P.; Turney, G.L.; Rowe, H. Electrical properties of wires of high permeability. *Nature* **1935**, *135*, 961. [CrossRef]
18. Harrison, E.P.; Turney, G.L.; Rowe, H.; Gollop, H. The electrical properties of high permeability wires carrying alternating current. *Proc. Roy. Soc.* **1936**, *157*, 451–479. [CrossRef]
19. Kurlyandskaya, G.V.; Buznikov, N.A.; Svalov, A.V. Giant Magnetoimpedance: 30 Years Since Rediscovery and Next Steps. *Phys. Met. Metallogr.* **2024**, *125* (Suppl. S1), S33–S61. [CrossRef]
20. Krempaský, J.; Šmejkal, L.; D'Souza, S.W.; Hajlaoui, M.; Springholz, G.; Uhlířová, K.; Alarab, F.; Constantinou, P.C.; Strocov, V.; Usanov, D.; et al. Altermagnetic lifting of Kramers spin degeneracy. *Nature* **2024**, *626*, 517–522. [CrossRef] [PubMed]

sensors

MDPI

Article

Permeability Sensors for Magnetic Steel Structural Health Monitoring

Evangelos V. Hristoforou

Laboratory of Electronic Sensors, Zografou Campus, National Technical University of Athens, 15780 Athens, Greece; hristoforou@ece.ntua.gr

Abstract: In this paper, magnetic permeability sensors able to perform structural health monitoring of magnetic steels, by means of determining residual strain and stress amplitude and gradient distribution, responsible for crack initiation, are presented. The good agreement between magnetic properties and residual strains and stresses is illustrated first resulting in the determination of the magnetic stress calibration (MASC) curves and the Universal MASC curve. Having determined differential magnetic permeability as a key magnetic property, able to measure and monitor residual strain and stress distribution in magnetic steels, the paper is devoted to the presentation of the permeability instruments and sensors developed in our lab. The classic single sheet testers and the electromagnetic yokes, are compared with new, low-power-consumption permeability sensors using the Hall effect and the anisotropic magnetoresistive (AMR) effect, discussing their advantages and disadvantages in magnetic steel structural health monitoring.

Keywords: structural health monitoring; magnetic steels; barkhausen noise; hall effect; anisotropic magnetoresistance

Academic Editor: Galina V. Kurlyandskaya

Received: 30 December 2024
Revised: 19 January 2025
Accepted: 19 January 2025
Published: 21 January 2025

Citation: Hristoforou, E.V. Permeability Sensors for Magnetic Steel Structural Health Monitoring. *Sensors* **2025**, *25*, 606. https://doi.org/10.3390/s25030606

1. Introduction

Magnetic steels cover a large range of structural and operational applications. Among them, transportation, energy, metallic buildings and bridges, as well as oil and gas are vital for the society and economy, requiring preventive maintenance. Apart from these, special scientific structures, like the steel tube rings used in the "Conseil Européen pour la Recherche Nucléaire" (CERN), the European Council for Nuclear Research and other important scientific infrastructure require special attention, particularly concerning the employed steel grades. In some cases, like ITER in southern France, stainless steel grades, like 316LN, being non-magnetic, are also used and need preventive maintenance. In general, a large variety of different steel grades are used in different specific applications. The preferable type of steel grade depends on its use: the grade, treatment and thickness of the used steel are dependent on the environment, like chemical ambient atmosphere, load, hydraulic stresses and low–high cycle fatigue, as well as environmental conditions like temperature, humidity and their gradients, in which it is used. The variety of these steels is so large, starting from single ferritic structures, low-carbon compounds, duplex steels, TRIP steels, maritime steels and so many other grades, that it is difficult to implement a universal method for maintenance and especially for preventive maintenance. For this reason, steel producers and manufacturers, like pipeline manufacturers, shipyards, heat exchanger producers, etc., follow specific procedures in alloying and post-manufacturing treatment, allowing for the proper and required structure and microstructure [1].

Steel failure may be critical and vital for several steel structures [2]. Steel failure in the energy cluster may result in disruptive damages. The domes of nuclear stations, suffering neutron fluence and therefore embrittlement, must be secure, far from being damaged, otherwise the public health of close and distant towns would be in danger, like in the Chernobyl accident. Environmental catastrophes may also affect nuclear stations, like in the recent Fukushima nuclear accident. Apart from that, steel failure in renewable, lignite and hydrogenation stations is also of importance, resulting in the disability of energy transmission. Similar problems appear in transportation. Railways and trains, suffering thermo-mechanical fatigue, in either a low or high cycle fatigue process, are in danger especially in large temperature gradients. For example, in north railway lines, sunshine causes elevated temperatures on the railways, so that at the day-to-night transition, the thermal gradient together with the mechanical fatigue may result in railway failure. After the privatization of railways, especially in northern countries with high thermo-mechanical fatigue, where the preventive maintenance depends on the profit of the railway operation, such accidents are rather frequent. Shipping is also a subject of steel failure, in the hull structure, as well as in shaft conditioning. Recently, shaft accidents that impact the ship's steering and propulsion ability are frequent, illustrating the need of preventive structural health monitoring and maintenance. The new era of electrification also requires special attention, not in terms of steel failure but in magnetic properties performance, requiring frequent operational conditioning monitoring with corresponding maintenance by means of localized magnetic annealing. The sector of oil and gas also requires special attention. The harsh environment of oil extraction, refineries and oil and gas transmission pipelines require special types of steels, which operate in a large variety of chemical environments at elevated temperatures. These conditions may result in unforeseen accidents, such as the Mexico gulf accident. On the other hand, steel structures like skyscrapers, metallic bridges and the like must also be protected against damage. Apart from these applications, the structural and conditional health monitoring in large scientific experiments like in CERN, ITER, etc. must be properly monitored in terms of structural and microstructural characterization. However, all these issues of structure and microstructure in all these steels grades can be summed up in one major parameter, affecting their structural health, namely the residual strains and stresses (named hereinafter stresses for simplicity reasons). These stresses are the accumulation of several different reasons for structural imperfections, like dislocations, precipitations. etc.: steel failure is dependent on the gradient of residual stresses [3]. These residual stresses are also subject to fatigue conditions, namely low and high cycle fatigue, concerning stresses above and below the yield point of the corresponding steel. In fact, laboratory tests under low and high cycle fatigue allow for the determination of the remaining lifetime of the corresponding steel structure. This challenge is rather difficult to be determined, being dependent on the specific steel under investigation and the environmental and stress conditions. However, the problem of the remaining lifetime can be actually solved, by knowing the residual strain and stress amplitude distribution. The processed signals can often be subject to uncertainties and/or inaccuracies. For this reason, recent scientific literature suggests using techniques based on fuzzy logic. In particular, the most recent research guidelines recommend grouping similar data in a fuzzy sense (through fuzzy divergence calculations that prove to be a measure of distance in a particular functional space) into specific groups to extract, for each of them, a representative characteristic of each grouping. Then, the classification of the signal can be performed by comparing the obtained fuzzy divergences with each representative data of each grouping [4].

The reference method to determine and monitor residual stresses is diffraction. The surface residual stresses are detected by X-ray diffraction, preferably following the Bragg–

Brentano set-up (XRD-BB) [5], while the bulk residual stresses are detected by neutron diffraction (ND) [6]. The spatial resolution of diffraction can be as low as 1 mm^2 using a high density of photons or neutrons. The methodology of the measurement of stresses is analyzed in the next chapter. Apart from these two classical methods, other techniques are also available, like synchrotron, offering even higher spatial precision. The currently used stress sensors in the field are strain gauges [7], while the drill hole method is also an alternative [8]. Strain gauges measure the local strain change after they are fixed in position. Their advantage is the straightforward measurement of strain changes, while the laborious methodology of fixing them on the steel surface, as well as their disability to monitor the distribution of strains, are their main disadvantages. Drill hole is a standardized method to monitor the stress tensor: after drilling a circular well on the steel surface, the circle is arranged to an ellipse, where the long and short diameters indicate the tensile and compressive stress history, respectively. Although the drill hole method is able to determine the history of stresses, it can only be used locally. The method is also somehow destructive.

For all these reasons, there is a great interest in determining and developing a non-destructive and contactless, if possible, method to provide stress and strain distribution measurement and monitoring along the surface and the bulk of steels. Such a method can be magnetic: it can be contactless and certainly non-destructive with the ability of scanning in several cases. Magnetic sensors can also be used in harsh environments, provided that the required power consumption permits their use. Our lab has studied such magnetic methods to correlate magnetic properties with residual stresses in magnetic steels for more than three decades. Apart from that, we have provided three different families of magnetic sensors, namely magnetic Barkhausen noise [9], differential permeability [10] and magnetoelastic sensors [11]. In this paper, after summarizing our methodology concerning the correlation of magnetic properties with residual stresses, we emphasize on the permeability sensors, namely the single sheet testers and the electromagnetic yokes, that serve as reference tools and lab equipment to determine stresses, as well as low consumption permeability sensors, like Hall and AMR sensors for in-field applications and measurements.

2. Magnetic Stress Calibration Curves

The use of any kind of instrument to monitor magnetic properties for the determination of residual stresses requires the comparison of residual stress and the magnetic property at the very same area-volume and direction, resulting in the determination of sensitivity, uncertainty, parametric dependence of uncertainty as well as the spatial resolution of the magnetic method. For this reason, the development and maintenance of magnetic steel samples, called hereinafter steel coupons, are necessarily required with well determined residual stresses on their surface and in their bulk.

Our group has initially taken the advantage of welding techniques to develop such stress coupons for calibration reasons. In fact, using autogenous steel welding of the same steel grade, it is possible to generate a steel coupon with a stress gradient across the weld (Autogenous welding means welding without any other additional compound to be added in the steel under welding.) For simplicity reasons, the stress coupons are usually flat. Such autogenous welding methods that can generate a stress gradient across the welding are the tungsten inert gas (TIG) method [12], the plasma welding method [13], the induction heating method [14], the electron beam welding [15], etc. The first three mentioned welding methods are widely used in the steel industry, while the fourth one is mainly used for research and development purposes. In all these welding techniques, the welding method is the same: the two parts of steel are molten and then pressed to be in touch or even mix each other. This way, during cooling, a stress gradient can be generated in the fusion zone (FZ) and the heat-affected zones (HAZ), provided that the two steel parts are forced to be flat. Practically, the base metal (BM) remains with a minimum amount of stresses. Otherwise, if the two components are not forced to be flat, the cooling welded steel is deformed to accumulate the heat gradient release.

Figure 1a illustrates such residual stress distribution in the fusion zone and heat-affected zones. Despite the phase transformations introduced inside the welded steel, the diffraction methods are capable of determining the residual strains existing within the FZ and HAZ areas [16]. All our steel coupons until now are based on such welding methods, despite issues like phase transformations and repeatability process. However, both phase transformation and sample repeatability are important and enlarge the uncertainty in inter-laboratory comparison tests. For this reason, we have started using another method, according to which phase transformation and sample repeatability are much better, allowing safer inter-laboratory comparison tests. This method is the local induction heating using a 6 mm straight, water-cooled conductor, transmitting a high-frequency current, generating eddy currents and therefore heat along the steel. The heated steel is set on top of a ceramic base and two vices are loosely fixing the ceramic and the steel under heat. Due to the relatively low thermal conductivity of steels, a temperature profile is generated in the steel, perpendicular to the orientation of the straight conductor direction, which can be regarded as a heat-affected zone (HAZ). The temperature distribution of the steel is monitored by an infrared camera. As soon as the maximum temperature reaches temperatures in the order of one-third of the melting temperature of the steel grade, thus allowing for no phase transformations, the two vices fix the ceramic-steel assembly and the whole assembly is inserted either in cold water or oil for quenching. This way, the temperature gradient is transferred into residual stresses, until the temperature in the steel is uniform (Figure 1b).

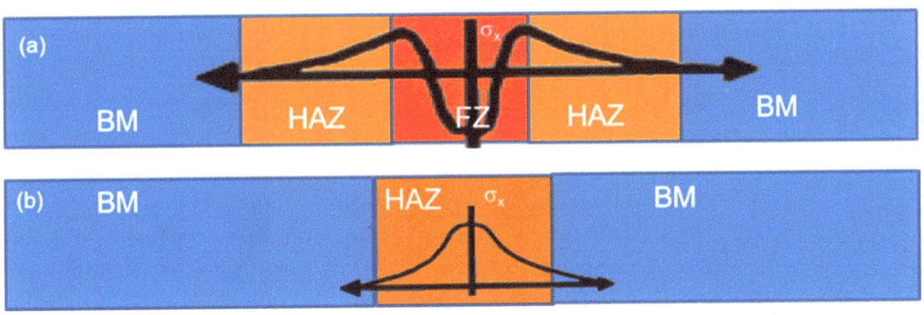

Figure 1. Stress distribution in steel coupons: (**a**) stresses in the fusion and heat-affected zones of welded samples; (**b**) stresses in inductively heated and quenched steel coupons.

The steel coupons used for the experiments of this paper are both welded and heated-quenched ones. These steel coupons must be kept in a safe place, without temperature, humidity and stress. Practically, they are stored in zero humidity champers, covered by soft blanquettes to avoid changes in localized residual stresses. Steel coupons of homogeneous stresses could also be prepared and stored, following heating, consequent quenching and then thermal recovery with different temperature and time intervals. However, this is not practically useful for in-field calibration reasons and therefore it is not followed as a best-practice. Then, these steel coupons undergo stress characterization. Thin, white indelible dots are used to mark different positions on top of the quenched steel, across the heat-affected zone, where the measurement of residual stresses is realized. Three lines of dots are designated on the surface of the steel, to have a number of areas with residual stresses for statistical reasons. They are determined in given directions, namely X and Y directions for surface stresses and X, Y and Z directions for bulk stresses. The surface strain measurement using the XRD-BB arrangement will be illustrated hereinafter. Pointing in one (let us say X) direction on an individual dot point of the steel coupon, the XRD response is determined, selecting the most stable 2theta angle with respect to sample tilting. We have experienced that the fourth peak of the XRD of ferrous steels is the most stable, which is close to 98°–99°. Then, the X-ray source and detector are fixed in this angle with a fourth decimal precision and rotate in an angle $\pm\psi$ from the vertical line, from −35° to +35° with respect to the sample, monitoring the change of the detected counts. Thus, the dependence of counts on $sin^2\psi$ determines the strain ε in this direction at the given dot, by the following formula:

$$\varepsilon_x = \frac{\frac{D_1 - D_n}{D_n}}{(1+n)sin^2\psi} \quad (1)$$

Positive inclination of this dependence corresponds to tensile stresses and negative inclination to compressive stresses. Then, the Young's modulus E, determined by the stress–strain curve of the steel, multiplied with ε_x, results in the residual stress σ_x:

$$\sigma_x = E\frac{\frac{D_1 - D_n}{D_n}}{(1+n)sin^2\psi} \quad (2)$$

This process is repeated in the Y-axis and then, the two stress components of this dot are determined. The next dot measurement follows, until the strains and stresses in all directions and dots of the steel are determined. The precision of the 2theta angle determination, the type of the X-ray source and the precision of the counts detected in different angles are the three main and decisive parameters for the quality of the dependence of counts on $sin^2\psi$. In case one of these parameters is not determined well, the dependence of counts on $sin^2\psi$ is not monotonic, resulting in large uncertainties in stress measurement. The counts are detected with a signal to noise ratio of 10^4 due to the time exposure in photons All this process is depicted in Figure 2. Neutron diffraction can detect the strains and stresses in the bulk of the steel, following a very similar procedure.

In all measurements of residual stresses, the initial measurement was the microstrain change, which was always multiplied with the Young's modulus E of the original compound, without taking into account the phase transformations or microstructural changes. This practically means that all measurements and experiments referred to microstrain changes. This ensures the validity of the measurements, avoiding any change of the Young's modulus E or any phase transformation and microstructural change.

Magnetic properties are measured on the surface and in the bulk of these steel coupons. These magnetic properties should be related to magnetic Barkhausen noise, magnetic permeability and magnetoelastic properties of the material, i.e., the dependence of the magnetostriction coefficient λ on residual stresses. These properties can be determined by measuring Barkhausen noise, magnetization hysteresis and magnetoacoustic emission measurements. Barkhausen noise can determine the amount of Barkhausen jumps by counting the number of pulses at each period of excitation, as well as the energy needed to overcome the obstacles of domain wall propagation, which is proportional to the integral of the Barkhausen jumps with respect to time. In practice, this energy is represented by the surface below the envelope of Barkhausen jumps. Magnetization loop offers the differential permeability dependence of the applied field and can be partially considered as the integral of the Barkhausen jumps sequence. In fact, the maximum differential permeability is the preferred magnetic property of the magnetization loop, being more sensitive than others, like coercivity, remanence magnetization and magnetic losses. The dependence of static magnetoelastic properties, namely the dependence of λ on stress, as well as the dependence of dynamic magnetoelastic properties, namely the amplitude and waveform of magnetoelastic waves generated in magnetic steels, that also offer significant information on stresses, is out of the scope of this paper.

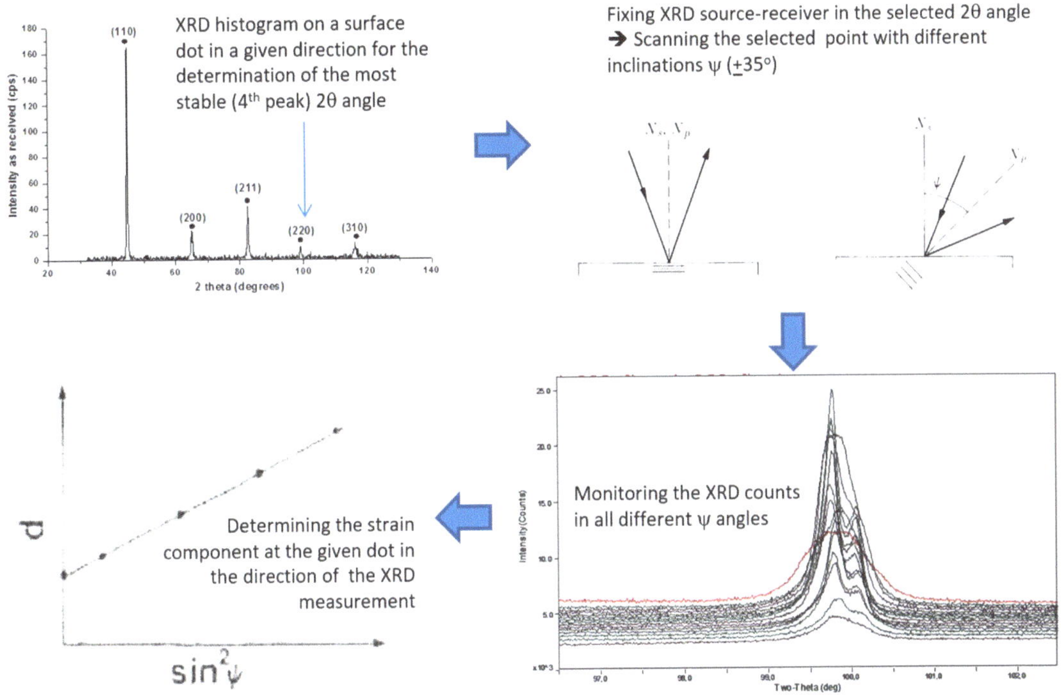

Figure 2. The methodology to determine the strain at a given dot of the under-test steel. At first, the XRD histogram determines the 2theta angle corresponding to the fourth (most stable) peak of the histogram (**top left**). Then, the X-ray source and detector are precisely fixed in this 2theta angle that rotates in an angle $\pm\psi$ (**top right**), allowing for the determination of counts with a signal to noise ratio of 10^4 (**bottom right**). Finally, the dependence of counts d on $sin^2\psi$ determines the strain ε and stress σ (**bottom left**).

Having had the tools to determine and monitor magnetic properties, the correlation between stresses and magnetic properties was realized, by means of measuring MBN and differential permeability at the same areas and directions, where residual stresses were measured (Figure 4). The reasons for good agreement are mainly related to the correlation of magnetic properties and stresses in the same direction, using a proper X-ray source and magnetic instrument. These measurements indicate that the MBN measurements are in good agreement with residual stresses. The output of the search coil related to the differential permeability measurements indicate less precise measurements, since the yoke was not calibrated, as described next in this paper.

We have initiated the measurements of Barkhausen noise and differential permeability using the MEB-2c probe of the Magnetics Laboratory, Gdansk, Poland (Figure 3). This probe uses an electromagnetic yoke to magnetize the steel under test, a search coil on the electromagnetic yoke to obtain a response proportional to the differential permeability of the steel under test, as well as a magnetic Barkhausen noise (MBN) search stylus set on the surface of the steel under test to provide the Barkhausen noise characteristics. The MBN stylus comprises of a short length and diameter coil, containing a thin ferrite rod for amplification of the MBN signal, which is weak and requires some magnetic and electronic pre-amplification.

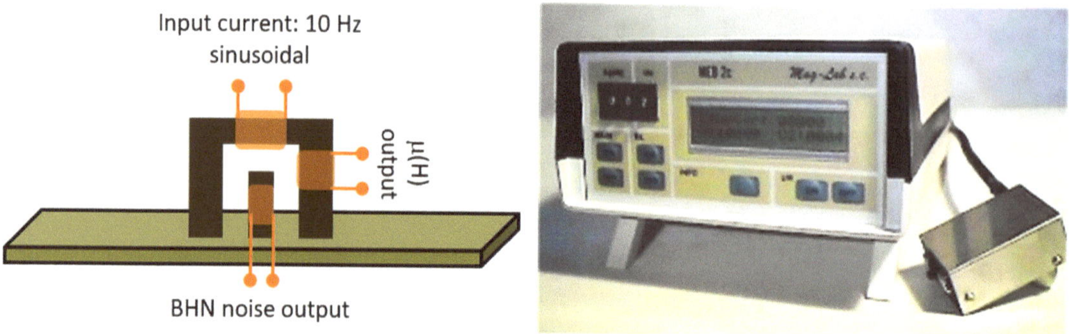

Figure 3. The MEB-2c probe used for the initial determination of magnetic properties. The schematic of the sensor (**left**). The actual probe with the electronics (**right**).

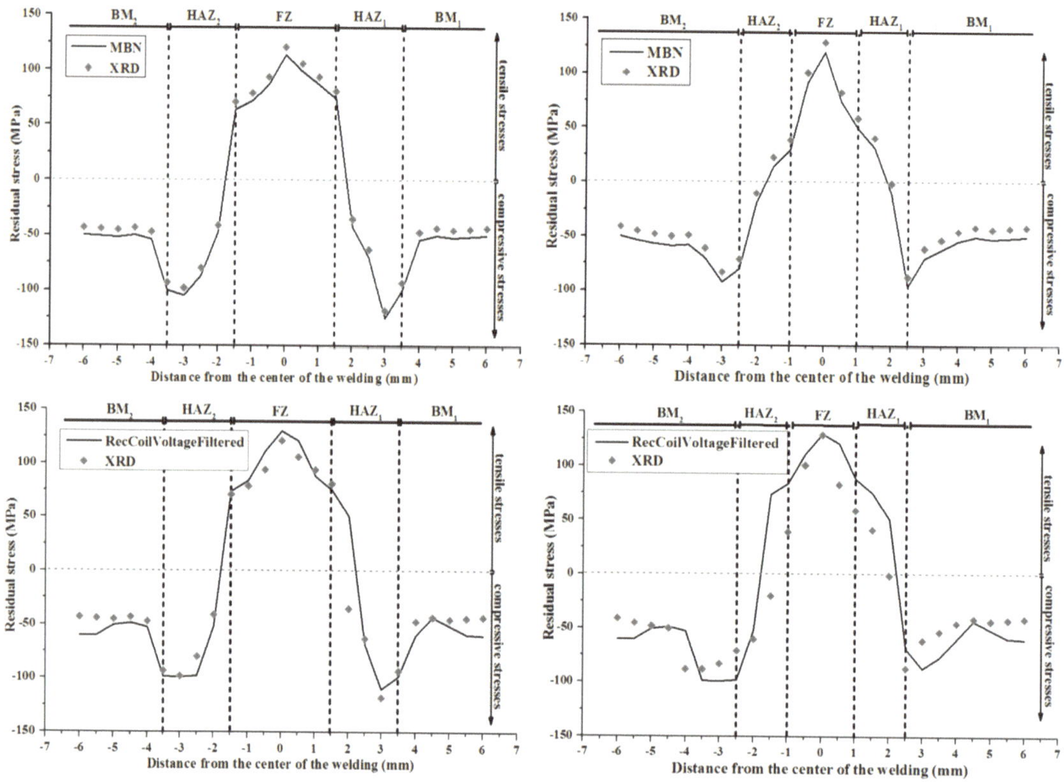

Figure 4. Indicative response of correlation between stresses and magnetic properties. **Top left**: correlation of MBN measurements with residual stresses in AISI 1008 steel, US Steel, Pittsburg, PA, USA, using TIG welding. **Top right**: correlation of MBN measurements with residual stresses in AISI 1008 steel using plasma welding. **Bottom left**: correlation of the differential permeability and stresses in AISI 1008 steel using TIG welding. **Bottom right**: correlation of the differential permeability and stresses in AISI 1008 steel using plasma welding. The larger uncertainty for the differential permeability response is due to the calibration of the electromagnetic yoke.

From these results, the pairs of stress and magnetic measurements resulted in the so-named magnetic stress calibration (MASC) curve. Until now, we have achieved 26 different types of MASC curves from different steel grades. All these curves are different to each other, in terms of residual stress amplitudes due to the corresponding yield point, as well as in magnetic properties due to the different MBN and permeability response. However, normalizing the X-axis with the corresponding yield point of the steel grade and the Y-axis with the maximum magnetic property (MBN or permeability), all MASC curves collapsed in on MASC curve, called Universal MASC curve [17]. Indicative results of the Universal MASC curve are illustrated in Figure 5. The Universal MASC curve is useful for the precise residual stress measurement in an unknown type of steel: measuring the stress–strain dependence of a dog-bone of the unknown type of steel, with simultaneous measurement of MBN or magnetic permeability, results in the determination of the yield point and the maximum amount of the magnetic property, respectively. Thus, multiplying the Universal MASC curve with these values results in the determination of the MASC curve of the unknown type of steel [18].

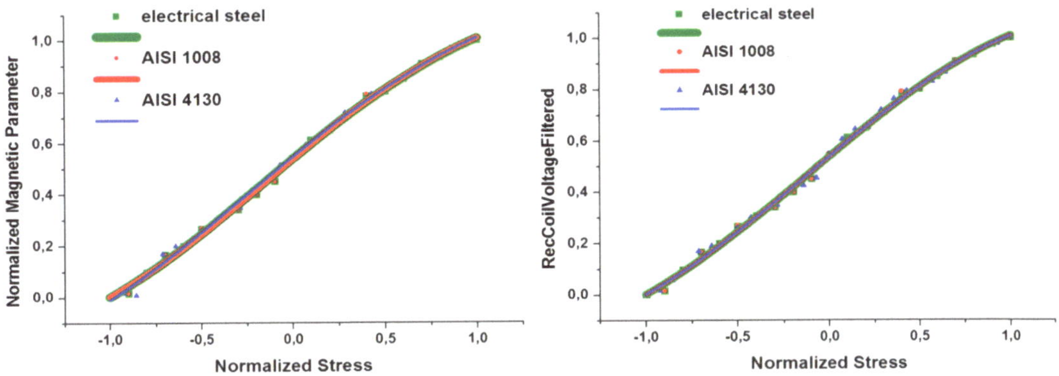

Figure 5. Universal MASC curves for MBN (**left**) and permeability measurements (**right**).

In order to provide robust stress distribution monitoring, we need to have precise sensors. MBN measurements, if not resulting from coils surrounding the under-test steel, are subject to uncertainties due to the not-perfect vertical orientation of the sensor with respect to the under-test steel. A slight deviation from their vertical position with respect to the surface of the steel under test, lower than a degree, results in significant changes of the MBN signal, more than 10%. Apart from that, MBN measurements suffer from the rather random, non-deterministic and non-repeatable MBN, requiring a sequence of measurements at each point to average the response and therefore result in a less uncertain measurement. Magnetic permeability measurements suffer less from this effect, being proportional to the integration of the MBN with respect to time or field. For this reason, the development of new types of robust permeability sensors that are able to provide precise information on the permeability of the under-test steel are required. The families of permeability-measuring instruments and sensors, developed in the laboratory, are presented next, illustrating their uncertainty levels and robust response. At the first page, the single sheet tester of the lab is presented, followed by the electromagnetic yoke sensors, both being principles of commonly accepted methods for permeability measurements in several labs around the world. Then, two other types of permeability sensors are presented based on the Hall effect and the anisotropic magnetoresistance (AMR) principle, offering several advantages related to in-field measurements.

3. Single Sheet Testers

The historical and reference instrumentation to monitor the magnetic permeability and the magnetization loop of soft magnetic materials has been the Epstein method [19], mainly dedicated to thin electric steel laminas. A large amount of these laminas is packed, arranged in a rectangular frame, thus allowing for a closed magnetic loop, permitting the application of Ampere's law when an excitation and search coil are arranged around them. In the Epstein topology, the excitation and search coils are coupled by the closed magnetic loop of the orthogonal frame generated by the thin steel laminas. The advantage of the Epstein method is the easily applicable equations of Ampere's law, resulting in a high signal to noise ratio in the order of 10^4 and in relatively low uncertainty, because of the large amount of magnetic mass involved in the measurement. This advantage is also responsible for the main disadvantage of the method: with the Epstein method, the average of the magnetization loop of all packed laminas is available, not permitting the measurement of each individual piece of steel. For this reason, the international steel community commonly accepted the ISO-ASTM A1036-04 (2020) standard entitled 'Standard Guide for Measuring Power Frequency Magnetic Properties of Flat-Rolled Electrical Steels Using Small Single Sheet Testers' [20], that allows for the determination of the dependence of the differential permeability and magnetization loop on the applied magnetic field using the so-called single sheet tester (SST).

We have adopted this standard and the experimental arrangement used in our lab is illustrated in Figure 6. With this arrangement, it is possible to measure the permeability and magnetization loop of different parts of a single steel sheet in low frequencies to avoid eddy currents in it. In fact, the excitation frequencies are from 0.1 Hz up to 10 Hz. In this arrangement, the excitation (or primary) and search (or secondary) coils surround the steel sheet under test and are directly coupled together. Our single sheet tester has been designed to accommodate single sheets with a thickness up to 0.5 mm and width up to 30 mm. Therefore, the constraints of the substrate of the primary and secondary coils were to have an internal free space of 1 mm and 31 mm external cross-section dimensions of 2 and 32 mm, respectively. The need to apply the Induction Law without compromises resulted in designing and developing a single-layer solenoid as the secondary coil. This coil should be short in order to monitor a small volume of the steel sheet under test. Therefore it has been designed to be 10 mm long, made from 0.05 mm enameled copper wire, thus allowing for 200 single-turn secondary solenoid. The primary coil should be longer than the secondary coil in order to apply a uniform field inside the secondary coil, so designed to be 30 mm long. Therefore, the substrate of the primary and secondary coils was 32 mm long with an internal 0.5 mm deep gap of 10 mm length in the middle of the substrate to accommodate the secondary coil, made by polymer powder using a 3D printer. The substrate was polished with alcohol in order to have a smooth surface allowing for no complexities of the secondary coil wire winding.

The low excitation frequency permits the use of multiple turns of the excitation coil. According to the needs of the biasing field, 0.3 mm up to 0.5 mm enameled copper wire were used for the development of the excitation coil. For 0.3 mm and 0.5 mm wire, 100 and 60 turns are accommodated in a single turn, respectively, thus requiring for three to four turns for the 0.3 mm wire and five to six turns for the 0.5 mm wire to make the excitation coil. Different excitation and search coils with their corresponding substrates can be developed in order to accommodate different cross-sections of the under-test steel.

The requirements of the ISO-ASTM A1036-04 (2020) standard [20] refer to the presence of two mirror-like, soft magnetic yokes that trap the magnetic flux allowing for the precise average length calculation on Ampere's law. There are some arguments on the type of the material of these yokes. A group of scientists and engineers wish to employ yokes with the maximum possible differential permeability to trap all magnetic lines (flux) from the single sheet. Another group wishes to use the same steel grade with the lamina under test. However, all agree to use yokes with a width larger than the width of the steel sheet under test. We have decided to use a couple of yokes made of amorphous ribbons with high differential permeability and magnetostriction close to zero to allow for magnetic flux trapping, without allowing magnetic lines to move outside the yoke. The width of the yokes is 50 mm, larger than the maximum width of the steel sheet under that which is 30 mm.

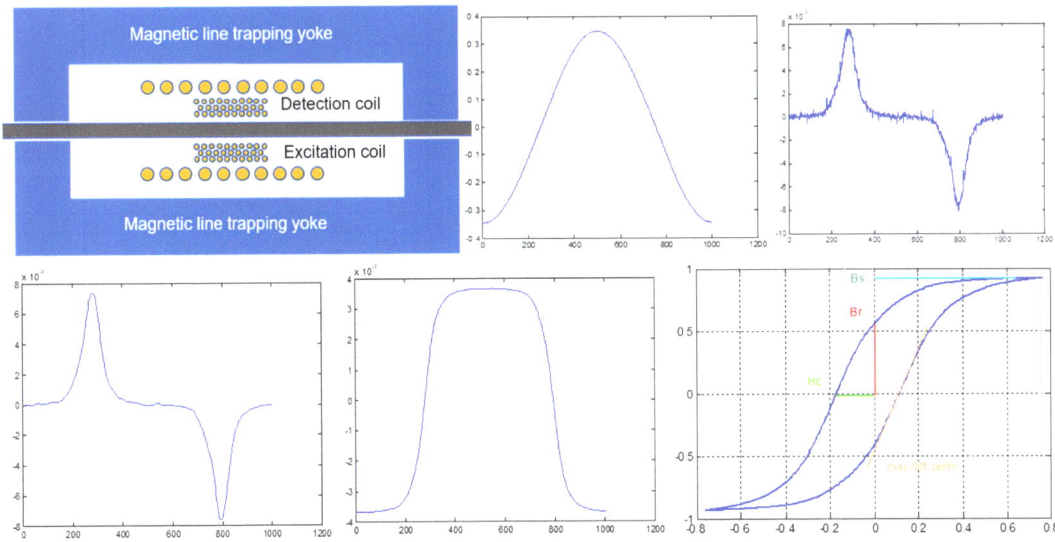

Figure 6. The schematic and waveforms of the single sheet tester of our lab. **Top left**: the schematic of the single sheet tester. **Top middle**: excitation current waveform proportional to the excitation field H. **Top right**: the voltage output of the secondary coil. **Bottom left**: the filtered voltage output of the secondary coil. **Bottom middle**: the integrated voltage output of the secondary coil, proportional to the magnetic flux B. **Bottom right**: the magnetization loop as a plot of B versus H.

The measurement of the applied field H can be performed by calculations based on the number of turns of the excitation coil and the amplitude of the transmitted current. However, we prefer to use Hall sensors at different and well-determined positions above the excitation coil. It is known that the dependence of the applied magnetic field follows a parabolic law on the distance from the surface of the steel [21]. Therefore, setting three Hall sensors on a substrate vertical to the steel under test, the parabolic projection of the field to the zero position results in the actual magnetic field H being applied on the single sheet. For this reason, four-pin Hall sensors from the market were arranged with a 5 mm distance from each other, with the closest Hall sensor been 1 mm from the surface of the primary coil. Thus, the magnetic field from each Hall sensor was measured for different excitation currents. The response of the Hall sensors for every excitation current had its own coefficients in the parabolic response. However, dividing these coefficients with the excitation current I resulted in collapsing all parabolic curves into a single one with an uncertainty of $\pm$100 ppm. Thus, the parabolic response of the magnetic field at different positions x is given by:

$$H(x,\ I) = aIx^2 + bIx + cI \tag{3}$$

where $c = 10\,\mathrm{m}^{-1} + 10^{-3}\mathrm{m}^{-1}$. Thus, the magnetic field on the surface (and therefore in the bulk) of the steel sheet is given by:

$$H(I) = cI \tag{4}$$

During measurements, the magnetic field was determined just by a measuring the excitation current I. Measuring B(H) in the arrangement of Figure 6 can be achieved by using Ampere's law:

$$\frac{N_e I(t)}{\sum \frac{1}{\mu_i(t)} \frac{l_i}{S_i}} = \Phi(t) \quad (5)$$

where N_e, $\mu_{i,}$ l_i, S_i and $\Phi(t)$ are the number of turns of the primary coil, the differential permeability at a given excitation field at a given part of the magnetic circuit, the average length of the magnetic part, the cross-section of the magnetic part and the flux passing through the magnetic circuit, respectively. So, for the case of the single sheet tester illustrated in Figure 6, Ampere's law is

$$\frac{N_e I(t)}{\frac{1}{\mu_{sss}(t)} \frac{l_{sss}}{S_{sss}} + 2\frac{1}{\mu_y(t)} \frac{l_y}{S_y}} = \Phi_{sss}(t) \quad (6)$$

where $\mu_{sss}(t)$, l_{sss}, S_{sss} are the differential magnetic permeability, the length and the cross-section of the single sheet tester, $\mu_y(t)$, l_y, S_y are the differential magnetic permeability, the length and the cross-section of the yokes and N_e is the number of turns of the primary coil. The lengths and cross-sections can be determined with relatively low uncertainty. The μ_y is a parameter that can be determined by using two pairs of yokes: each one of the yokes is fed by a primary (excitation) and a secondary (search) coil, having the same characteristics like the single sheet tester in terms of number of turns, length and diameter of the enameled copper wire. Then, these two yokes are set opposite each other with one exciting the other measuring the output signal. Ampere's law for this case is

$$\frac{N_e I(t)}{2\frac{1}{\mu_y(t)} \frac{l_y}{S_y}} = \Phi_y(t) \quad (7)$$

Hence,

$$\mu_y(t) = 2\frac{l_y}{S_y} \frac{\Phi_y(t)}{N_e I(t)} \quad (8)$$

but $\Phi(t)$ is related to the voltage output of the secondary (search) coil:

$$V(t) = -N_o \frac{d\Phi}{dt} \quad (9)$$

Therefore,

$$\Phi(t) = -\frac{1}{N_o} \int V_y(t) dt \quad (10)$$

So,

$$\mu_y(t) = 2\frac{l_y}{S_y} \frac{\int V_y(t) dt}{N_e N_o I(t)} \quad (11)$$

Or

$$\mu_y(I) = 2\frac{l_y}{S_y} \frac{\left(\int V_y(t) dt \middle| I\right)}{N_e N_o I} \quad (12)$$

Thus, the μ_{sss}, which is the actual unknown parameter, using Equation (6), is given by

$$\frac{N_e I}{\frac{1}{\mu_{sss}(I)} \frac{l_{sss}}{S_{sss}} + 2\frac{1}{2\frac{l_y}{S_y} \frac{\left(\int V_y(t) dt \middle| I\right)}{N_e N_o I}} \frac{l_y}{S_y}} = \Phi_{sss}(I) = -\frac{1}{N_o} \int \left(V_{sss}(t) dt \middle| I\right) \quad (13)$$

Or

$$\mu_{sss}(I) = \frac{1}{N_o N_e I} \frac{l_{sss}}{S_{sss}} \frac{\left(\int V_y(t)dt\middle|I\right)\left(\int V_{sss}(t)dt\middle|I\right)}{\left(\int V_y(t)dt\middle|I\right) - \left(\int V_{sss}(t)dt\middle|I\right)} \tag{14}$$

All parameters at the right part of Equation (14) are known or can be known. Therefore, $\mu_e(I)$ or the $\mu_{sss}(H)$ can be determined, since $H = cI$. This is the software logic to precisely determine $\mu_{sss}(H)$. Apart from that, there is also another hardware solution. According to this, two secondary coils are arranged symmetrically around the axis of the primary coil, while one is empty and the other includes the steel under test. The two coils are connected in series opposition and their output voltage is proportional to the susceptibility of the steel sheet, instead of the permeability, like in the previous arrangement. The errors introduced by the fact that the cross-section of the secondary coil(s) is not the same with the steel sheet is rather negligible since almost all magnetic flux is included in the steel sheet. This way, a steel coupon with uniform distribution of residual stresses can be calibrated in terms of magnetic properties. The nature of this measurement methodology, i.e., the fact that a piece of steel is cut to be characterized, indicates that the single sheet tester cannot be used for field measurements.

Despite its disability of being used in the field, SST can be used for several lab measurements, determining the quality of steels, monitoring frequency dependence, determine minor loops and therefore determine the first-order reversal curve (FORC). Knowing the characteristics of the magnetization process, a number of parameters can be tested this way. Furthermore, the SST arrangement can also be used for MBN measurements in a way that the uncertainties caused by the precision of the vertical arrangement of the MBN search coil are missing in the SST arrangement: the search coil is almost impossible to move in an angle with respect to the sample orientation. Having the disadvantage of not being a really localized measurement, the SST arrangement used for MBN measurements is the proper MBN set-up allowing for small measurement uncertainties.

4. Electromagnetic Yokes

The electromagnetic yoke (EMY) employs some principles of the single sheet tester standard and is capable of being used in this field, since it can scan surfaces without any need of cutting the steel under test. The Π shape of these electromagnetic yokes allows for the yoke magnetization using a low-frequency current. Typical arrangements of electromagnetic yokes are illustrated in Figure 7, where the schematic of such a typical yoke is demonstrated with an excitation coil transmitting a low-frequency sinusoidal current, magnetizing the yoke. The search coil is also on the Π-shaped yoke, where the proximity with the steel under test offers a closed magnetic loop, magnetizing the steel under test, thus allowing the calculation of its permeability by using Ampere's law. An important issue of this measurement is the distance between the yoke and the under-test steel. This gap may be of significant importance, being decisive for the uncertainty of the permeability measurement. For this reason, a contactless sensor for low air-gap determination, namely time-of-flight ultrasonic or laser transceiver, can be used.

At first, the permeability dependence on magnetic field is the same as for the case of the calibration of yokes of the single sheet tester. Thus, Equation (12) can be used to determine the permeability dependence on current and excitation field. Then, following the corresponding procedure for the single sheet tester, Ampere's law for the case of a zero gap between the yoke and steel under test gives

$$\mu_s(I) = \frac{1}{N_o N_e I} \frac{l_s}{S_s} \frac{\left(\int V_y(t)dt\middle|I\right)\left(\int V_s(t)dt\middle|I\right)}{\left(\int V_y(t)dt\middle|I\right) - \left(\int V_s(t)dt\middle|I\right)} \tag{15}$$

where μ_s is the permeability of the steel under test and $V_s(t)$ is the voltage output of th search coil of the yoke during measurement of the steel under test. Taking the air ga into account, Equation (15) is more complicated, but can be analytically solved, too. I case the electromagnetic yoke is to be used in curly surfaces, its legs are of circular shap allowing for continuous touching on the surface of the steel under test. However, in thi case, the precision of measurement is not as good as for the case of flat yoke legs, becaus magnetic flux escapes from the curly surface of the yoke in a rather unpredictable wa In electromagnetic yokes, the question of the kind of the yoke's material is also presen The two different schools are also present: use of either high-permeability material, o material of the same steel grade as the steel under test. In fact, in our group, we used highl permeable amorphous ribbon yoke, while high-quality electronics can clearly distinguis the change of the signal due to the presence of the steel under test. Electronics refer t current power amplifiers, like IGBT transistors and high-quality signal amplifiers togethe with 16-bit A to D converters, controlled by a proper micro-controller, like ESP32. Apar from that, the measurement of the ambient temperature is also required, since permeabilit is strongly dependent on temperature. Thus, both lift-off between yoke and steel under tes as well as the ambient temperature are two significant parameters, affecting the uncertaint of measurements. Precise measurements of permeability require measurements of bot parameters (lift-off and ambient temperature) in order to correct the readout of the searc coil of the yoke. This is important when in-field measurements are considered.

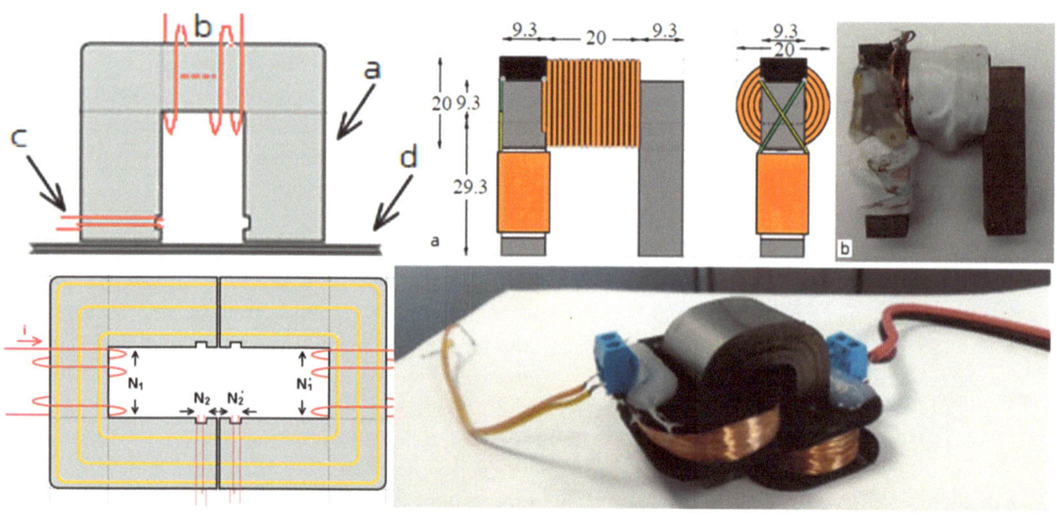

Figure 7. Typical electromagnetic yoke arrangements. **Top left**: schematic of the electromagneti yoke (a), with the excitation coil (b), search coil (c) and steel under test (d). **Top right (a)**: exac geometrical configuration of an ARMCO yoke. **Top right (b)**: the actual ARMCO electromagneti yoke. **Bottom left:** calibration methodology of the permeability of the electromagnetic yoke. **Bottom right**: near-zero magnetostriction amorphous ribbon yoke with the excitation and detection coils able to operate on flat steel surfaces.

Although electromagnetic yokes are used in steel non-destructive tests, they suffer from several disadvantages. A certain disadvantage is the large area in which the yoke is able to measure permeability, while the other is the time needed for each individual measurement, being at least equal to the period of the excitation current. In case of 0.1 Hz the time needed for a single measurement in a single direction is more than 10 s, which is very long for specific applications. This way, when electromagnetic yokes need to be used in the field, their excitation frequency is larger, up to 10 Hz, to allow for faster measurements

Such a frequency increase results in eddy currents involved in the output signal which is not favorable for a proper quasi-dc permeability measurement.

The way of using electromagnetic yokes in real environments depends on the specific in-field applications. In case that electric energy for feeding the electronics of the yoke and particularly the current amplifier of the excitation coil are acceptable for field measurements, electromagnetic yokes can be used, offering a large amount of information on magnetic properties. However, there are cases where energy requirements do not permit the use of the electromagnetic yokes. One reason may be that such energy is not readily available in the field, while the other is that the sensor must consume energy up to a certain threshold for anti-explosive reasons. For both cases, electromagnetic yokes cannot be efficient. Bearing in mind that at least two A rms are required for a proper permeability measurement, under at least 5 V power supply, the total energy consumption of the yoke is larger than 10 W, thus making its use impossible in cases of a restricted amount of energy demand.

However, electromagnetic yokes can offer significant information in quality control labs, where the speed of measurement and energy consumption are not an issue, because they offer information that may approach the information and the quality of information provided by the single sheet testers. As an example, the use of these yokes in ductile to brittle transition temperature in quality control labs of the steel industry is of vital importance, since they can offer detailed information of FORC analysis of magnetization loops. For some particular applications of steel structures, like heat exchangers, the use of electromagnetic yokes is preferable with respect to single sheet testers, due to shape reasons and the disability of cutting them for quality control.

For all these reasons, we have developed new families of sensors, having a much smaller size, much lower energy consumption, having a time response in the order of ms or even μs, allowing for fast, reliable and safe measurements, with power consumption in the order of mW. These families of sensors are based on the Hall effect and the Anistropic Magnetoresistance (AMR) effect and are presented next.

5. Hall Sensors

The Hall effect is well-known for sensors and transducers [22]. They offer a linear dependence of the magnetic field component vertical to the surface of the sensing element, being able to have a large variety of span from mT up to T, with a sensitivity in the order of $\mu THz^{-1/2}$. This sensitivity requires the magnetic biasing field to properly use the advantages of Hall sensors in magnetic permeability measurements. Therefore, the design of our permeability sensors based on the Hall effect use the principle of magnetic bias. Suppose a permanent magnet yoke, i.e., a yoke comprised of a permanent magnet, magnetizing two magnetized soft magnetic bars, as depicted in Figure 8, comprising the so-called permanent magnet yoke. In the absence of a magnetic steel under test, magnetic lines follow a large path from one soft magnetic pole to the other (left). As soon as the magnetic steel is under the permanent magnet yoke at a certain distance from the soft magnetic bars, the magnetic lines tend to be more vertical towards the surface of the magnetic steel under test (middle). The larger the permeability of the steel under test, the more vertical the magnetic lines on the surface of the steel, provided that the distance between the soft magnetic bars and the steel under test remains the same. Therefore, the magnetic field at the edge of the soft magnetic bar increases with the permeability of the steel under test.

At the same time, the magnetic field far from the edge of the soft magnetic bar decreases with the permeability of the magnetic steel. It is clear that for this family of sensors, the measurement of the distance between the soft magnetic bar and the steel under test is of vital importance. Similarly, the ambient temperature is also important, affecting both, the performance of the Hall sensor and the permeability of the steel under test.

Therefore, a time-of-flight sensor and an ambient temperature sensor are used between the two soft magnetic bars to adjust the Hall sensor response at the proper levels, thus affecting dramatically the uncertainty of the measurement. In Figure 8, the actual permanent magnet yoke with the Hall sensors at the bottom of the soft magnetic bars can be seen. The packaging of the system, including a 16-bit A to D converter, together with an ESP32 microcontroller, inside a 3D-printed package, can also be observed (right). This sensor can also measure magnetic anisotropy by rotating it on top of the steel under test. Such a measurement is illustrated in Figure 9.

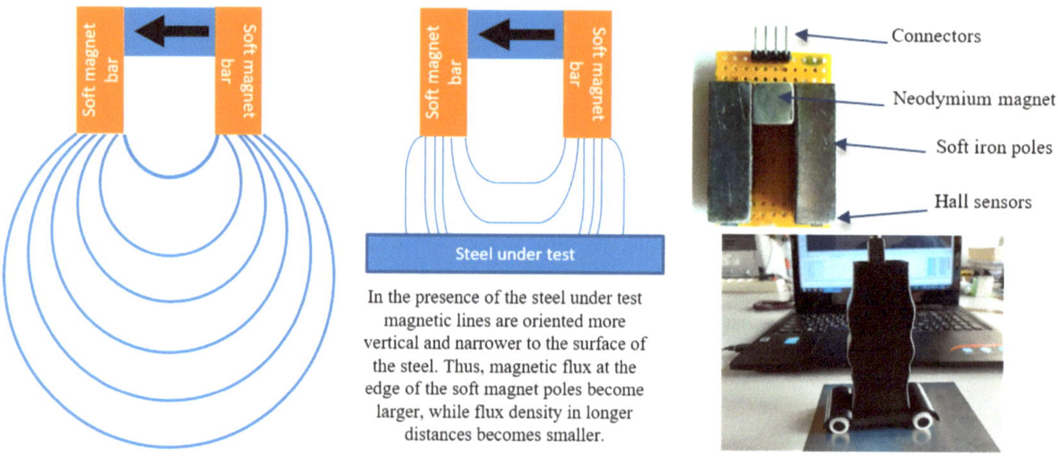

Figure 8. The main principle of our family of permeability sensors based on the Hall effect. **Left**: magnetic lines in the absence of the steel under test. **Middle**: magnetic lines in the presence of the steel under test. **Right**: the actual permanent magnet yoke with the Hall sensors at the bottom of the soft magnetic bars and the packaging of the Hall sensor arrangement, hosting four wheels allowing for permeability distribution monitoring.

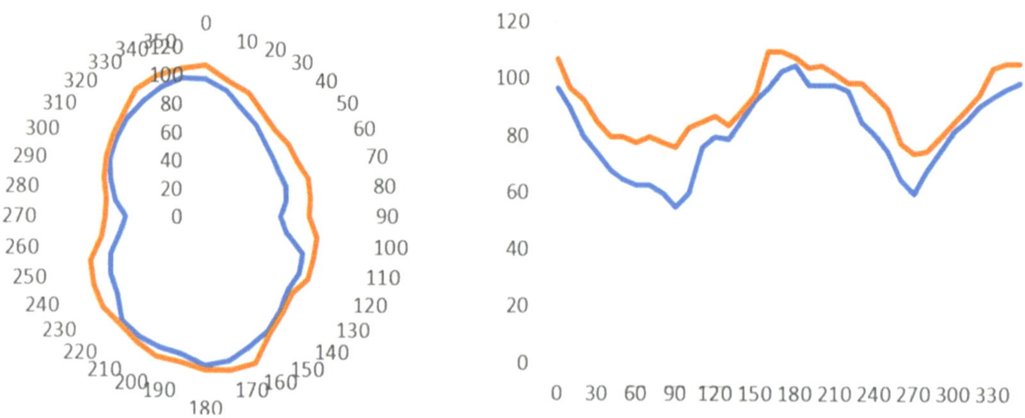

Figure 9. Magnetic anisotropy measurements, obtained by rotating the sensor on top of the steel under test in different areas of non-oriented electric steel. **Left**: Polar anisotropy response. **Right**: Cartesian anisotropy response.

Apart from this type of Hall sensor arrangement, there is another possible magnetization means, comprised of a permanent magnet in the form of a cylinder, being vertical to the surface of the steel under test, with its magnetization along the axis of the cylinder. Thus, following the same principle like in Figure 8, in the absence of steel under test, the magnetic lines getting out of the cylinder follow a large path until entering back to the other pole of the permanent magnetic cylinder. In the presence of the steel under test, the magnetic lines are more vertical on the surface of the steel, dependent on the local permeability and magnetic anisotropy of the steel, while they escape from the steel at a distance also dependent on the local permeability and magnetic anisotropy of the steel.

Therefore, arranging a ring of Hall sensors, radially oriented and surrounding the edge of the permanent magnetic cylinder, results in the measurement of the local magnetic permeability of the steel on a ring, formulated by the diameter of the permanent magnetic cylinder. The distribution of permeability is the indication of the magnetic anisotropy of this region of the steel under test. A typical response of such anisotropy is indicated in Figure 10, using a 12 mm × 12 mm Si substrate, where 24 Hall sensors radially developed with a 15° pitch, having a 10 mm diameter permanent magnetic bar in the middle of the arrangement, are used.

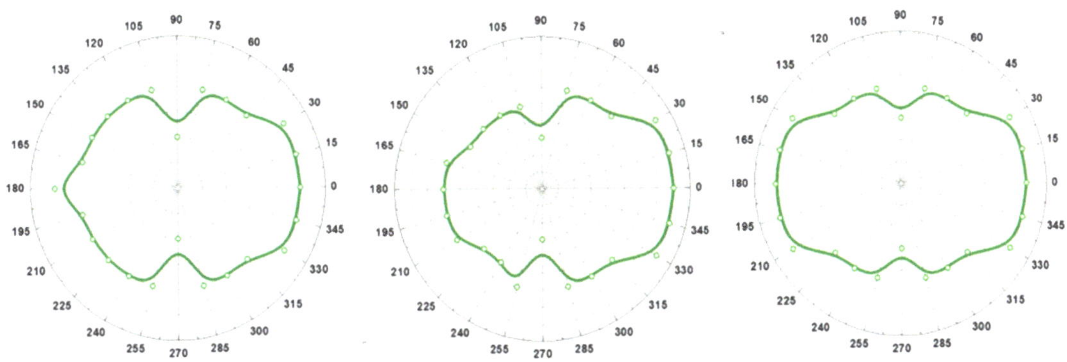

Figure 10. Polar magnetic anisotropy (arbitrary units) in the heat-affected zones (HAZ) of welded AISI 1008 flat steel samples. **Left**: TIG-welded sample. **Middle**: Plasma-welded sample. **Right**: Induction heating sample.

The advantages of this sensor arrangement open a new era in magnetic permeability measurements in steels. In the first place, they can achieve a fast measurement. Bearing in mind that the time response of the Hall sensor, including the A to D conversion and signal transmission through the microcontroller of the sensor is less than 1 ms, and the fact that the size of the Hall sensors is in the order of 0.1 mm × 0.1 mm, the speed of measurement is in the order of 0.1 m/s or 360 m/h with a spatial resolution of 0.1 mm. This speed is sufficient for several steel production and manufacturing lines. Furthermore, it is incomparably higher than the speed of electromagnetic yokes. Furthermore, using the permanent magnetic cylinder arrangement, the magnetic anisotropy is able to be monitored with the same speed and spatial resolution. Apart from that, the power consumption of the permanent magnet yoke is in the order of 50 mA, including the ESP32 microcontroller, allowing for the possibility of passing the anti-explosive ATEX specifications [23]. However, the permanent yoke arrangement suffers from the same size issues, like the electromagnetic yoke, due to the similar topology of the Π-shaped yoke. In order to improve sensitivity, energy consumption and other properties, another family of permeability sensors has been developed, based on the AMR effect, presented in the next chapter.

6. AMR Sensors

AMR sensors are more sensitive than the Hall sensors. In fact, they can have a sensitivity in the order of 1 $nTHz^{-1/2}$ and a span of several hundreds of μT. This way, they do not need a biasing magnetic field in the order of several tens of mT to operate as the

Hall sensors do. Instead, they can operate with the biasing field of the Earth's field, being in the order of some tens of μT. This way, they do not need permanent magnets to measure permeability. Although they are used as sensing elements in electromagnetic and magnetic yokes to monitor cracks present on the surface of the steel, they can also be used without these yokes to monitor the magnetic permeability in given surface areas of the steel under test, having a size in the order of the size of the AMR sensor.

The main principle of operation is the following: suppose that the steel under test has a residual stress on its surface. If this material is positive magnetostrictive in the direction of measurement and has a tensile stress field in this direction, then its permeability increases [24]. Bearing in mind that the Earth's field is parallel to the surface of the steel and having such a local permeability increase, the Earth's field (magnetic lines) penetrates the steel under test in the region of the increased permeability. Earth's field has three components: H_x corresponding to the in-plane direction along measurement (X-axis), H_y corresponding to the in-plane direction orthogonal to the measurement axis (Y-axis) and H_z corresponding to the direction vertical to the plane of measurement (Z-axis). Then, H_z will decrease (the Z-axis is considered positive for amplitudes from the surface of the steel under test upwards) and therefore H_x will slightly decrease, because the magnetic lines of the X-axis are trapped inside the steel. Consequently, the H_y will slightly increase, because it has to behave the opposite of the X-axis, because the sign of magnetostriction should be opposite to the sign of magnetostriction of the X-axis [25].

If the steel is positive magnetostrictive in the direction of measurement and has a compressive stress field in this direction, then its permeability decreases. Thus, following the same analysis as above, H_z will increase and therefore H_x will slightly increase, while H_y will slightly decrease. If the steel is negative magnetostrictive in the direction of measurement and has a tensile stress field in this direction, then its permeability decreases. Following the same analysis as above, H_z will increase and therefore H_x will slightly increase, while H_y will slightly decrease. If the steel is negative magnetostrictive in the direction of measurement and has a compressive stress field in this direction, then its permeability increases. Thus, H_z will decrease and therefore H_x will slightly decrease while H_y will slightly increase. All these are depicted in Table 1. A schematic describing this process is illustrated in Figure 11.

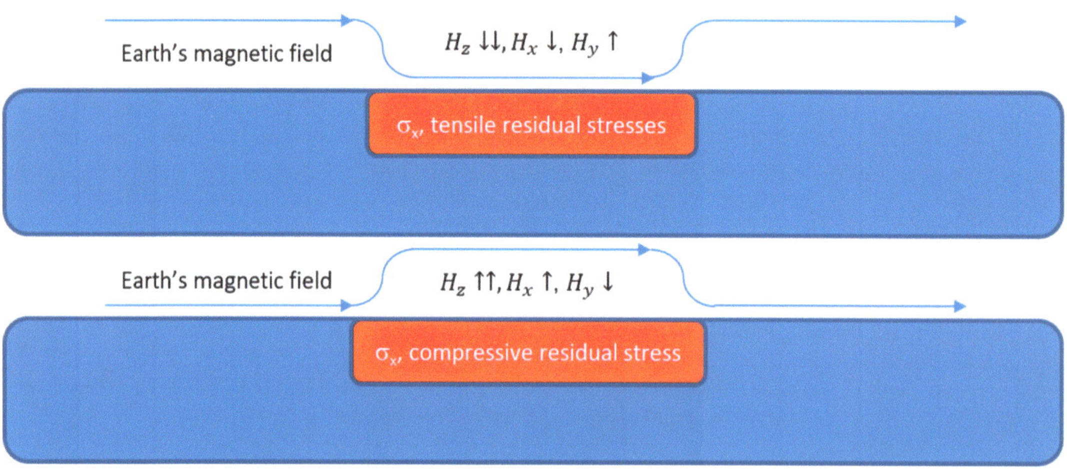

Figure 11. Representation of the magnetic lines of Earth's field for tensile and compressive stresses in positive magnetostrictive steels. The opposite behavior occurs in negative magnetostrictive materials.

Table 1. Change of permeability and AMR sensor response due to stresses and magnetostriction.

λ	σ	μ	H_z	H_x	H_y
+	+	↑	↓↓	↓	↑
+	−	↓	↑↑	↑	↓
−	+	↓	↑↑	↑	↓
−	−	↑	↓↓	↓	↑

A critical issue arising from this analysis is how to distinguish positive and negative magnetostrictive materials in known and unknown steel grades in order to select the proper sign of residual stresses at the area of measurement. Regardless of the steel grade, a magnetostriction constant is not universally either positive or negative. It may be positive or negative in given crystallographic directions, but it is also dependent on the amount of magnetostriction in given coordinates: in case λ_x in the X-axis is positive at a given region, then λ_y in the Y-axis is negative in most cases at the same region. Until now, this issue has remained unsolved, and there is ongoing research in order to find a solution. However, since the gradient of stress is the important parameter, one should count on the absolute value of stress which appears to be symmetric with respect to the Y-axis of the MASC curve. Therefore, AMR scanning may provide the absolute value of stress along the measurement, which tells about the stress gradient. Apart from that, the areas of severe probability of suffering a nano-crack generation are those demonstrating a gradient from one sign of stress to another, when stresses are close to the yield point of the steel under test. Therefore, the absolute stress value and the gradient stress are the two decisive parameters for the determination of the position of crack initiation.

The AMR sensor used in our group is the ST microelectronics LSM30-3D sensor. It has been selected due to the sensitivity of 1 nTHz$^{-1/2}$, the three-dimensional measurement of field and the ambient temperature sensor hosted on the integrated circuit, which is vital for more precise measurements. Two types of AMR sensors have been developed, one with a biasing permanent magnet and another one without permanent magnet yoke. A typical response of the AMR sensor on non-oriented electric steel which suffered induction heating and consequent quenching, therefore giving rise to residual stresses and strains, is illustrated in Figure 12. In this response, the coincidence of the sign of the AMR response in X and Z axes can be seen, as well as the opposite sign in the Y-axis direction. This response refers to the AMR sensor with the permanent magnet yoke. The same response without the yoke is illustrated in Figure 12 demonstrating better sensitivity, allowing for the monitoring of smaller residual stresses present along the length of the measurement.

In case the AMR sensor needs a lift-off for measurement, like a (non-magnetic) coating, a time-of-flight sensor, either ultrasonic or laser, can be used in order to correct the readout response. Otherwise, concerning naked steels under test, the polymeric sensor packaging is attaching the steel under test, taking into account the precise distance between the AMR sensor and the outer part of the sensor packaging, which is 0.5 mm $\pm$ 0.01 mm. The excitation and detection electronics are driven by the ESP32 microcontroller, used in all our Hall and AMR sensors arrangements. The main advantage of the AMR sensor is the ability to monitor three-dimensional fields with a sensitivity better than the Hall sensor. Another advantage of the AMR sensor is the fast measurement, less than 1 ms, including the time needed for the microcontroller. Bearing in mind the size of the AMR sensor being 0.1 mm $\times$ 0.1 mm, similar to the Hall sensor, it can scan surfaces as fast as the Hall sensor. The power consumption is also similar to the Hall sensor, with the most amount of power required for the microcontroller. This way, the AMR sensor can also be used in harsh environments as with the Hall sensor, provided that both pass the specific anti-explosive ATEX requirements.

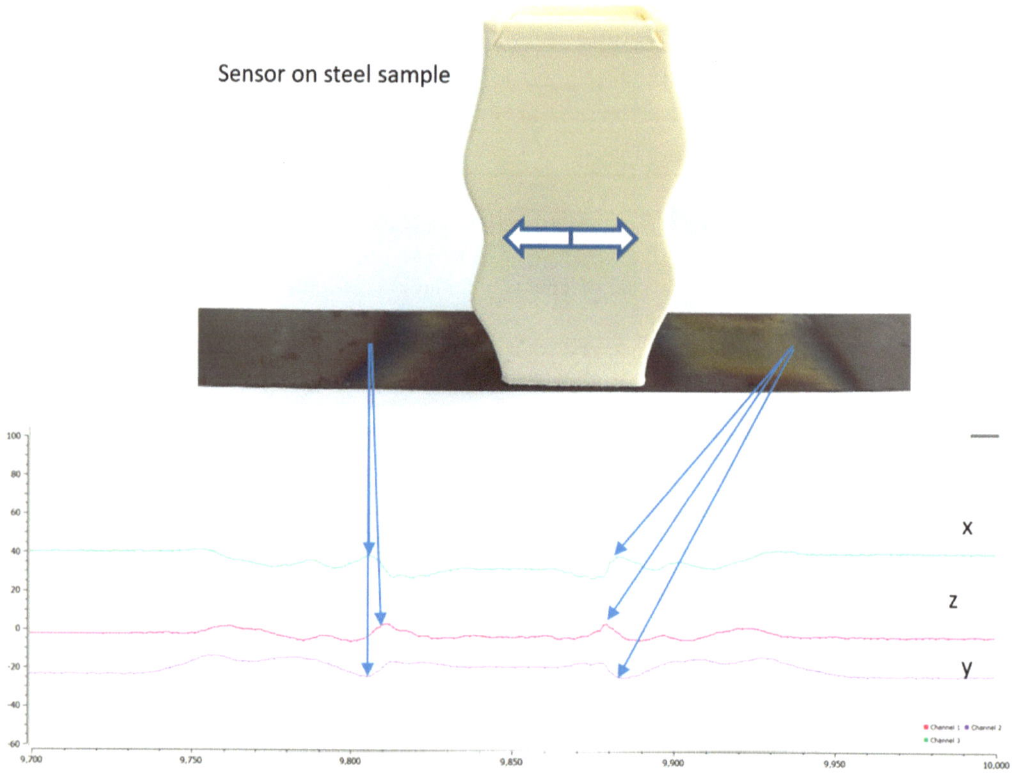

Figure 12. The AMR sensor used with biasing permanent magnet yoke, able to show the changes of field in the X, Y and Z axes, demonstrating the presence of stresses in a non-oriented steel after local induction heating and consequent quenching. **Top**: The sensing element from ST microelectronics, together with its packaging on top of a non-oriented steel. **Bottom**: Response of the field change on the three directions X, Y and Z.

Magnetic field sensors can in general be used for such measurements, like GMR [26], TMR [27], Spin Valves [28], and GMI [29–32] sensors. All these sensors can take the place of AMR sensors with similar results. Furthermore, other types of magnetic sensors can also be employed with the possibility of measuring residual stresses [33–35], using the magnetoelastic coupling theory [36,37].

Bearing in mind the advantages of the AMR sensor with respect to the rest of the sensors, we have provided the dependence of the magnetic component H_z on the residual stresses in two different types of steels, namely non-oriented electric steel and 42CrMo4 steel grade, which is suitable for maritime applications. The two samples have been prepared by using localized RF induction heating and consequent quenching in water. The samples were measured in three different lines across the area of heating and quenching. The results are given in Figure 13, illustrating a sigmoid response, reaching values of residual stresses closer to the yield point with respect to the stress values obtained with the autogenous welding process.

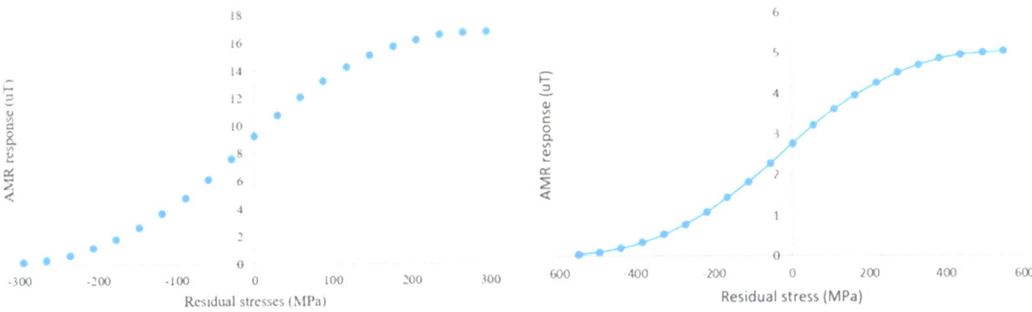

Figure 13. MASC curves of non-oriented electric steel (**left**) and 42CrMo4 maritime steel grade (**right**), using RF induction heating and consequent quenching.

7. Discussion

The four different families of instruments and sensors, able to measure and monitor the differential magnetic permeability of steels, are important in the steel industry, in both production and manufacturing, as well as in structural and operational steel applications. At first, the single sheet tester (SST) arrangement is useful for quality control laboratories in the steel industry. As an example, our SST instrument has been used in the quality control laboratory of Corinth Pipe Works (CPW) in Greece to monitor the ductile to brittle transition temperature (DBTT) in magnetic steels for pipeline manufacturing. The classic method of monitoring such transition temperature is to drop a weight from a distance and determine the energy required for breaking the steel. CPW accepted the use of our magnetic technology in their factory to monitor the permeability of such samples. Indeed, the experimental set-up was developed the quality control facilities of CPW, to compare the magnetic technique with the classical, standardized response of the drop weight facility of the factory. Next to this drop weight instrument, there is a cooling pool to drop the temperature of the steel under test. As soon as the temperature reaches the required level, the piece of steel is transferred quickly from the cooling pool to the drop weight instrument to drop the weight and then to determine the energy required for breaking the steel sample. As soon as the temperature of the steel is below DBTT, the breaking energy decreases. The SST instrumentation was set next to the above-mentioned cooling pool, and the magnetic permeability of the steel under test was determined at each temperature level. Figure 14 illustrates the set-up and the achieved results. According to these results, the ductile to brittle transition temperature window was within the region measured by the standard method. However, the time required to determine DBTT in one sample using the standard method is more than 8 h, while the magnetic method requires no more than 10 min for one sample, without any kind of special and laborious sample preparation. This application of the permeability sensors to determine DBTT, as well as to predict the crack initiation of magnetic steels are the two examples according to which the sensors and the methodology of using them to predict the crack initiation are clear indications of the superiority of these sensors with respect to the current state of the art.

The electromagnetic yoke has been used for the same reason, i.e., quality control procedures in testing magnetic steels, before the development of our low-power-consumption Hall and AMR sensors.

As already mentioned in previous chapters, there are two parameters affecting the performance of the permeability sensors. The first is the lift-off effect, causing uncertainties in measurements: an increase of the lift-off results in a decreased amplitude of the under-determination magnetic permeability, illustrating an incorrect amount of residual strain and stress at the area of measurement, according to the sigmoid response of the MASC curves presented in Section 2. The second is the ambient temperature during measuremen The dependence of permeability on temperature is more or less decreasing and paraboli Permeability decreases with temperature, affecting the determination of residual stresse on the under-test steel too. The response of all permeability sensors is independent o the specific and mentioned application, like energy, transportation, oil and gas, stee structures, etc. The only parameters affecting their response are the lift-off effect an ambient temperature.

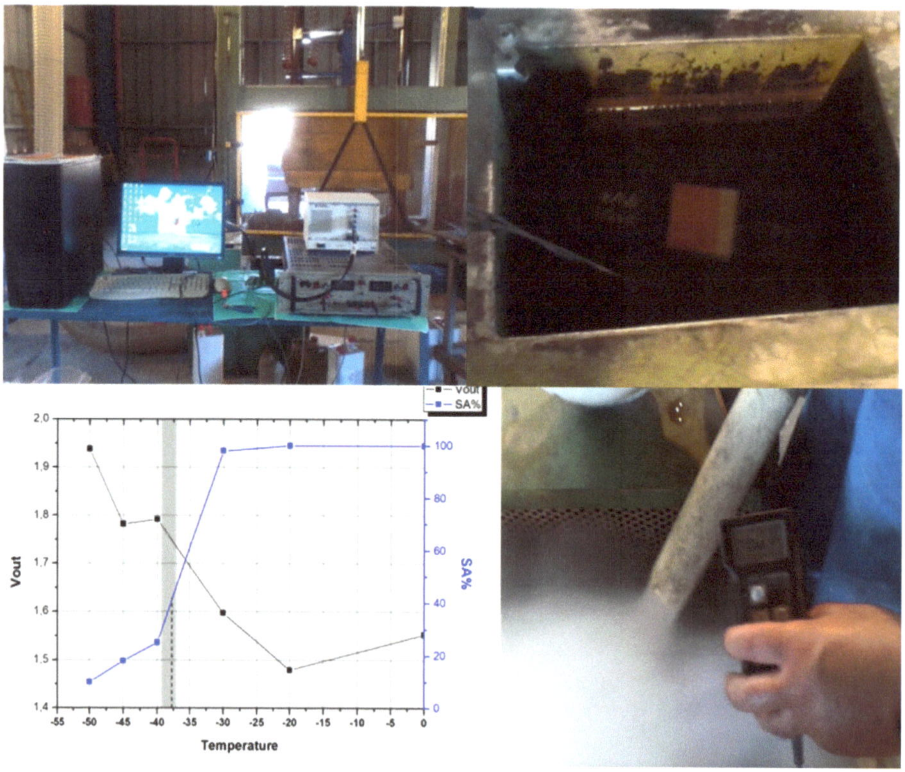

Figure 14. The SST of our lab used for the DBTT determination. Top left: the SST of our lab (behinc SST a part of the standard Drop Weight test can be seen). Top right: the cooling pool where th electro-mechanical part of the SST instrument was firmly set. Bottom right: manual measurement o temperature for security reasons. Bottom left: comparison of the magnetic measurements with th drop weight tests. The agreement between the two methods can be seen.

The steel applications foreseen by our laboratory are many and different. Apar from steel manufacturers (namely pipeline, shipyards, steel coil producers, heat exchange manufacturers, etc.) where the inspection of the incoming raw materials and the fina product are needed, the steel end-users that need such measurements are divided into five main categories:

- Transportation, namely on-board stress tests in ships, railways and train wagons, as well as electric steels for the largely developing industry of electric motors for e-motior applications, etc.
- Energy, namely heat exchangers in several energy production activities, test of domes in nuclear stations, tests of the carrying steel bar in underwater current transmis sion, etc.

- Oil and gas, namely quality control of the front-end tools to cut the ores, tests in vessels under pressure and pipelines of refineries, oil and gas transmission pipelines, etc.
- Several steel structures, like steel bridges, metallic buildings, special facilities like the 27 km ring and the foreseen 100 km new ring at CERN, etc.

The comparison of the four different families of sensors is depicted in Table 2, allowing for the determination of the advantages and disadvantages of each one. The Hall and AMR sensors can be used in harsh environments, like oil and gas transmission lines, vessels under pressure, etc., during operation and not only in the shut-down periods. Of course, it is impossible for them to be used in neutron fluence environments, like in nuclear stations. For this particular reason, new types of Hall and AMR sensors should be developed, made of semiconductors able to withstand high energies. Examples of these materials are the garnets that have been extensively used in space applications, having proven their resistance in high-energy particles. Apart from that, the AMR sensor can be used to develop a pen-like sensor for some specific applications: for example, surveyors in ships, energy stations, oil and gas applications and other steel structures may have such a pen, calibrated against residual stresses and strains, transferring data to their cell phone or tablet or laptop in order to see by their own eyes the validity of measurements and evaluate the quality control procedures, based on our magnetic steel health-monitoring technology.

Table 2. Comparison of the four families of permeability sensors.

Characteristics Sensors	Use in the Field	Consumption (W)	Speed of Tests (mm/s)	Sensitivity (ppm)	Uncertainty (%)	Cost of Use (k€)
SST	No	>100	0.01	10	0.1	50–180
EMY	Yes	>50	0.01	100	0.5	30–60
Hall	Yes	~0.05	100	1000	±2	0.5–1
AMR	Yes	~0.05	100	1000	±1	0.5–1

Another important issue is the development of technology for the structural health monitoring of non-magnetic ferrous steels, like the austenitic 304, 316, 316L and 316LN steel grades. In fact, all these steels are non-magnetic until the end of their life: as soon as they suffer the development of nano-cracks, they undergo a phase transformation from austenitic to martensitic phase, which is magnetic. To demonstrate a feeling on such phase transformation, one can think of the following experiment: cutting an austenitic steel without cooling results in a magnetic layer at the vicinity of cutting, which corresponds to the martensitic phase. Thus, the measurement of a small magnetic substance, by the Hall or the AMR sensor, is an indication of the beginning of the end of life of the austenitic steel. As the sensitivity of the AMR sensor is superior to the Hall sensor, it is very logical to use AMR sensors for the austenitic steel health monitoring. Bearing in mind the vast amount of applications of austenitic steels, our team has initiated work in this subject, mainly targeting applications in ITER, south France.

One of the most significant future outcomes of magnetic steel health monitoring is the remaining lifetime prediction, based on the monitoring of the differential permeability change with time. The most straightforward method to foresee the position and the time of failure of such a steel grade is the projection of a continuous time-dependent function towards the yield point of the steel. The time-based projection is separated into micro-time, mid-time and macro-time functions of permeability change, representing the accumulation of residual and hydraulic stresses. For example, a sudden and short approach to the yield point due to a high hydraulic stress is not necessarily the preamble of a steel failure: if it happened once and then the magnetic properties are changing slightly, the steel in question does not suffer possible failure. Instead, a stress gradient changing from −20% up to 20%

of the yield point in a space of a few mm will result in steel failure, in a time window determined by the projection of the actual data toward the yield point. Another approach apart from following a continuous projection of the magnetic permeability towards the yield point is the realization of high and low cycle fatigue studies (i.e., fatigue studies less or higher than the yield stress point, respectively), to have the experimental evidence of the permeability levels approaching the yield point.

Having determined the position and the time window for the steel failure, we are also able to fix the issue of residual stress by using a localized heat induction treatment. Transmitting a sinusoidal current at a frequency bandwidth from 10 to 30 kHz and a current amplitude in the order of 30 A results in eddy currents in the near-surface of the sample. Bearing in mind the low thermal conductivity of steel, it is expected that a temperature profile, with maximum temperature below one-third of the melting point of the corresponding steel grade, provides stress annihilation.

Future work includes experiments under low and high cycle fatigue under various thermodynamic and electromagnetic conditions.

The presented sensors and the technology to provide the correlation of the sensor response with the actual residual stresses in the tested steels is the basis, i.e., the working documents of the Quality Management System of our laboratory, towards its accreditation according to the ISO 17025 Standard.

8. Conclusions

In this work, after presenting the dependence of magnetic properties, like MBN and differential magnetic permeability, on residual stresses, the sensors measuring magnetic permeability were presented. These sensors are suitable for laboratory use, like the single sheet tester instrument or the electromagnetic yoke, as well as for in-field measurements, like sensors based on the Hall effect and the AMR principle. The advantages and disadvantages of these sensors have been discussed, offering applications in steel manufacturing and corresponding use in steel structures.

Funding: This research received no external funding.

Informed Consent Statement: Informed consent was obtained from all subjects involved in the study.

Data Availability Statement: Data are available at request by the author.

Acknowledgments: Acknowledgements are due to Aphrodite Ktena, a close collaborator in the field of magnetism and magnetic properties of steels since 1993; Peter Svec, a close collaborator in residual stress monitoring; Xenia Vourna, post-doc of our laboratory in the field of structural characterization of steels and determination of MASC curves; Spyros Angelopoulos, post-doc in our laboratory for his work in the field of magnetic sensors; and Tatiana Damatopoulou, post-doc in our laboratory working on the accreditation of our laboratory according to the ISO 17025 for steel health monitoring.

Conflicts of Interest: The authors declare no conflict of interest.

References

1. Thelning, K.E. *Steel and Its Heat Treatment*; Butterworth-Heinemann: Oxford, UK, 2013.
2. Birat, J.-P. Society, Materials, and the Environment: The Case of Steel. *Metals* **2020**, *10*, 331. [CrossRef]
3. Huff, M. Residual Stresses in Deposited Thin-Film Material Layers for Micro- and Nano-Systems Manufacturing. *Micromachines* **2022**, *13*, 2084. [CrossRef]
4. Versaci, M.; Angiulli, G.; La Foresta, F.; Laganà, F.; Palumbo, A. Intuitionistic fuzzy divergence for evaluating the mechanical stress state of steel plates subject to bi-axial loads. *Integr. Comput.-Aided Eng.* **2024**, *31*, 363–379. [CrossRef]
5. Luo, Q.; Yang, S. Uncertainty of the X-ray diffraction (XRD) sin2 ψ technique in measuring residual stresses of physical vapor deposition (PVD) hard coatings. *Coatings* **2017**, *7*, 128. [CrossRef]

6. Mo, F.; Sun, G.; Li, J.; Zhang, C.; Wang, H.; Chen, Y.; Liu, Z.; Yang, Z.; Li, H.; Yang, Z.; et al. Recent Progress of Residual Stress Distribution and Structural Evolution in Materials and Components by Neutron Diffraction Measurement at RSND. *Quantum Beam Sci.* **2018**, *2*, 15. [CrossRef]
7. Zhao, Y.; Liu, Y.; Li, Y.; Hao, Q. Development and application of resistance strain force sensors. *Sensors* **2020**, *20*, 5826. [CrossRef]
8. Ammar, M.M.; Shirinzadeh, B.; Lai, K.Z.; Wei, W. On the sensing and calibration of residual stresses measurements in the incremental hole-drilling method. *Sensors* **2021**, *21*, 7447. [CrossRef]
9. Maciusowicz, M.; Psuj, G. Classification of Grain-Oriented Electrical Steel Sheets by Magnetic Barkhausen Noise Using Time-Frequency Analysis and Selected Machine Learning Algorithms. *Appl. Sci.* **2022**, *12*, 12469. [CrossRef]
10. Szewczyk, R. Generalization of the model of magnetoelastic effect: 3D mechanical stress dependence of magnetic permeability tensor in soft magnetic materials. *Materials* **2020**, *13*, 4070. [CrossRef]
11. Augustyniak, B.; Chmielewski, M.; Piotrowski, L.; Sablik, M.J. Designing a magnetoacoustic emission measurement configuration for measurement of creep damage in power plant boiler tubes. *J. Appl. Phys.* **2002**, *91*, 8897–8899. [CrossRef]
12. Jeyaprakash, N.; Haile, A.; Arunprasath, M. The parameters and equipments used in TIG welding: A review. *Int. J. Eng. Sci. (IJES)* **2015**, *4*, 11–20.
13. Sahoo, A.; Tripathy, S. Development in plasma arc welding process: A review. *Mater. Today Proc.* **2021**, *41*, 363–368. [CrossRef]
14. Ahmed, T.J.; Stavrov, D.; Bersee, H.E.N.; Beukers, A. Induction welding of thermoplastic composites—An overview. *Compos. Part A Appl. Sci. Manuf.* **2006**, *37*, 1638–1651. [CrossRef]
15. Węglowski, M.S.; Błacha, S.; Phillips, A. Electron beam welding–Techniques and trends–Review. *Vacuum* **2016**, *130*, 72–92. [CrossRef]
16. Leggatt, R.H. Residual stresses in welded structures. *Int. J. Press. Vessel. Pip.* **2008**, *85*, 144–151. [CrossRef]
17. Hristoforou, E.; Vourna, P.; Ktena, A.; Svec, P. On the universality of the dependence of magnetic parameters on residual stresses in steels. *IEEE Trans. Magn.* **2015**, *52*, 6201106. [CrossRef]
18. Mangiorou, E.; Damatopoulou, T.V.; Angelopoulos, S.; Kalkanis, K.; Vourna, P.; Ktena, A.; Hristoforou, E. Revisiting the universality law in magnetically detected residual stresses in steels. *AIP Adv.* **2024**, *14*, 025126. [CrossRef]
19. Wilson, J.W.; Zhou, L.; Davis, C.L.; Peyton, A.J. High temperature magnetic characterisation of structural steels using Epstein frame. *Meas. Sci. Technol.* **2021**, *32*, 125601. [CrossRef]
20. *ISO-ASTM A1036-04*; Standard Guide for Measuring Power Frequency Magnetic Properties of Flat-Rolled Electrical Steels Using Small Single Sheet Testers. ASTM International: West Conshohocken, PA, USA, 2020.
21. Ušák, E. A New Approach to the Evaluation of Magnetic Parameters for Non-Destructive Inspection of Steel Degradation. *J. Electr. Eng.* **2010**, *61*, 100–103.
22. Popović, R.S. Hall-effect devices. *Sens. Actuators* **1989**, *17*, 39–53. [CrossRef]
23. Leroux, P. New regulations and rules for ATEX directives. *IEEE Ind. Appl. Mag.* **2006**, *13*, 43–51. [CrossRef]
24. Sablik, M.J.; Jiles, D.C. Coupled magnetoelastic theory of magnetic and magnetostrictive hysteresis. *IEEE Trans. Magn.* **1993**, *29*, 2113–2123. [CrossRef]
25. Jiles, D.C. Theory of the magnetomechanical effect. *J. Phys. D Appl. Phys.* **1995**, *28*, 1537. [CrossRef]
26. Cubells-Beltrán, M.D.; Reig, C.; Madrenas, J.; De Marcellis, A.; Santos, J.; Cardoso, S.; Freitas, P.P. Integration of GMR sensors with different technologies. *Sensors* **2016**, *16*, 939. [CrossRef]
27. Jin, Z.; Mohd Noor Sam, M.A.I.; Oogane, M.; Ando, Y. Serial MTJ-based TMR sensors in bridge configuration for detection of fractured steel bar in magnetic flux leakage testing. *Sensors* **2021**, *21*, 668. [CrossRef]
28. Svalov, A.V.; Savin, P.A.; Kurlyandskaya, G.V.; Gutierrez, J.; Barandiaran, J.M.; Vas, V.O. Spin-valve structures with Co-Tb-based multilayers. *IEEE Trans. Magn.* **2002**, *38*, 2782–2784. [CrossRef]
29. Gonzalez-Legarreta, L.; Corte-Leon, P.; Zhukova, V.; Ipatov, M.; Blanco, J.M.; Gonzalez, J.; Zhukov, A. Optimization of magnetic properties and GMI effect of Thin Co-rich Microwires for GMI Microsensors. *Sensors* **2020**, *20*, 1558. [CrossRef]
30. Mohri, K. Review on recent advances in the field of amorphous metal sensors and transducers. *IEEE Trans. Magn.* **1984**, *20*, 942–947. [CrossRef]
31. Panina, L.V.; Mohri, K. Effect of magnetic structure on giant magneto-impedance in Co-rich amorphous alloys. *J. Magn. Magn. Mater.* **1996**, *157–158*, 137–140. [CrossRef]
32. Kurlyandskaya, G.V.; Garcia-Arribas, A.; Fernandez, E.; Svalov, A.V. Nanostructured Magnetoimpedance Multilayers. *IEEE Trans. Magn.* **2012**, *48*, 1375–1380. [CrossRef]
33. Alves, F.; Bensalah, A.D. New 1D–2D magnetic sensors for applied electromagnetic engineering. *J. Mater. Process. Technol.* **2007**, *181*, 194–198. [CrossRef]
34. Cobeno, A.F.; Zhukov, A.P.; Pina, E.; Blanco, J.M.; Gonzalez, J.; Barandiaran, J.M. Sensitive magnetoelastic properties o amorphous ribbon for magnetoelastic sensors. *J. Magn. Magn. Mater.* **2000**, *215–216*, 743–745. [CrossRef]
35. Barandiaran, J.M.; Gutierrez, J. Magnetoelastic sensors based on soft amorphous magnetic alloys. *Sens. Actuators A Phys.* **1997**, *59*, 38–42. [CrossRef]

36. Barandiarán, J.M.; Gutiérrez, J.; García-Arribas, A. Magneto-elasticity in amorphous ferromagnets: Basic principles and applica-tions. *Phys. Status Solidi A* **2011**, *208*, 2258–2264. [CrossRef]
37. Garcia, D.; Munoz, J.L.; Kurlyandskaya, G.; Vazquez, M.; Ali, M.; Gibbs, M.R.J. Magnetic domains and transverse induced anisotropy in magnetically soft CoFeB amorphous thin films. *IEEE Trans. Magn.* **1998**, *34*, 1153–1155. [CrossRef]

sensors

MDPI

Article

Effects of Magnetostatic Interactions in FeNi-Based Multilayered Magnetoimpedance Elements

Grigory Yu. Melnikov [1], Sergey V. Komogortsev [2,3], Andrey V. Svalov [1], Alexander A. Gorchakovskiy [2], Irina G. Vazhenina [2,3] and Galina V. Kurlyandskaya [1,*]

1 Institute of Natural Sciences and Mathematics, Ural Federal University, 620002 Ekaterinburg, Russia; grigory.melnikov@urfu.ru (G.Y.M.); andrey.svalov@urfu.ru (A.V.S.)
2 Kirensky Institute of Physics, Federal Research Center SB RAS, 660036 Krasnoyarsk, Russia; komogor@iph.krasn.ru (S.V.K.); vigetch@list.ru (A.A.G.) irina-vazhenina@mail.ru (I.G.V.)
3 Polytechnic School, Siberian Federal University, 660041 Krasnoyarsk, Russia
* Correspondence: galinakurlyandskaya@urfu.ru

Abstract: Multilayered $[Cu(3\ nm)/FeNi(100\ nm)]_5/Cu(150\ nm)/FeNi(10\ nm)/Cu(150\ nm)/FeNi(10\ nm)/Cu(150\ nm)/[Cu(3\ nm)/FeNi(100\ nm)]_5$ structures were obtained by using the magnetron sputtering technique in the external in-plane magnetic field. From these, multilayer magnetoimpedance elements were fabricated in the shape of elongated stripes using the lift-off lithographic process. In order to obtain maximum magnetoimpedance (MI) sensitivity with respect to the external magnetic field, the short side of the rectangular element was oriented along the direction of the technological magnetic field applied during the multilayered structure deposition. MI sensitivity was defined as the change of the total impedance or its real part per unit of the magnetic field. The design of the elements (multilayered structure, shape of the element, etc.) contributed to the dynamic and static magnetic properties. The magnetostatic properties of the MI elements, including analysis of the magnetic domain structure, indicated the crucial importance of magnetostatic interactions between FeNi magnetic layers in the analyzed $[Cu(3\ nm)/FeNi(100\ nm)]_5$ multilayers. In addition, the uniformity of the magnetic parameters was defined by the advanced technique of the local measurements of the ferromagnetic resonance field. Dynamic methods allowed investigation of the elements at different thicknesses by varying the frequency of the electromagnetic excitation. The maximum sensitivity of 40%/Oe with respect to the applied field in the range of the fields of 3 Oe to 5 Oe is promising for different applications.

Keywords: magnetic multilayers; permalloy; magnetic properties; magnetic domains; magnetostatic interactions; ferromagnetic resonance; magnetoimpedance; magnetic field sensors

Citation: Melnikov, G.Y.; Komogortsev, S.V.; Svalov, A.V.; Gorchakovskiy, A.A.; Vazhenina, I.G.; Kurlyandskaya, G.V. Effects of Magnetostatic Interactions in FeNi-Based Multilayered Magnetoimpedance Elements. *Sensors* **2024**, *24*, 6308. https://doi.org/10.3390/s24196308

Academic Editor: Arcady Zhukov

Received: 3 September 2024
Revised: 26 September 2024
Accepted: 27 September 2024
Published: 29 September 2024

1. Introduction

Magnetic soft multilayered structures are relevant in numerous technological applications. Modern electronics are developing in the direction of patterned multilayered films with different shaped electronic components [1–4]. One of the perspective applications is the fabrication of sensitive elements for detecting small magnetic fields up to the signals corresponding to the stray fields of magnetic labels or biomagnetic signals from living systems [5–8]. Magnetoimpedance (MI) is one of the most promising effects on which sensitive magnetic detectors can be based. Its advantage is its highest sensitivity with respect to external applied magnetic fields observed in a low field in the order of a few Oersted at room temperature [9–12].

The MI effect is the change of the ferromagnetic conductor impedance (Z) when high-frequency current flows through it and an external magnetic field is applied. This is a classical electrodynamic effect related to the dependence of the skin penetration depth (δ) on the dynamic magnetic permeability (μ) [13,14]. The skin penetration depth can be estimated as $\delta = (\pi f \sigma \mu)^{-0.5}$, where f is the driving current frequency, and σ is the conductivity of the material. As can be seen, frequency is a significant parameter; the lower frequency is more acceptable for technical applications. On the other hand, low frequency leads to a higher skin penetration depth; consequently, the thickness of the material should be higher. A high sensitivity of MI elements is observed for a frequency of the order of tens of MHz at a thickness of about half of the micron of the magnetic layers.

Permalloy ($Fe_{20}Ni_{80}$) is a magnetically soft material used in most sensor applications [14–17]. However, at a thickness of about 100 nm transition, an in "transcritical" state is observed. This state of permalloy is characterized by an increase in the coercivity H_c and particular features of the shape of the magnetic hysteresis loop, stripe magnetic domains formation, and appearance of "rotational anisotropy" [18,19]. To avoid this effect the concept of nanostructuring was proposed [18–20]. Nanostructuring is the separation of magnetic layers of permalloy by a conductive material [21,22]. Copper is frequently a useful material due to availability, high conductivity, and ease of etching in the course of the standard techniques of electronic component production. The thickness of the cooper spacers in MI multilayered structures has been analyzed previously and, for the present studies, we just took into account existing references [19–23]. However, the analysis of magnetostatic interactions between magnetic FeNi layers in $[Cu(3\ nm)/FeNi(100\ nm)]_5$ multilayered structures is absent in the literature.

Based on a previous research, reports of high magnetoimpedance effects are observed in symmetric structures, with two layers of $[Cu(3\ nm)/FeNi(100\ nm)]_5$ separated by 500 nm Cu [22–24]. The MI element requires a stripe shape for measurements in "microstripe" lines and induced magnetic anisotropy perpendicular to the short side of the stripe. In this case the highest magnetoimpedance is observed when a high frequency current and magnetic field are directed along the long side of the element [24,25]. Although magnetoimpedance is high for this type of structure, it is still lower in comparison with theoretical predictions One reason for the low MI effect is the inhomogeneity of the magnetic properties in the FeNi layers, i.e., the magnetic anisotropy field, the dispersion of induced magnetic anisotropy axes, and magnetization. Also, other complicated contributions (for instance, magnetostatic interactions between magnetic layers) can be essential. The asymmetry of the magnetic properties of the magnetic layers is due to a long deposition time and the presence of a thick Cu layer in the middle. Previously [26], we have shown that a 0.5 μm thick Cu lead has a rather high average grain size up to 50 nm in comparison with the typical size of 12–25 nm for a thin FeNi film of 100 nm. The grain size and texture of the Cu lead usually contribute to the structural peculiarities of the FeNi layer immediately above the Cu lead and, consequently, to the properties of the other layers of the top multilayered structure. As a result, the structure and the magnetic properties of the top and bottom multilayers differ from each other and the asymmetry of their properties become reflected in the decrease in MI. In a previous work [27], we suggested that the problem of magnetic property asymmetries related to thick Cu leads [26] can be partially solved using the nanostructuring of FeNi layers by Cu sublayers. However, this complex technological step requires further investigation.

Understanding magnetic properties with inhomogeneities like multilayered structures is an actual task. Static methods such as vibrating sample magnetometry (VSM) and Kerr microscopy give signals from a whole sample or a thin surface area (about 20 nm for a Kerr microscope). Indirect dynamic magnetic methods such as ferromagnetic resonance (FMR) and MI effect related to dynamic magnetic permeability allow set investigation thickness (skin penetration depth) by variation of frequency. Resonance spectra for FMR are analyzed by widely used Sul–Smith equations [28,29], while, for MI effects, a more complicated system using the Landau–Lifshitz equation with special boundary conditions is calculated [30–32]. These theoretical approximations estimate magnetic anisotropy fields, dispersion of induced magnetic anisotropy axes, magnetization, etc.

The measurements of the impedance Z and its active (R) and reactive (X) components ($Z(f) = R(f) - iX(f)$) were widely discussed in the literature [33,34]. However, for practical applications in certain conditions devices with active part R control working principle may have an advantage [35].

In this work, FeNi/Cu-based multilayered MI rectangular elements separated by Cu lead were obtained by a standard lift-off lithographic process. Elements were investigated by static (VSM, Kerr microscopy) and dynamic (FMR, MI effect) methods in order to estimate magnetic anisotropy field, dispersion of induced magnetic anisotropy axes, and magnetization at different thicknesses, revealing the crucial role of the magnetostatic interactions between FeNi layers in $[Cu(3\ nm)/FeNi(100\ nm)]_5$ multilayered structures.

2. Experiment

Multilayered $[Cu(3\ nm)/FeNi(100\ nm)]_5$/Cu(150 nm)/FeNi(10 nm)/Cu(150 nm)/FeNi(10 nm)/Cu(150 nm)/$[Cu(3\ nm)/FeNi(100\ nm)]_5$ MI elements were obtained by dc magnetron sputtering onto Corning glass substrates at room temperature. Parameters of deposition were next, with a background pressure of 3×10^{-7} mbar and a working argon pressure of 3.8×10^{-3} mbar. The thickness of the layers was defined by the deposition time using previously calibrated rates. For fabrication, a batch of the magnetoimpedance sensitive elements of a standard optical lift-off lithography was employed [36]. The obtained magnetoimpedance elements were configured with an open magnetic flux [26], i.e., the rectangular elements consisted of a number of layers of the same width and length (0.5 mm × 10.0 mm) organized as a vertical structure (Figure 1a).

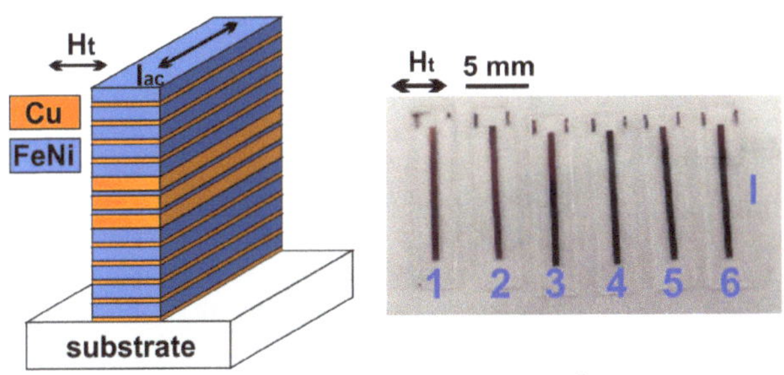

(a) (b)

Figure 1. Scheme of multilayered $[Cu(3\ nm)/FeNi(100\ nm)]_5$/Cu (150 nm)/[FeNi (3 nm)/Cu(150 nm)/[FeNi (3 nm)/Cu(150 nm)/$[FeNi(100\ nm)/Cu(3\ nm)\ ]_5$ element in magnetoimpedance geometry. Ht—direction of the application of technological magnetic field during multilayered structure deposition. Iac—direction of the flow of the high frequency not alternating current during magnetoimpedance applications. Note that the structures are shown in their real scale (**a**). Photograph of 1–6 lithographic MI element arrangements; l is a length of 10 mm (**b**).

Two batches of twelve elements (batch I and batch II) were arbitrarily selected for characterization by different techniques. Therefore, the denomination of the elements included the batch number (I or II), and position (from 1 to 6) (Figure 1b). MI element thicknesses for magnetic properties estimation were measured by sharp step and analyzed with a Dektak 150 stylus profilometer (Veeco, Somerset, NJ, USA). The thickness of all samples was (1.20 ± 0.04) μm. During the deposition process, an in-plane constant technological magnetic field Ht = 100 Oe was applied along the short side of the MI elements in order to induce a transverse uniaxial in-plane magnetic anisotropy.

Static magnetic properties measurements were carried out by the means of a 7407 VSM vibrating sample magnetometer (Lake Shore Cryotronics, London, UK) and a magneto-optical Kerr effect (MOKE) using a specialized optical microscope Evico (Evico, Dresden, Germany). The latter equipment was also used for the magnetic domain structure observation in different external magnetic fields applied in the plane of the MI element.

A rectangular multilayered MI sensitive element was placed into a "microstripe" line, which was contacted by highly conductive silver painting on the short sides. A uniform constant external magnetic field (H) up to 100 Oe was created by a pair of Helmholtz coils. It was applied along the long side of the rectangular element and therefore the longitudinal magnetoimpedance configuration was employed. The alternating current flowed parallel to the external magnetic field, providing the highest sensitivity of the MI ratio.

The total impedance was measured by a network analyzer (Agilent E8358A) in a frequency range of 0.1–400 MHz, with 1 mA amplitude of the excitation current across the multilayered element. The calibration and mathematical subtraction of the test fixture contribution procedures were performed in accordance with previously well-described procedures. The MI ratio and MI ratio sensitivity for the resistive component of impedance were calculated as follows:

$$\Delta R/R(H) = \frac{R(H) - R(H_{max})}{R(H_{max})} \cdot 100\% \quad (1)$$

where R(H) and $R(H_{max})$ are the resistances corresponding to the external magnetic fields H and H_{max}, respectively. The magnetic field sensitivity of the MI ratio, i.e., the change of the real component of the impedance ratio per unit of the external magnetic field, was determined by the following expression:

$$s(\Delta R/R) = \frac{\Delta R/R}{\Delta H} \quad (2)$$

where $\Delta H = 0.1$ Oe is the increment for an external magnetic field.

The ferromagnetic resonance (FMR) of the MI multilayered structures was studied on the basis of the measurements of absorption spectra by a home-made scanning FMR spectrometer (Kirensky Institute of Physics, Krasnoyarsk, Russia) using a microstrip resonator with a hole 0.8 mm in diameter [37]. This means that high frequency properties were measured from the local area of about 1 mm in diameter. FMR spectra were measured at a fixed frequency 1.3 GHz and at room temperature. The measurements were carried out with the direction of the external constant magnetic field Hc hanging in the film plane, i.e., in in-plane configuration (the angle between the magnetization M and the applied constant field was variable) (Figure 2).

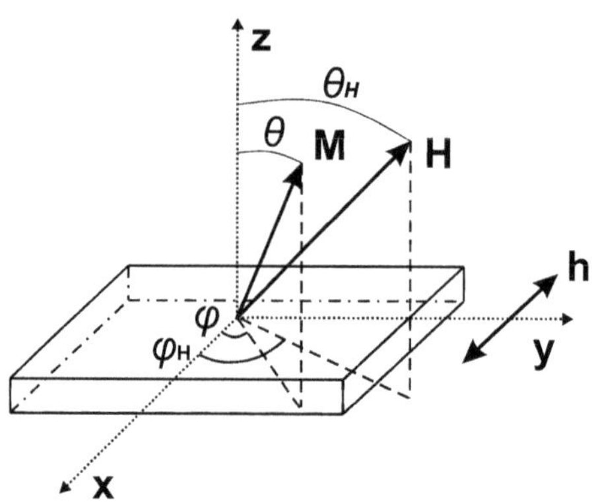

Figure 2. Scheme of FMR measurements. Here, M is the magnetization vector, H is an external constant magnetic field, and h is a microwave rf field. For definition of all angles, see also the main text.

In all configurations, the radio frequency alternating magnetic field (ac) magnetic field h was perpendicular to the external constant magnetic field, i.e., h~⊥ H. The longitudinal MI effect employed in this work (according to the literature it is the most useful configuration for applications) also dealt with the same configuration of the external magnetic field and the direction of the flow of the alternating current. FMR is an indirect method for determining the magnetic characteristic of magnetic thin film. It is based on the investigation of resonance field depending on the angle at the in-plane (θ_H = 90°) or out-of-plane (φ_H = const) configuration.

The resonance field at fixed θ_H and φ_H angles for homogeneous film could be defined by an equation of motion of magnetization:

$$\left(\frac{\omega}{\gamma}\right)^2 = \frac{1}{M_{eff}{}^2 \cdot sin^2(\theta)} \cdot \left(\frac{\partial^2 E}{\partial \theta^2} \cdot \frac{\partial^2 E}{\partial \varphi^2} - \left(\frac{\partial^2 E}{\partial \varphi \partial \theta}\right)^2\right) \quad (3)$$

Taking into account the magnetization equilibrium state [27–29].

$$\frac{\partial E}{\partial \varphi} = \frac{\partial E}{\partial \theta} = 0 \quad (4)$$

where $\gamma = 1.758 \times 10^7$ Hz/Oe is the gyromagnetic ratio; ω is the resonance frequency; E is the free energy density. The equation of free energy density includes the following: E_H is the energy of Zeeman; E_{nz} is the demagnetization field perpendicular to the plane; E_u is the induced uniaxial magnetic anisotropy in-plane ($K_u = H_a \times M_{eff}/2$); E_{nxy} is the energy of demagnetization in-plane of the film [38–40].

$$\begin{gathered} E = E_H + E_{nz} + E_u + E_{nxy} \\ E_H = -M_{eff} \cdot H \cdot (\sin(\theta) \cdot \sin(\theta_H) \cdot \cos(\varphi - \varphi_H) + \cos(\theta) \cdot \cos(\theta_H)) \\ E_{nz} = \frac{N_z}{2} \cdot M_{eff}{}^2 \cdot cos^2(\theta) \\ E_u = -K_u \cdot sin^2(\theta) \cdot cos^2(\varphi - \varphi_0) \\ E_{nxy} = \frac{M_{eff}{}^2}{2}(N_x \cdot cos^2(\varphi) + N_y \cdot sin^2(\varphi)) \cdot sin^2(\theta) \end{gathered} \quad (5)$$

where θ is theangle between the z-axis and magnetization; θ_H is the angle between the z-axis and the external magnetic field; H is the external magnetic field; M_{eff} is the effective magnetization; φ is the angle between the x-axis and magnetization; φ_H is the angle between the x-axis and the external magnetic field; φ_0 is the angle between the x-axis and the magnetic anisotropy axes; N_x, N_y, and N_z are the demagnetization factors; K_u is the constant of induced uniaxial magnetic anisotropy ($K_u = H_a \times M_{eff}/2$).

In this work, the in-plane configuration of FMR (θ_H = 90°) was measured. Therefore, the equilibrium direction of magnetization lay in-plane (θ = 90°) due to the influence of the demagnetization field being perpendicular to the plane. Varied parameters were effective magnetization M_{eff} and magnetic anisotropy field H_a. The parameters M_{eff} and H_a were varied until the experimental curve matched the approximation curve with an accuracy of 5% or less. Local measurements of ferromagnetic resonance field were made using a homemade scanning FMR spectrometer and a microstrip resonator with a hole 0.8 mm in diameter. This device allowed to measure the heterogeneities of the magnetic properties along the long side of MI elements.

The measurements of MI effects and FMR spectra were carried out at a high frequency, so it was important to take into account skin effect. The advantage of skin effect for magnetodynamic methods is the opportunity to measure magnetic properties of a local area of the sample with a thickness equal to the skin depth. Skin penetration depth can be estimated by Equation [41]:

$$\delta = c \cdot \sqrt{\frac{1}{4\pi^2 \cdot f \cdot \mu(f) \cdot \sigma}} \quad (6)$$

where σ is the conductivity, f is the current frequency, $\mu(f)$ is the transverse magnetic permeability at resonance frequency, and c is the speed of light.

Let us estimate the skin penetration depth for frequencies at which measurements were carried out. For this, evaluate transverse magnetic permeability for MI elements investigated in this paper (6): effective magnetization M_{eff} = 750 G, thickness d = 1 µm, permalloy conductivity $\sigma = 3 \times 10^{16}$ s^{-1}, Gilbert damping parameter κ = 0.02, magnetic anisotropy field Ha = 7 Oe, ψ -deviation of the magnetic anisotropy axis from the transverse direction [31].

$$\mu = 1 + \frac{\gamma 4\pi M_{eff}\left(\gamma 4\pi M_{eff} + \omega_1 - i\kappa\omega\right)sin^2\theta}{\left(\gamma 4\pi M_{eff} + \omega_1 - i\kappa\omega\right)(\omega_2 - i\kappa\omega) - \omega^2}, \quad (7)$$

$$\omega_1 = \gamma\left[H_a cos^2(\theta - \psi) + Hsin\theta\right]$$

$$\omega_2 = \gamma[H_a cos(2(\theta - \psi)) + Hsin\theta]$$

$$H_a sin(\theta - \psi)cos(\theta - \psi) = Hcos\theta$$

In the case of the MI effect, when the frequency of the current changes from 100 MHz to 400 MHz, the minimum skin depth varies from 600 nm to 300 nm and is observed in the field of magnetic anisotropy H_a = 7 Oe. With a further increase in frequency to 1.3 GHz in the case of FMR, the skin penetration depth in the resonance field from 15 Oe to 30 Oe is about 200 nm. Since, at the measurements of the MI effect and FMR, the penetration of the exciting electromagnetic field occurs from two surfaces of multilayered element structures, the signal actually comes from a thickness equal to twice the skin penetration depth.

3. Results and Discussion

3.1. Static Magnetic Properties

Static magnetic properties were investigated by a VSM magnetometer and a magneto optical Kerr microscope. The VSM magnetometer allowed to investigate the magnetic properties of the whole sample, including quantitative estimates of magnetization. The Kerr microscopy was associated with the properties of the surface layer in the case of FeNi and visualized the magnetic domain structure.

According to the VSM magnetic hysteresis loops, the MI elements had the coercive force value of about Hc = 1 Oe, and the anisotropy field was estimated as a field close to saturation Ha = 5 Oe. Although, during the deposition the technical magnetic field was applied along the short side of the elements, the easy magnetization axis (EMA) was along long side of the element. This behavior related to the shape anisotropy, which created an additional demagnetizing field directed along the short side of the MI elements. However, the axis of uniaxial magnetic anisotropy along the short side was confirmed by the Kerr microscope, which showed that the magnetization process occurred by magnetic domain wall displacement in this direction (Figure 3). To estimate the shape of the MI elements for calculating the demagnetizing field, an infinitely long prism was confirmed, with a thickness of t = 500 nm and width of w = 500 µm, for which the demagnetizing factors were N_x = 0.027; $N_y \approx 0$; $N_z \approx 12.539$ (N_x, along the short side; N_y, along the long side; N_z, perpendicular to the plane) [38,39].

This choice was due to the following factors. The length of the elements exceeded the width by several orders of magnitude, so the demagnetizing factor along the long side could be neglected, which corresponded to a prism with infinite length. In a multilayer structure separated by interlayers of Cu, the demagnetizing fields would be less than in a non-separated structure, i.e., the greater the thickness of the interlayers, the smaller these fields. In this case, it was assumed that the magnetostatic interaction between two multilayers separated by a thick Cu layer was negligible, and the magnetostatic interaction between the FeNi layers in each multilayer was the same as if they were not separated by Cu interlayers. Then, to estimate the demagnetizing fields, we could take the demagnetization factor for a thin film with a thickness of t = 500 nm equal to the thickness of one FeNi multilayer. It is worth noting that, for bodies of a non-ellipsoidal shape, the demagnetization factor depended in a complex way not only on the shape but also on the magnetic properties of the material, the magnetic state of the body, the distribution of magnetization in the sample, and the coordinates of the observation point. Therefore, for the estimation, we considered the state of magnetic saturation, and, by demagnetizing factor, we meant its average value over the volume (magnetometric demagnetizing factor) [38,39]. So, the estimation for the demagnetization field resulted in the value of about Hd = 22 Oe for the saturation state of the permalloy, with Ms = 820 Gs.

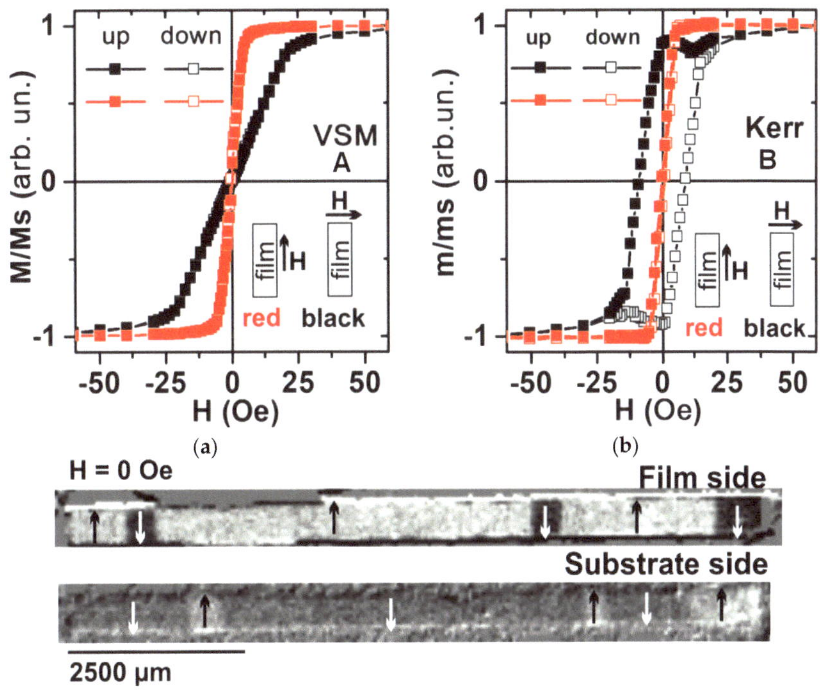

(a) (b)

(c)

Figure 3. VSM (**a**) and magneto-optical Kerr (**b**) in-plane magnetic hysteresis loops measured along the long (red curves) and short (black curves) sides of MI elements. Magnetic domain images obtained from both sides of the MI element A at zero magnetic field (**c**). Here "up" is for the measurements increasing in the external field starting from the saturation in the maximum negative field and "down" is for the measurements decreasing in the external field starting from the saturation in the maximum positive external field.

The magnetization process along the long side of the MI elements according to the Kerr microscopy occurred by magnetization rotation for the anisotropy field of about Ha = 7 Oe. The VSM and Kerr microscopy magnetic hysteresis loops for magnetization along the long side were similar. Thus, the magnetization processes mainly occurring by magnetization rotation were very close for the surface layer and the whole sample (Figure 3a,b, "red"). However, the magnetization process along the short side of the MI element was different for the whole sample (Figure 3a, "black") and the surface layer (Figure 3b, "black"), which pointed out the strong effects of magnetostatic interaction between FeNi layers and shape anisotropy.

First of all, at H = 0 Oe, according to the VSM measurements, the whole element had zero magnetization. However, according to the Kerr microscopy data, the individual magnetic FeNi layer was close to magnetic saturation along the short side (Figure 3a,b, "black"). This was possibly due to the "antiparallel" ordering of magnetic moments in the adjacent FeNi layers leading to a closed magnetic flux state.

It was confirmed by the magnetic domain images of the free FeNi layer and the FeNi layer closest to the glass substrate at zero magnetic field that they were almost magnetically saturated having an opposite direction of magnetization (Figure 3c). The above observed behavior indicated not only an induced magnetic anisotropy but also a magnetostatic interaction between layers played an important role in the formation of effective transverse magnetic anisotropy with the axis oriented along the short side, which was necessary fo the high sensitivity of the magnetoimpedance effect. Secondly, the magnetic hysteresi loop corresponding to the surface layer measured by Kerr microscopy had a "rectangular shape. However, it described a rather complex and "non-classical" magnetic hysteresi process. Re-magnetization began when the external magnetic field was still directed alon the current magnetization vector of the layer (Figure 4a).

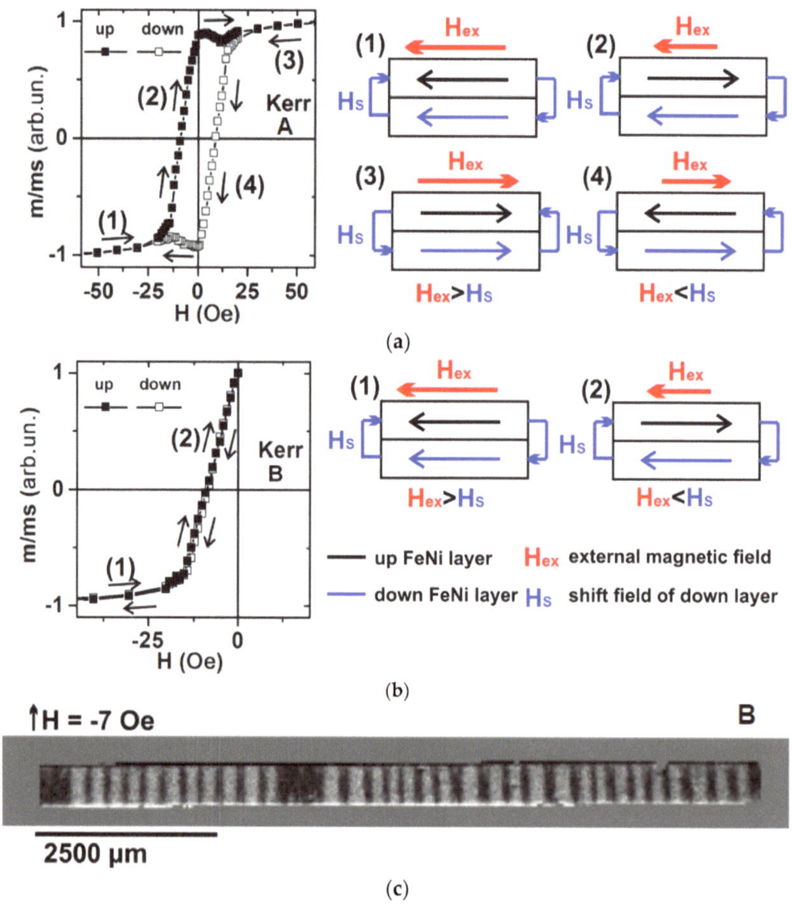

Figure 4. Kerr microscopy surface magnetic hysteresis loops measured along the short side of M elements with the scheme of magnetization of elements: (**a**) magnetization from −100 Oe to 100 O and back; (**b**) magnetization from −100 Oe to 0 Oe and back; (**c**) image of magnetic domains of M element corresponding to the external magnetic field H_{ext} = 7 Oe. Here "up" is for the measurement increasing in the external field starting from the saturation in the maximum negative field an "down" is for the measurements decreasing in the external field starting from the saturation in th maximum positive external field. Orientation of the external magnetic field is indicated by the re arrow; orientation of the magnetization of the top layer is indicated by the black arrow; orientation o the magnetization of the bottom layer is indicated by the blue arrow.

This feature was due to magnetic shift field *Hs*, which the FeNi bottom layer create as a result of the magnetostatic interaction, which became very important in the analyze [Cu(3 nm)/FeNi(100 nm)]$_5$ multilayered structure. The whole magnetization process ca be described as follows (Figure 4a). At point (1), the magnetizations of the top and botto FeNi layers were directed along the direction of the external magnetic field Hex. At poin

(2), when Hex became lower than Hs, the upper layer began with the re-magnetization process. At point (3), the value of the external field Hex became sufficient in order to change the orientation of the FeNi bottom layer, and the magnetization of the upper and bottom layers became directed along the direction of the external field. For the opposite direction of the external magnetic field, the above-described process was repeated, (point (4)). In the case under consideration, the magnetic field Hex = 7 Oe at which the magnetic moment equaled zero did not define the coercive force as for classic hysteresis; this value was close to the value of the shift field Hs. For the measurement of the coercive force using the magneto-optical Kerr microscopy, it was necessary to measure the magnetic hysteresis loop without the re-magnetization of the FeNi bottom layer. Figure 4b shows such a process. In this case, the coercivity Hc = 1 Oe had the same value as for the VSM measurements.

3.2. Dynamic Magnetic Properties

Figure 5 shows the experimental (points) and theoretical (line) dependences of the resonance fields for the MI elements denominated by A and B. The theoretical approximation was carried out with the point that the films had magnetic anisotropy along the short side of the elements according to the Kerr microscopy ($\varphi_0 = 0°$) (see also Figures 3 and 4). Demagnetization factors were chosen as follows: $N_x \approx 0.027$; $N_y \approx 0$; $N_z \approx 12.539$ (N_x, along the short side; N_y, along the long side; N_z, perpendicular to the plane). In this case, approximation could be reached for the next varied parameters (M_{eff} and H_a), i.e., for element A M_{eff} = 740 Gs, H_a = 10 Oe; for element B M_{eff} = 780 Gs, Ha = 14 Oe. This also confirmed the presence of induced magnetic anisotropy along the short side. The minimum of the resonance field was observed at the direction of the field being parallel to the long side of the element. This pointed out the dominant effect of the demagnetization field over the induced magnetic anisotropy at the magnetic saturation state. The Effective magnetization value was in an agreement with the values of magnetization carried out by VSM and Kerr microscopy, but the magnetic anisotropy was higher.

If we selected the demagnetizing factor N_x = 0.022, a good approximation could be reached for the next varied parameters (Meff and H_a), i.e., for element A M_{eff} = 740 Gs, H_a = 7 Oe; for element B M_{eff} = 780 Gs, H_a= 10 Oe. So, the value of the magnetic anisotropy field corresponded to the results of the Kerr microscopy and VSM magnetometry, while the effective magnetization remained the same. Thus, FMR can be used to estimate the demagnetization factor from other known quantities. The demagnetizing factor describing the demagnetization field could be lower due to the presence of Cu layers, which decreased the magnetostatic interaction between the FeNi layers with increasing thicknesses.

An interesting feature was observed at an angle when the external field was applied along the short side of the element ($\varphi = 180° \pm 5°$). There was a local minimum (Figure 5b). This complicated behavior could be explained by the following suggestions. The first suggestion is that, at a frequency of 1.3 GHz, the resonant field of about 32 Oe was not sufficient for magnetization saturation due to the influence of the demagnetization field (Figure 3a, "black"). Thus, the inhomogeneity of the resonance field was present along the element. The second suggestion is that the magnetostatic interaction between the permalloy layers (see the explanation in Section 3.1) made an additional contribution.

As the scanning FMR spectrometer consisted of a microstrip resonator with a hole of 0.8 mm in diameter, it allowed to measure the heterogeneities of the magnetic properties along the long side of the MI elements, with ane accuracy about the size of the hole. It can be seen in Figure 5c that, at the edges of the elements, the values of M_{eff} decreased and H_a increased. It could be supposed that this behavior was connected with the complicated distribution of the demagnetization fields at the edges.

The magnetoimpedance effect is another magnetodynamic effect connected with FMR. The difference between them is that energy absorption does not have resonant behavior and it is measured in the external field range near the field of saturation [42–44]. Energy absorption is related to the resistive component of impedance. Therefore, as the next step, we considered the real part of the MI study. The magnetoimpedance effect was measured in the "microstripe" line, in which the high frequency current flowed through the element. The current created an rf-magnetic field (h) like in the FMR measurements, and the resistance (energy absorption) was measured with the changing external static magnetic field (H). The parameters M_{eff} and H_a shown in Figure 5c were determined from the angular dependence of the FMR field taken from a certain region of the MI strip element. The position for each measurement is shown using the gray circles (Figure 6).

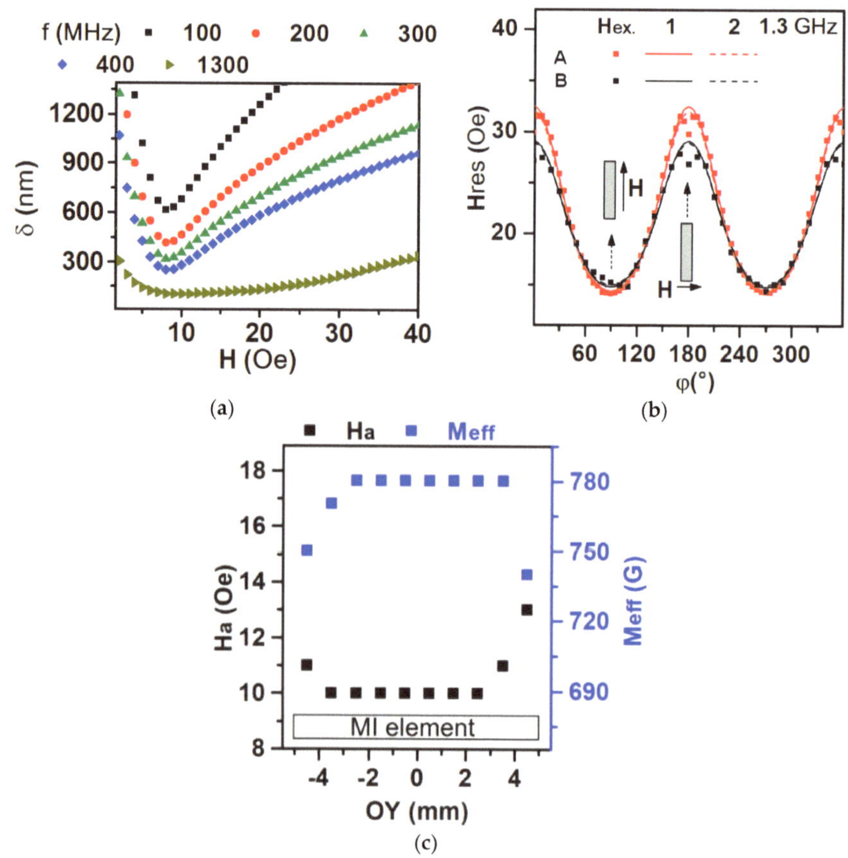

Figure 5. (**a**) Dependence of the skin penetration depth on the value of the magnetic field at different frequencies of the exciting electromagnetic field for the f values in the range of the experimental studies. (**b**) The angular dependence of the resonant field for MI elements A and B at a frequency of 1.3 GHz. Lines 1 and 2 are theoretical calculations of N_x = 0.027 and N_x = 0.022, respectively, points of the experiment. (**c**) Distribution of H_a and M_{eff} values along the long side of MI element A.

This angular dependence was calculated using the Stoner–Wohlfarth model, an essential assumption of which is the homogeneity of magnetization, and the uniform rotation of magnetization within the sample. For the sample as a whole, the resonance fields (15 ÷ 32 Oe) corresponded to the saturated state of the MI strip element (see Figure 3a). In the case of applying the external magnetic field along the long axis of the MI element, the orientational homogeneity of magnetization was significantly disrupted near the edges of the strip (Figure 6a, where the regions of inhomogeneous magnetization are colored blue and pink). The deviations of M_{eff} and H_a values near the strip edges (Figure 5c) were exactly related to this. It was evident that the region of non-uniform magnetization orientation extended approximately 1 mm from each edge of the stripe. For the point in the center of the strip (from which the angular dependence shown in Figure 5b was taken), the magnetization could be considered uniform in the angular range $\phi > 10 \div 90$ deg, but would be non-uniform at $\phi = 0 \div 10$ deg (in Figure 6b, it corresponded with the angle between the field and the long axis of the MI element, i.e., 90 deg − ϕ = 80 deg). In this angular range, the angular dependence of the resonant field (Figure 6b) exhibited a deviation from the line corresponding to the Stoner–Wohlfarth model. In the region between the strip edges (1 mm long), Figure 5c exhibits remarkable homogeneity of the absolute values of M_{eff} and H_a along the strip length. This means that, when detecting an external field by an element of a given geometry using MI, the direction of the external field along the long axis of the element is preferable, and the length of the element significantly exceeds the length of the edge effect by ~1 mm.

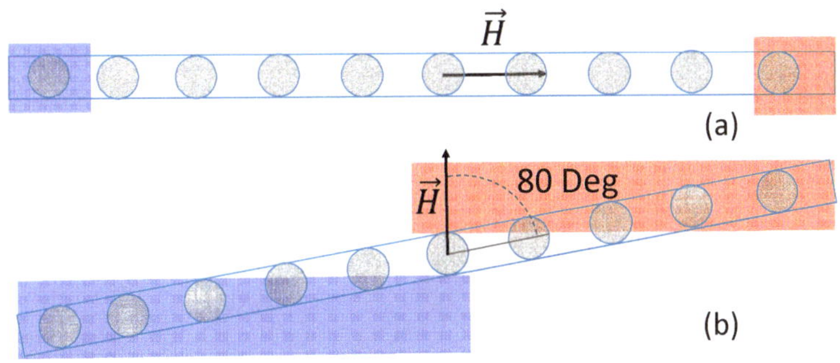

Figure 6. Schematic representation of the geometry of local FMR measurements for MI strip element: external magnetic field is applied along the long (**a**) and short (**b**) sides of the element. See also the main text.

The frequency behavior of magnetoimpedance was analyzed as a frequency dependence of the maximum value of the MI ratio ($\Delta R/R_{max}$) taken from $\Delta R/R(H)$ dependencies for each frequency. For the frequency of the 253 MHz value, the maximum of the magnetoimpedance effect was observed (Figure 7a). MI can be used as a method for the definition of an effective magnetic anisotropy field and an effective magnetic anisotropy direction.

The magnetic anisotropy field corresponded to the maximum MI ratio in the field dependence of the MI ratio for the real part of the impedance. The "two peaks" shape of the MI ratio field dependence corresponded with the magnetic anisotropy along the short side of the MI element [24]. At the frequency of 253 MHz, the magnetic anisotropy field H_a was close to 5 Oe and was directed along the short side of the elements (Figure 7). For the values of skin penetration depth in the range from 600 nm to 300 nm, the corresponding frequency range was from 85 MHz to 400 MHz (Figure 5a), where the magnetic anisotropy field was changed in the range from 5 Oe to 7 Oe (Figure 7c,d). This effect was connected with the inhomogeneity of the magnetic anisotropy field and the dispersion of the EMA over the volume of the elements. As the frequency increased, the skin penetration depth decreased, and the MI signal came from a smaller area near the surface. For example, in paper [31], theoretical calculations without the contribution of the inhomogeneity of the magnetic anisotropy gave non-shifted MI ratio curves.

This possibly pointed out that the surface of elements had a higher magnetic anisotropy field and dispersion of the EMA axis, which was also in accordance with the VSM measurements of H_a = 5 Oe for the whole sample (signal from about one μm thickness) corresponding to the MI measurements for a skin-penetration depth of 600 nm (85 MHz). From the skin-penetration depth of 300 nm (400 MHz), the MI measurements were H_a = 7 Oe corresponding to Kerr microscopy with H_a = 7 Oe, with a signal from thickness at about 20 nm and FMR with Ha: 7 Oe (A) and 10 Oe (B) for signal corresponding to a skin-penetration depth of 180 nm.

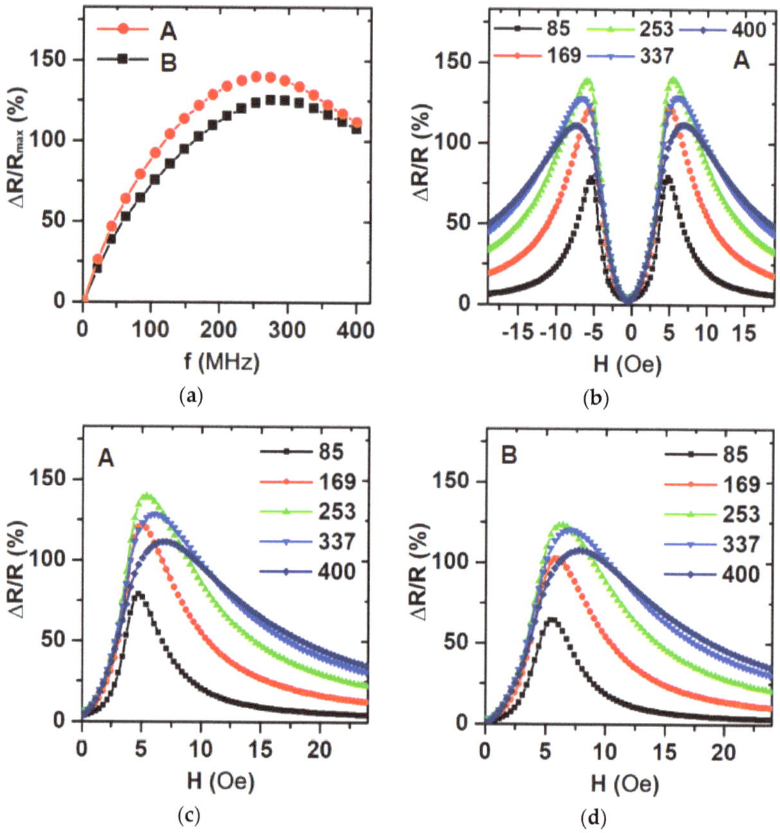

Figure 7. Frequency dependence of maximum MI ratio for real part of the impedance ratio (**a**); field dependence of MI ratio for resistance: (**b**,**c**) A, (**d**) B. Numbers in the legend correspond to the value of the frequency of the exciting current.

The method of optical-lift off lithography allowed to obtain big batches of MI elements with good repeatable properties, which is important for industrial mass production [27]. Fabricated MI elements showed a maximum MI ratio close to 130% at a frequency of 253 MHz. The maximum sensitivity of 40%/Oe at a frequency of 253 MHz in the range of the external fields from 3 Oe to 5 Oe is quite satisfactory for different applications [45–47]. In addition, it is worth mentioning that the MI properties of the elements of the batch (Figure 1a) were rather close to each other with an accuracy of at least 10%. This is a reasonable result, indicating the possibility of the application of a developed fabrication process for the mass production of MI elements under consideration.

As it mentioned above, in some cases, the operational properties for the real component of the impedance are preferable. In our previous studies [28], the magnetoimpedance effect was analyzed for MI elements of similar types but with a focus on total impedance variation $\Delta Z/Z$. The maximum MI values ($\Delta Z/Z_{max}$) of the magnetoimpedance ratio was close to 105% for 169 MHz frequency, i.e., the maximum for total impedance variation was lower in comparison with the real component changes. The maximum sensitivity of 30%/Oe for $\Delta Z/Z$ was also lower in comparison with $\Delta R/R$ values of about 40%/Oe. Thus, real component detection indeed may have an advantage in certain conditions.

Although we mainly considered the MI effect as a candidate for the possible sensor application of a designed sensitive element, these data can be useful for different electronic components where high magnetic permeability is desired. Properties of the FeNi rectangular elements with different width-to-length ratio and permalloy components with other geometries are currently under special interest in view of their applications as electronic components in many modern devices designed for technological applications [48–51]. Even more, to some extent, the obtained result can be used for the analysis of magnetostatic interactions in more complex MI composites when size and shape effects become very important or we are dealing with arrays of the elements [51–54].

4. Conclusions

Rectangular multilayered magnetoimpedance elements $[Cu(3\ nm)/FeNi(100\ nm)]_5/Cu(150\ nm)/FeNi(10\ nm)/Cu(150\ nm)/FeNi(10\ nm)/Cu(150\ nm)/[Cu(3\ nm)/FeNi(100\ nm)]_5$ were obtained by rf-sputtering and standard lift-off lithographic processes as batches of 12 long $(0.5 \times 10.0\ (mm^2))$ elements. In MI elements, there is competition between transverse magnetic anisotropy, which consist of induced magnetic anisotropy, and magnetostatic interaction between magnetic layers and longitudinal magnetic anisotropy due to shape magnetic anisotropy.

Magnetostatic properties of MI elements including analysis of the magnetic domain structure indicate the crucial importance of magnetostatic interactions in the analyzed $[Cu(3\ nm)/FeNi(100\ nm)]_5$ multilayered structures. Local parameters of ferromagnetic resonance were measured along the stripe lengths, showing that the deviations of the effective magnetization and the anisotropy field values near the strip edges are related to the orientational inhomogeneity of the magnetization of the MI strip element (being approximately 1 mm from each of the strip edges). The center of the strip can be considered uniform in the angular range $\phi > 10° \div 90°$, but at $\phi = 0° \div 10°$, the angular dependence of the resonant field exhibits a deviation from the Stoner–Wohlfarth model. In the region between the strip edges (central 8 mm long part), remarkable homogeneity of the absolute values of M_{eff} and H_a along the strip length is exhibited, i.e., the detection of an external magnetic field by an element of a given geometry, using magnetoimpedance, is preferable along the long axis of the element, and the length of the element should significantly exceed 2 mm, corresponding with the sum of the length of the edges of the inhomogeneities.

An obtained linear MI response with respect to the external magnetic field is not near the zero magnetic field, but has a bias effect. Previously, there were attempts to add an antiferromagnetic material to the FeNi multilayer in order to shift the linear magnetic field range to a near zero magnetic field [55]. However, technologically, the usage of a double element configuration might be easier. In any case, the direction to use biasing in MI planar systems is an interesting task to develop in the future.

The MI effect allows to investigate inhomogeneities of magnetic properties within multilayered structures at different thicknesses by varying skin penetration depth with current frequency. The skin penetration depth at FMR is fixed by resonance condition, but this method provides a remarkable quantitative estimation of magnetic inhomogeneities. The maximum sensitivity with respect to the external magnetic field being 40%/Oe at a frequency of 253 MHz in the range of the external fields of 3 Oe to 5 Oe is quite satisfactory for different applications, keeping in mind the advantages of active component R detection in many electronic devices.

Author Contributions: G.V.K., A.V.S., S.V.K. and G.V.K.; methodology, A.A.G., I.G.V. and A.V.S.; software, G.Y.M. and I.G.V.; validation, S.V.K., I.G.V. and G.V.K.; formal analysis, G.Y.M. and A.A.G.; investigation, A.V.S., I.G.V. and G.Y.M.; data curation, S.V.K. and G.V.K.; writing—original draft preparation, G.Y.M., G.V.K., I.G.V. and S.V.K.; writing—review and editing, G.V.K., I.G.V., G.Y.M. and S.V.K.; visualization, G.Y.M. and S.V.K.; supervision, G.V.K. All authors discussed the results and implications and commented on the manuscript at all stages. All authors have read and agreed to the published version of the manuscript.

Funding: This research was financially supported by the Ministry of Science and Higher Education of the Russian Federation, project no FEUZ-2023-0020.

Institutional Review Board Statement: This work did not involve humans or animals and therefore it did not require the Institutional Review Board Statement and approval.

Informed Consent Statement: Not applicable.

Data Availability Statement: Data available from the corresponding author upon reasonable request.

Acknowledgments: The authors acknowledge the possibility to use the Krasnoyarsk Regional Center of Research Equipment of the Federal Research Center "Krasnoyarsk Science Center SB RAS" for ferromagnetic resonance studies. We thank V.N. Lepalovskij for useful discussions and special support.

Conflicts of Interest: The authors declare no conflicts of interest.

References

1. Yang, Y.; Liu, T.; Bi, L.; Deng, L. Recent advances in development of magnetic garnet thin films for applications in spintronics and photonics. *J. Alloys Compd.* **2021**, *860*, 158235. [CrossRef]
2. Yachmeneva, A.E.; Pushkareva, S.S.; Reznikc, R.R.; Khabibullina, R.A.; Ponomareva, D.S. Arsenides-and related III-V materials-based multilayered structures for terahertz applications: Various designs and growth technology. *Prog. Cryst. Growth Charact. Mater.* **2020**, *66*, 100485. [CrossRef]
3. Pan, L.; Xie, H.; Cheng, X.; Zhao, C.; Feng, H.; Cao, D.; Wang, J.; Liu, Q. Tuning the ferromagnetic resonance frequency of soft magnetic film by patterned permalloy micro-stripes with stripe-domain. *J. Magn. Magn. Mater.* **2018**, *457*, 46–51. [CrossRef]
4. Melzer, M.; Kaltenbrunner, M.; Makarov, D.; Karnaushenko, D.; Karnaushenko, D.; Sekitani, T.; Someya, T.; Schmidt, O.G. Imperceptible magnetoelectronics. *Nat. Commun.* **2015**, *6*, 6080. [CrossRef]
5. Nabaei, V.; Chandrawati, R.; Heidari, H. Magnetic biosensors: Modelling and simulation. *Biosens. Bioelectron.* **2018**, *103*, 69–86. [CrossRef]
6. Semirov, A.V.; Derevyanko, M.S.; Bukreev, D.A.; Moiseev, A.A.; Kudryavtsev, V.O.; Safronov, A.P. Magnetoimpedance of cobalt-based amorphous ribbons/polymer composites. *J. Magn. Magn. Mater.* **2016**, *415*, 97–101. [CrossRef]
7. Sobczak-Kupiec, A.; Venkatesan, J.; AlAnezi, A.A.; Walczyk, D.; Farooqi, A.; Malina, D.; Hosseinie, S.H.; Tyliszczak, B. Magnetic nanomaterials and sensors for biological detection. *Nanomed. Nanotechnol. Biol. Med.* **2016**, *12*, 2459–2473. [CrossRef]
8. Ren, C.; Bayin, Q.; Feng, S.; Fu, Y.; Ma, X.; Guo, J. Biomarkers detection with magnetoresistance-based sensors. *Biosens. Bioelectron.* **2020**, *165*, 112340. [CrossRef]
9. Wang, T.; Zhou, Y.; Lei, C.; Luo, J.; Xie, S.; Pu, H. Magnetic impedance biosensor: A review. *Biosens. Bioelectron.* **2017**, *90*, 418–435. [CrossRef]
10. Wang, T.; Chen, Y.Y.; Wang, B.C.; He, Y.; Li, H.Y.; Liu, M.; Rao, J.J.; Wu, Z.Z.; Xie, S.R.; Luo, J. A giant magnetoimpedance-based separable-type method for supersensitive detection of 10 magnetic beads at high frequency. *Sens. Actuators A Phys.* **2019**, *300*, 111656. [CrossRef]
11. Buznikov, N.A.; Safronov, A.P.; Orue, I.; Golubeva, E.V.; Lepalovskij, V.N.; Svalov, A.V.; Chlenova, A.A.; Kurlyandskaya, G.V. Modelling of magnetoimpedance response of thin film sensitive element in the presence of ferrogel: Next step toward development of biosensor for in-tissue embedded magnetic nanoparticles detection. *Biosens. Bioelectron.* **2018**, *117*, 366–472. [CrossRef] [PubMed]
12. Fodil, K.; Denoual, M.; Dolabdjian, C.; Harnois, M.; Senez, V. Dynamic sensing of magnetic nanoparticles in microchannel using GMI technology. *IEEE Trans. Magn.* **2013**, *49*, 93–96. [CrossRef]
13. Correa, M.A.; Bohn, F.; Chesman, C.; da Silva, R.B.; Viegas, A.D.C.; Sommer, R.L. Tailoring the magnetoimpedance effect of NiFe/Ag multilayer. *J. Phys. D Appl. Phys.* **2010**, *43*, 295004–295007. [CrossRef]
14. Antonov, A.S.; Gadetskii, S.N.; Granovskii, A.B.; D'yachkov, A.L.; Paramonov, V.P.; Perov, N.S.; Prokoshin, A.F.; Usov, N.A.; Lagar'kov, A.N. Giant magnetoimpedance in amorphous and manocrystalline multilayers. *Fiz. Met. Metalloved.* **1997**, *83*, 60–71.
15. Tumanski, S. *Handbook of Magnetic Measurements*; CRC Press: Boca Raton, FL, USA, 2011; Volume 719. [CrossRef]
16. Hida, R.; Falub, C.V.; Perraudeau, S.; Morin, C.; Favier, S.; Mazel, Y.; Saghi, Z.; Michel, J.P. Nanolaminated FeCoB/FeCo and FeCoB/NiFe soft magnetic thin films with tailored magnetic properties deposited by magnetron sputtering. *J. Magn. Magn. Mater.* **2018**, *453*, 211–219. [CrossRef]
17. Naumova, L.I.; Milyaev, M.A.; Zavornitsyn, R.S.; Pavlova, A.Y.; Maksimova, I.K.; Krinitsina, T.P.; Chernyshova, T.A.; Proglyado, V.V.; Ustinov, V.V. High-sensitive sensing elements based on spin valves with antiferromagnetic interlayer coupling. *Phys. Met. Metallogr.* **2019**, *120*, 653–659. [CrossRef]
18. Sugita, Y.; Fujiwara, H.; Sato, T. Critical thickness and perpendicular anisotropy of evaporated permalloy films with stripe domains. *Appl. Phys. Lett.* **1967**, *10*, 229–231. [CrossRef]
19. Svalov, A.V.; Aseguinolaza, I.R.; Garcia-Arribas, A.; Orue, I.; Barandiaran, J.M.; Alonso, J.; Fernández-Gubieda, M.L.; Kurlyandskaya, G.V. Structure and magnetic properties of thin Permalloy films near the "transcritical" state. *IEEE Trans. Magn.* **2010**, *46*, 333–336. [CrossRef]
20. Egelhoff, W.F.; Bonevich, J.; Pong, P.; Beauchamp, C.R.; Stafford, G.R.; Unguris, J.; McMichael, R.D. 400-fold reduction in saturation field by interlayering. *J. Appl. Phys.* **2009**, *105*, 013921. [CrossRef]

21. Vas'kovskii, V.O.; Savin, P.A.; Volchkov, S.O.; Lepalovskii, V.N.; Bukreev, D.A.; Buchkevich, A.A. Nanostructuring effects in soft magnetic films and film elements with magnetic impedance. *Tech. Phys.* **2013**, *58*, 105–110. [CrossRef]
22. Svalov, A.V.; Gonzalez Asensio, B.; Chlenova, A.A.; Savin, P.A.; Larranaga, A.; Gonzalez, J.M.; Kurlyandskaya, G.V. Study of the effect of the deposition rate and seed layers on structure and magnetic properties of magnetron sputtered FeNi films. *Vacuum* **2015**, *119*, 245–249. [CrossRef]
23. Kurlyandskaya, G.V.; Svalov, A.V.; Fernandez, E.; Garcia-Arribas, A.; Barandiaran, J.M. FeNi-based magnetic layered nanostructures: Magnetic properties and giant magnetoimpedance. *J. Appl. Phys.* **2010**, *107*, 09C502. [CrossRef]
24. Cos, D.; Barandiarán, J.M.; García-Arribas, A.; Vas'kovskiy, V.O.; Kurlyandskaya, G.V. Longitudinal and Transverse Magnetoimpedance in FeNi/Cu/FeNi Multilayers with Longitudinal and Transverse Anisotropy. *IEEE Trans. Magn.* **2008**, *44*, 3863–3866. [CrossRef]
25. Kikuchi, H.; Sumida, C.; Nakai, T.; Hashi, S.; Ishiyama, K. Effects of dc bias current on behaviors and sensitivity of thin-film magnetoimpedance element. *IEEE Trans. Magn.* **2017**, *53*, 4003704. [CrossRef]
26. Kurlyandskaya, G.V.; Chlenova, A.A.; Fernández, E.; Lodewijk, K.J. FeNi-based flat magnetoimpedance nanostructures with open magnetic flux: New topological approaches. *J. Magn. Magn. Mater.* **2015**, *383*, 220–225. [CrossRef]
27. Melnikov, G.Y.; Vazhenina, I.G.; Iskhakov, R.S.; Boev, N.M.; Komogortsev, S.V.; Svalov, A.V.; Kurlyandskaya, G.V. Magnetic Properties of FeNi/Cu-Based Lithographic Rectangular Multilayered Elements for Magnetoimpedance Applications. *Sensors* **2023**, *23*, 6165. [CrossRef] [PubMed]
28. Komogortsev, S.V.; Vazhenina, I.G.; Kleshnina, S.A.; Iskhakov, R.S.; Lepalovskij, V.N.; Pasynkova, A.A.; Svalov, A.V. Advanced Characterization of FeNi-Based Films for the Development of Magnetic Field Sensors with Tailored Functional Parameters. *Sensors* **2022**, *22*, 3324. [CrossRef]
29. Vazhenina, I.G.; Iskhakov, R.S.; Yakovchuk, V. Characteristics of Angular Dependences of Parameters of Ferromagnetic and Spin-Wave Resonance Spectra of Magnetic Films. *Phys. Met. Metallogr.* **2022**, *123*, 1084–1090. [CrossRef]
30. Antonov, A.S.; Iakubov, I.T. The high-frequency magneto-impedance of a sandwich with transverse magnetic anisotropy. *J. Phys. D Appl. Phys.* **1999**, *32*, 1204–1208. [CrossRef]
31. Buznikov, N.A.; Kurlyandskaya, G.V. Modeling of magnetoimpedance effect in nanostructured multilayered films. *J. Phys. Conf. Ser.* **2019**, *1389*, 012132. [CrossRef]
32. Buznikov, N.A.; Antonov, A.S. A model for asymmetric magnetoimpedance effect in multilayered bimagnetic films. *J. Magn. Magn. Mater.* **2016**, *420*, 51–55. [CrossRef]
33. Delooze, P.; Panina, L.V.; Mapps, D.J.; Ueno, K.; Sano, H. Effect of transverse magnetic field on thin-film magneto impedance and application to magnetic recording. *J. Magn. Magn. Mater.* **2004**, *272–276*, 2266–2268. [CrossRef]
34. Ripka, P.; Zaveta, K. Magnetic Sensors: Principles and Applications. In *Handbook of Magnetic Materials*; Buschow, K.H.J., Ed.; Elsevier: Amsterdam, The Netherlands, 2009; Volume 18, pp. 347–420.
35. de Cos, D.; García-Arribas, A.; Barandiarán, J.M. Analysis of magnetoimpedance measurements at high frequency using a microstrip transmission line. *Sens. Actuators A* **2004**, *115*, 368–375. [CrossRef]
36. Moritz, H. Optical single layer lift-off process. *IEEE Trans. Electron Devices* **1985**, *32*, 672–676. [CrossRef]
37. Belyaev, B.A.; Gorchakovskii, A.A.; Boev, N.M.; Izotov, A.I.; Shabanov, D.A. Ferromagnetic Resonance Scanning Spectrometer. Patent RU 2747100C1, 26 April 2021. Available online: https://new.fips.ru/registers-doc-view/fips_servlet?DB=RUPAT&DocNumber=2747100&TypeFile=html (accessed on 2 September 2024).
38. Chen, D.-X.; Pardo, E.; Sanche, A. Demagnetizing factors for rectangular prisms. *IEEE Trans. Magn.* **2005**, *41*, 2077–2088. [CrossRef]
39. Aharoni, A. Demagnetizing factors for rectangular ferromagnetic prisms. *J. Appl. Phys.* **1998**, *83*, 3432–3434. [CrossRef]
40. Shcherbinin, S.V.; Svalov, A.V.; Melnikov, G.Y.; Kurlyandskaya, G.V. Angular Dependence of the Ferromagnetic Resonance Parameters of $[Ti/FeNi]_6/Ti/Cu/Ti/[FeNi/Ti]_6$ Nanostructured Multilayered Elements in the Wide Frequency Range. *Nanomaterials* **2020**, *10*, 433. [CrossRef]
41. Antonov, A.; Gadetsky, S.; Granovsky, A.; D'yatckov, A.; Sedova, M.; Perov, N.; Usov, N.; Furmanova, T.; Lagar'kov, A. High-frequency giant magneto-impedance in multilayered magnetic films. *Phys. A Stat. Mech. Its Appl.* **1997**, *241*, 414–419. [CrossRef]
42. Jayasekara, W.P.; Bain, J.A.; Kryder, M.H. High frequency initial permeability of NiFe and FeAlN. *IEEE Trans. Magn.* **1998**, *34*, 1438–1440. [CrossRef]
43. Yelon, A.; Menard, D.; Britel, M.; Ciureanu, P. Calculations of giant magnetoimpedance and of ferromagnetic resonance response are rigorously equivalent. *Appl. Phys. Lett.* **1996**, *69*, 3084–3085. [CrossRef]
44. Valenzuela, R.; Zamorano, R.; Alvarez, G.; Gutierrez, M.P.; Montiel, H. Magnetoimpedance, ferromagnetic resonance, and low field microwave absorption in amorphous ferromagnets. *J. Non-Cryst. Solids* **2007**, *353*, 768–772. [CrossRef]
45. Grosz, A.; Haji-Sheikh, M.J.; Mukhopadhyay, S.C. *High Sensitivity Magnetometers*; Springer: Berlin/Heidelberg, Germany, 2017; Volume 19. [CrossRef]
46. Buznikov, N.A.; Svalov, A.V.; Kurlyandskaya, G.V. Influence of the parameters of permalloy-based multilayer film structures on the sensitivity of magnetic impedance effect. *Phys. Met. Metallogr.* **2021**, *122*, 223–229. [CrossRef]
47. Feng, Z.; Zhi, S.; Wei, M.; Zhou, Y.; Liu, C.; Lei, C. An integrated three-dimensional micro-solenoid giant magnetoimpedance sensing system based on MEMS technology. *Sens. Actuators A Phys.* **2019**, *299*, 111640. [CrossRef]

48. Solovev, P.N.; Belyaev, B.A.; Boev, N.M.; Skomorokhov, G.V.; Izotov, A.V. Magnetic anisotropy and ferromagnetic resonance in inhomogeneous demagnetizing fields near edges of thin magnetic films. *J. Phys. Cond. Matt.* **2024**, *36*, 195803. [CrossRef] [PubMed]
49. Malagò, P.; Lumetti, S.; Holzmann, D.; Ortner, M.; Roshanghias, A. Magnetic field sensors for non-invasive current monitoring in wire-bond-less power modules. *Proceedings* **2024**, *97*, 100. [CrossRef]
50. Zare, M.; Jamilpanah, L.; Barough, V.; Sadeghi, A.; Ghanaatshoar, M.; Mohseni, M. Role of electrospun fibers coated on magnetoimpedance effect of Co-based ribbons. *Appl. Phys. A Solids Surf.* **2024**, *130*, 90. [CrossRef]
51. Zhang, S.; Jiang, Q.; Yanfeng, Y. Design of a novel magnetic induction switch with a permalloy film and a trans-impedance amplifier circuit. *Inventions* **2024**, *9*, 4. [CrossRef]
52. Amalou, F.; Gijs, M.A.M. Giant magnetoimpedance in trilayer structures of patterned magnetic amorphous ribbons. *Appl. Phys. Lett.* **2002**, *81*, 1654–1656. [CrossRef]
53. Mu, Y.; Zhu, W.; Zhang, M.; Yi, Y. High-sensitivity giant magneto-impedance microsensor with grooved grating patterned magnetic films. *IEEE Trans. Magn.* **2024**, *24*, 25422–25429. [CrossRef]
54. Karnaushenko, D.; Karnaushenko, D.D.; Makarov, D.; Baunack, S.; Schäfer, R.; Schmidt, O.G. Self-assembled on-chip-integrated giant magneto-impedance sensorics. *Adv. Mater.* **2015**, *27*, 6582–6589. [CrossRef]
55. García, C.; Florez, J.M.; Vargas, P.; Ross, C.A. Effect of the exchange bias coupling strength on the magnetoimpedance of IrMn/NiFe films. *J. Appl. Phys.* **2011**, *109*, 07D735. [CrossRef]

sensors

MDPI

Article

Analysis of Relationship between Microwave Magnetic Properties and Magnetic Structure of Permalloy Films

Nikita A. Buznikov *, Andrey N. Lagarkov, Sergey A. Maklakov, Sergey S. Maklakov, Alexey V. Osipov, Konstantin N. Rozanov and Polina A. Zezyulina

Institute for Theoretical and Applied Electromagnetics, Russian Academy of Sciences, Moscow 125412, Russia; n.a.buznikov@gmail.com (A.N.L.); sergeymaklakov@yandex.ru (S.A.M.); squirrel498@gmail.com (S.S.M.); avosipov@mail.ru (A.V.O.); k.rozanov@yandex.ru (K.N.R.); zez-p@yandex.ru (P.A.Z.)

* Correspondence: n_buznikov@mail.ru

Abstract: Changes in the microwave permeability of permalloy films with an increase in the film thickness are studied. Measurement data on the evolution of microwave permeability with film thickness are analyzed in the framework of a model for the film with a regular stripe domain structure and out-of-plane magnetic anisotropy. A correlation between the microwave magnetic properties and magnetic structure of permalloy films is established. It is demonstrated that the observed decrease in the ferromagnetic resonance frequency and the static permeability with a growth in the film thickness can ascribed to the appearance of perpendicular anisotropy and the formation of a stripe domain structure. The calculated dependences of the ferromagnetic resonance frequency and static permeability on the film thickness are in reasonable agreement with the measurement results. Based on the analysis of these dependences, the domain width in the permalloy films is estimated. It is found that for thick permalloy films, the domain width is of the order of the film thickness. The results obtained may be useful for high-frequency applications of soft magnetic films.

Keywords: permalloy films; microwave permeability; ferromagnetic resonance; out-of-plane anisotropy; stripe domain structure

Citation: Buznikov, N.A.; Lagarkov, A.N.; Maklakov, S.A.; Maklakov, S.S.; Osipov, A.V.; Rozanov, K.N.; Zezyulina, P.A. Analysis of Relationship between Microwave Magnetic Properties and Magnetic Structure of Permalloy Films. *Sensors* **2024**, *24*, 6165. https://doi.org/10.3390/s24196165

Academic Editor: Evangelos Hristoforou

Received: 2 August 2024
Revised: 16 September 2024
Accepted: 22 September 2024
Published: 24 September 2024

1. Introduction

In recent decades, thin soft magnetic films have been studied extensively due to their possible use in high-frequency applications, such as magnetic inductors [1–10], patch antennas [7,8], micro-transformers [3], magnetoelastic sensors [2,9], tunable microwave filters [5,6,10], electromagnetic shielding [11], etc. In these applications, high values of microwave permeability are required. The microwave permeability of a material may be estimated based on the static permeability μ_s and the ferromagnetic resonance frequency f_{res}. To achieve high values of the microwave permeability, both μ_s and f_{res} should be as high as possible.

For bulk magnetic materials, the product of these values is limited by the Snoek law [12]. Soft magnetic films are promising for obtaining high microwave permeability, since the product of the static permeability and ferromagnetic resonance frequency in the films may exceed the Snoek limit [13–15]. For thin magnetic films, the microwave permeability is evaluated by using the Acher parameter:

$$k_A = \frac{(\mu_s - 1) f_{res}^2}{(\gamma 4\pi M_0)^2}, \tag{1}$$

where M_0 is the saturation magnetization of the material and $\gamma \approx 3$ GHz/kOe is the gyromagnetic ratio.

For a uniformly magnetized film with in-plane magnetic anisotropy, k_A tends to unity. In real magnetic films, the Acher parameter may decrease ($k_A < 1$) as a result of deviations from the uniform magnetic structure, in particular, due to the appearance of out-of-plane magnetic anisotropy and a domain structure [15,16]. The model for a magnetic film having a stripe domain structure and an arbitrary angle between the anisotropy axis and the film plane predicts a decline in microwave permeability [17]. This prediction is in a qualitative agreement with the results of the measurements of microwave permeability in permalloy [18] and cobalt films [19] with different thicknesses.

Permalloy films attract much attention due to their low coercivity, high static permeability and low magnetocrystalline anisotropy, resulting in high microwave permeability [20–25]. The excellent soft magnetic properties of permalloy films are very promising for sensors applications, in particular, for the development of giant magnetoimpedance sensors [26–29]. However, out-of-plane anisotropy can appear in permalloy films in cases when the film thickness exceeds some critical value [20,30–38]. The appearance of out-of-plane anisotropy may be related to columnar structure formation as well as magnetocrystalline and magnetoelastic anisotropy. The presence of out-of-plane anisotropy in a film can lead to a transition into a transcritical state. The transcritical state is characterized by a specific hysteresis loop (so-called transcritical loop) [39], enhanced coercivity and the formation of a stripe domain structure in the film. As a result, the microwave permeability of a film is significantly reduced after its transition into a transcritical state.

This paper deals with a quantitative comparison of the results predicted by the model in [17] with measurement data for the microwave permeability of permalloy films. Based on the microwave permeability data, changes in the magnetic structure of permalloy films appearing with an increase in the film thickness are analyzed. It is found that the observed decrease in the static permeability and the ferromagnetic resonance frequency is related to the appearance of perpendicular magnetic anisotropy and formation of a stripe domain structure in the films. The width of the domains in the films is estimated by the analysis of the dependence of the Acher parameter on the film thickness.

2. Model

In this section, we briefly describe the main results obtained by using a previously proposed model [17]. It was assumed that the film has uniaxial anisotropy, and the anisotropy axis makes the angle ψ with the film plane. It was also assumed that the film has a regular stripe domain structure. The motion of the domain walls was neglected. The geometry and angles used in the model are presented schematically in Figure 1.

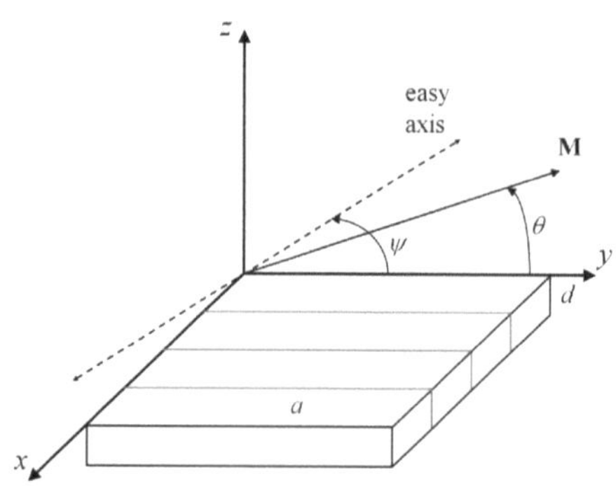

Figure 1. A sketch of the geometry used for analysis. A film of thickness d is in the x–y plane, and the anisotropy axis and equilibrium magnetization vector **M** are in the y–z plane. The angle between the anisotropy axis and the film plane is ψ, and the width of the domains is a.

Note that in general case, the stripe domain structure arises in films as a result of competition between perpendicular magnetic anisotropy, exchange interactions and magnetostatic energy. It is well known that the stripe domain structure exists if the film thickness exceeds some critical value [40–42]. This means that there is no stripe domain structure in very thin ferromagnetic films.

For films with out-of-plane anisotropy, the effect of the stripe domain structure on the demagnetizing factor N_z in the direction transverse to the film plane is described in [43–45].

$$N_z = \frac{32(1+Q^{-1})^{1/2}}{\pi q} \times \sum_{j=1}^{\infty} \frac{1}{(2j-1)^3} \times \frac{1}{1+(1+Q^{-1})^{1/2}\coth[(2j-1)q]} \quad (2)$$

Here, $q = (\pi d/2a)(1 + Q^{-1})^{1/2}$; d is the film thickness; a is the domain width; $Q = H_a/4\pi M_0$; H_a is the anisotropy field. For very thin films, for which $a/d >> 1$, the demagnetizing factor tends to 4π as in the case of the single-domain film. The value of N_z decreases monotonically with the domain aspect ratio a/d, and at $q >> 1$, the demagnetizing factor N_z is expressed as in [17]:

$$N_z \approx \frac{56\zeta(3)}{\pi^2} \times \frac{a/d}{1+(1+Q^{-1})^{1/2}}, \quad (3)$$

where $\zeta(3) \approx 1.20$ is the Riemann ζ function of 3.

The equilibrium magnetization angle θ with respect to the film plane was found by minimizing the free energy that consists of anisotropy and demagnetizing energy. In general case, the magnetization angle θ is given by

$$\tan 2\theta = \frac{(4\pi Q/N_z)\sin 2\psi}{1+(4\pi Q/N_z)\cos 2\psi}. \quad (4)$$

It should be noted that for soft magnetic films with relatively wide domains, where $4\pi Q/N_z << 1$, the equilibrium magnetization angle θ is small [17]. In this case, the magnetization deviates slightly from the film plane, and Equation (4) is rewritten as

$$\theta \approx (2\pi Q/N_z)\sin 2\psi\,[1-(4\pi Q/N_z)\cos 2\psi] \quad (5)$$

The ferromagnetic resonance frequency $f_{\rm res}$ was found by solving the linearized Landau–Lifshitz–Gilbert equation. In the framework of the model, the resonance frequency is defined as the frequency where the real part of the permeability $\mu' = 1$. The resonance frequency $f_{\rm res}$ depends on the anisotropy axis deviation angle ψ, the factor Q and the domain aspect ratio a/d and is expressed as

$$\begin{aligned} f_{\rm res}^2/(\gamma 4\pi M_0)^2 = {} & (N_z/4\pi)^2\sin^2\theta + Q(N_z/4\pi)[\cos^2\psi - 2\sin\psi\sin\theta\cos(\psi-\theta)] \\ & +Q^2\cos^2(\psi-\theta)\,. \end{aligned} \quad (6)$$

For soft magnetic films, for which $Q << 1$, with relatively wide domains, when $\theta << 1$, Equation (6) is simplified as follows:

$$f_{\rm res}^2/(\gamma 4\pi M_0)^2 \approx Q\cos^2\psi\,[(N_z/4\pi)+Q]. \quad (7)$$

It follows from Equation (7) that the ferromagnetic resonance frequency $f_{\rm res}$ decreases monotonically with a growth in the anisotropy axis angle ψ and with a decrease in the domain aspect ratio a/d [17].

The value of the static permeability $\mu_{\rm s}$ is found from the general expression for permeability at zero frequency. In general cases, the static permeability is written as

$$\mu_{\rm s} = 1+(\gamma 4\pi M_0)^2[(N_z/4\pi)\cos 2\theta + Q\cos 2(\psi-\theta)]/f_{\rm res}^2. \quad (8)$$

From Equations (1), (6) and (8), the Acher parameter $k_{\rm A}$ is given by

$$k_{\rm A} = (N_z/4\pi)\cos 2\theta + Q\cos 2(\psi-\theta). \quad (9)$$

If the deviation of the equilibrium magnetization angle is small, $\theta << 1$, the Ache parameter is simplified as

$$k_A \approx (N_z/4\pi) + Q\cos 2\psi. \quad (1C$$

Since $Q << 1$, the Acher parameter depends slightly on the deviation of the anisotrop axis from the film plane and is mainly governed by the domain aspect ratio a/d.

3. Experimental

Thin permalloy films with a nominal composition of $Ni_{80}Fe_{20}$ on a polyethylene tereph thalate (PET) substrate were obtained by DC magnetron sputtering at room temperatur The films were sputtered under an Ar flow at 0.67 Pa pressure. A flexible substrate c 12 μm thickness was fixed on a rotating barrel-type substrate holder. The barrel height wa 20 cm, while the diameter was 20 cm. A planar extended magnetron sputtering source wa used, and the sputtering power per target area was 10 W/cm^2. The sputtering target-to substrate distance was 5 cm. The film thickness was controlled by a ZYGO New View 730 laser interferometer.

The microwave permeability measurements were carried out in a 7/3 coaxial mea surement line. A scheme of the measurement setup can be found elsewhere [46,47]. Fo the measurement, the strips were cut from a film and wound into a hollow cylindrica roll. The inner and outer diameters of the wound stripes were 3 and 7 mm. The frequenc dependence of microwave permeability was measured in the range of 0.1 to 5 GHz. The direction of the cut film stripes was chosen to obtain the maximum response in permeabilit with the easy magnetization axis being across the strip. The microwave permeability wa measured in the absence of an external magnetic field.

4. Results and Discussion

The measured frequency dependence of microwave permeability for permalloy film with different thicknesses is shown in Figure 2. It follows from Figure 2 that both the static permeability μ_s and ferromagnetic resonance frequency f_{res} drop sharply if the film thickness exceeds 220 nm. Note that the static permeability was obtained by approximating the experimental data on the real part of the permeability to the low frequency range. The resonance frequency decreases from 1.7 to 0.6 GHz when the film thickness increases from 80 to 1760 nm. The static permeability drops by a factor of four with an increase in the film thickness. A decrease in both the resonance frequency and static permeability leads to a significant decline in the Acher parameter given by Equation (1).

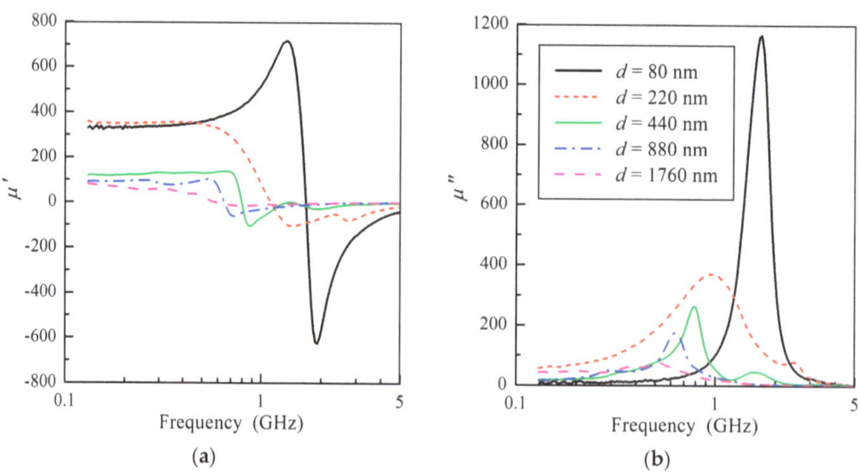

Figure 2. The measured real (**a**) and imaginary (**b**) parts of microwave permeability versus frequency at different values of film thickness d.

A qualitative explanation for the observed evolution of the ferromagnetic resonance frequency and static permeability with the thickness of permalloy film is as follows. For a film with a thickness of 80 nm, the anisotropy axis is located in the film plane. With an increase in the film thickness, the anisotropy axis deviates from the film plane, and perpendicular anisotropy appears. An increase in perpendicular magnetic anisotropy may be related to columnar microstructure formation and a magnetoelastic effect [37]. A growth in perpendicular anisotropy and a corresponding increase in the anisotropy axis angle ψ lead to a decrease in the ferromagnetic resonance frequency f_{res} in accordance with Equation (7). It should be noted that the appearance of perpendicular magnetic anisotropy is confirmed by the observed transition to transcritical hysteresis loops in thick permalloy films [18].

Furthermore, an increase in the film thickness is accompanied by a decrease in the static permeability (see Figure 2). This indicates that the in-plane anisotropy field also increases with the film thickness. A change in in-plane anisotropy may be related to the magnetoelastic effect, which contributes greatly to the measured microwave permeability of the films [47]. Note that although the magnetostriction constant in $Ni_{80}Fe_{20}$ alloy has a low value [48], the presence of magnetoelastic effects in $Ni_{80}Fe_{20}$ thin films obtained by DC magnetron sputtering was reported previously [18,37]. An increase in in-plane anisotropy was also observed in hysteresis loops [18]. An analysis of changes in internal stresses in permalloy films leading to an increase in anisotropy is beyond the scope of this paper. However, it should be noted that the magnetoelastic origin of the increase in anisotropy is confirmed by data on microwave permeability obtained for amorphous films with very low magnetostriction [49]. In these films, a decrease in the microwave permeability was not observed even for relatively thick films due to a low magnetoelastic effect. For further analysis, we introduce the in-plane H_{in} and perpendicular H_{per} anisotropy fields. Taking into account that $\theta << 1$, these fields can be expressed in terms of the effective anisotropy field H_a and the anisotropy axis angle ψ:

$$H_{\text{in}} = H_a \cos\psi \cos(\theta - \psi) \approx H_a \cos^2 \psi, \tag{11}$$

$$H_{\text{per}} = H_a \sin\psi \cos(\theta - \psi) \approx H_a \sin\psi \cos\psi. \tag{12}$$

Figure 3 shows the values of the anisotropy axis angle ψ and anisotropy fields H_a, H_{in} and H_{per} obtained as a result of fitting the measured data on microwave permeability. The effective anisotropy field H_a and the anisotropy axis angle ψ are obtained from calculations of the resonance frequency f_{res} and the static permeability μ_s by using Equations (2), (4), (7) and (8) and from a comparison of the calculated and measured values of f_{res} and μ_s. It follows from Figure 3 that the effective anisotropy field H_a increases sharply when the film thickness exceeds 220 nm and the transition into a transcritical state appears. In this case, the perpendicular anisotropy field H_{per} becomes higher than the in-plane anisotropy field H_{in}.

A comparison of the measured and calculated dependences of the ferromagnetic resonance frequency f_{res} and the static permeability μ_s on the film thickness is shown in Figure 4. For calculations, we use the value of the saturation magnetization M_0 = 805 G and the values of the effective anisotropy field H_a and anisotropy axis angle ψ presented in Figure 3. The resonance frequency and the static permeability are found by using Equations (7) and (8), respectively. It follows from Figure 4 that the calculated dependence of the resonance frequency is in good agreement with the results of the measurements, and the discrepancy between the calculated and measured values of static permeability does not exceed 10%.

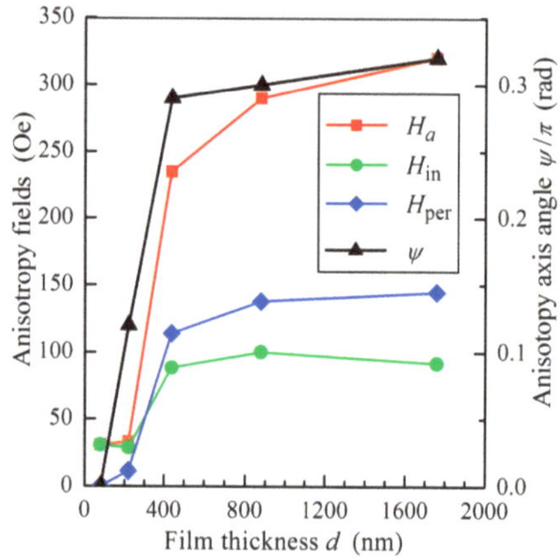

Figure 3. The variations in the anisotropy axis angle ψ, effective anisotropy field H_a, in-plane anisotropy field $H_{\rm in}$ and perpendicular anisotropy field $H_{\rm per}$ (symbols) versus the film thickness d used in the analysis. The lines are guides for eyes.

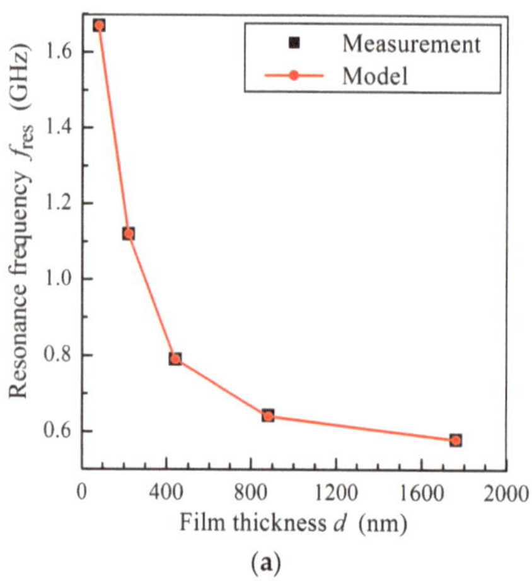

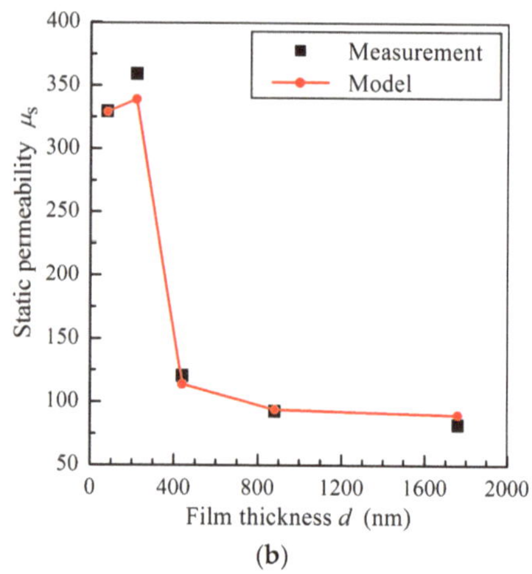

Figure 4. The ferromagnetic resonance frequency $f_{\rm res}$ (**a**) and the static permeability $\mu_{\rm s}$ (**b**) as functions of film thickness d. The lines are guides for eyes.

Figure 5 shows the dependence of the Acher parameter $k_{\rm A}$ on the film thickness restored from the results of measurements and calculated using the model. The Acher parameter reduces from 1 to 0.08 with a rise in the permalloy film thickness from 80 to 440 nm. The sharp drop in $k_{\rm A}$ is attributed to the transformation in the film magnetic structure with an increase in film thickness leading to the appearance of perpendicular anisotropy and a stripe domain structure.

As mentioned above, in soft magnetic films with out-of-plane anisotropy, the Acher parameter depends mainly on the value of the demagnetized factor N_z (see Equation (10)). According to Equation (2), the value of N_z is governed by the domain aspect ratio a/d. Therefore, we can estimate the domain width a by using the obtained dependence of the Acher parameter on the film thickness.

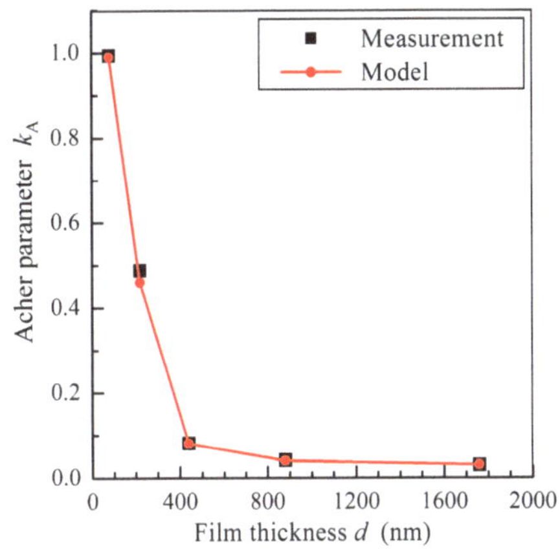

Figure 5. The Acher parameter k_A as a function of the film thickness d. The line is a guide for eyes.

Shown in Figure 6 is the dependence of the domain width on the film thickness calculated by Equations (2), (4) and (9). The domain width decreases sharply when the film thickness exceeds 220 nm and the transition into a transcritical state appears. For permalloy films with thicknesses higher than 440 nm, the domain width is of the order of 0.5 μm. The calculated domain width correlates with the stripe domain structure period observed for sputtered permalloy films [50,51]. Note that the domain aspect ratio a/d for permalloy films under study is not too low and is about 0.4 for a film with a thickness of 1760 nm, as follows from the data presented in Figure 6.

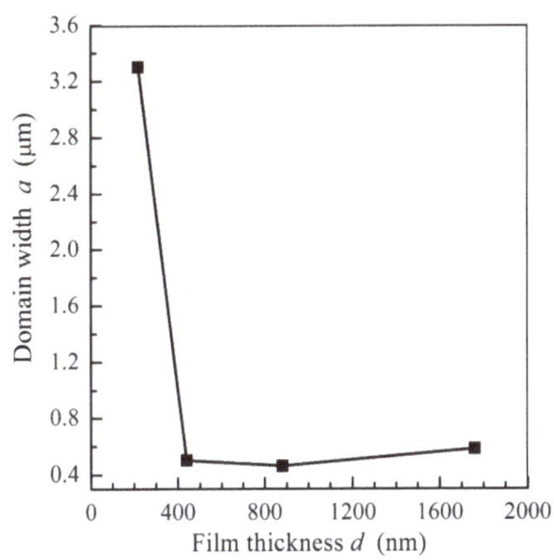

Figure 6. The calculated dependence of the domain width a on the film thickness d (squares). The line is a guide for eyes.

For not-too-narrow domains, when $(a/d)Q^{1/2} >> 1$, we can obtain an expression for the domain aspect ratio in explicit form. Combining Equations (3) and (10), we have

$$a/d \approx \frac{\pi^3}{16.8} \times \frac{k_A - H_a \cos 2\psi / 4\pi M_0}{1 + (1 + 4\pi M_0 / H_a)^{1/2}} \quad (13)$$

Note that calculations by means of Equation (13) give the same values for the domain width a, as shown in Figure 6, when the permalloy film thickness is higher than 220 nm.

The decrease in the Acher parameter with the domain aspect ratio a/d in films with out-of-plane anisotropy can be explained as follows [17]. With a decrease in a/d, the dynamic demagnetizing fields at the domain walls increase, which restrains variations in magnetization in the film plane. As a result, magnetization variation in the direction transverse to the film plane becomes more preferable, which leads to a decrease in microwave permeability.

We assume above that there is the stripe domain structure in the permalloy films under study. In general, a more complex micromagnetic configuration may arise in soft magnetic films. However, a comparison of the measurement data on microwave permeability with the results of the model does not contradict the assumption of the existence of a stripe domain structure in the permalloy films.

To conclude this section, note that a decrease in microwave permeability with a growth in film thickness was also studied for cobalt films [19]. For these films, a slow decrease in the resonance frequency and the static permeability was observed, and the Acher parameter reduced gradually from 0.70 to 0.26 when the film thickness increased from 20 to 160 nm. The measured hysteresis loops exhibited a shape close to the transcritical one for cobalt films with a thickness of 470 nm. As in the case of permalloy films, a decrease in the microwave permeability of cobalt films is attributed to the appearance of out-of-plane anisotropy. However, the slower decrease in the Acher parameter in cobalt films is due to the fact that the domain aspect ratio a/d still remains quite high, even for the thickest films.

5. Conclusions

The microwave permeability of a thin film can be estimated based on the saturation magnetization of the material by using the Acher parameter, described in Equation (1). It tends to the limiting value of unity for a single-domain film with in-plane magnetic anisotropy. The appearance of out-of-plane anisotropy and a domain structure results in a decrease in the Acher parameter. The analysis of the Acher parameter can be used to study the relationship between the microwave magnetic properties and magnetic structure of films.

We studied changes in the magnetic structure of permalloy films with an increase in the film thickness based on the measured microwave permeability. The measurement data were analyzed by means of a model for film with a regular stripe domain structure and out-of-plane magnetic anisotropy [17]. The observed decrease in the ferromagnetic resonance frequency with a growth in the film thickness was attributed to the appearance of perpendicular anisotropy. Both the in-plane and perpendicular anisotropy increase with the film thickness due to the magnetoelastic effect. As a result, the static permeability drops sharply when the transition into a transcritical state appears. It should be noted that the ferromagnetic resonance frequency also decreases with a growth in the film thickness despite the increase in the in-plane anisotropy field. This is due to the fact that the resonance frequency depends on the perpendicular anisotropy field and decreases with the domain aspect ratio a/d (see Equation (7)).

It is predicted that for soft magnetic films with out-of-plane anisotropy, the value of the Acher parameter is mainly governed by the ratio of the domain width and film thickness. This allows one to estimate the domain width in the film by using measurement data on the microwave magnetic properties. It is found that for relatively thick permalloy films, the domain width is of the order of the film thickness. The formation of a stripe domain structure results in a sharp decrease in the Acher parameter for thick films due to the influence of dynamic demagnetizing fields at the domain walls.

Author Contributions: Conceptualization, N.A.B., S.S.M. and A.V.O.; methodology, N.A.B.; formal analysis, N.A.B.; investigation, S.A.M., A.V.O. and P.A.Z.; resources, S.A.M.; writing—original draft preparation, N.A.B.; writing—review and editing, N.A.B., S.S.M., A.V.O. and K.N.R.; supervision, A.N.L., S.S.M. and K.N.R. All authors have read and agreed to the published version of the manuscript.

Funding: This research received no external funding.

Institutional Review Board Statement: Not applicable.

Informed Consent Statement: Not applicable.

Data Availability Statement: Data available from the corresponding author upon reasonable request.

Conflicts of Interest: The authors declare no conflicts of interest.

References

1. Yamaguchi, M.; Suezawa, K.; Takahashi, Y.; Arai, K.I.; Kikuchi, S.; Shimada, Y.; Tanabe, S.; Ito, K. Magnetic thin-film inductors for RF-integrated circuits. *J. Magn. Magn. Mater.* **2000**, *215*, 807–810. [CrossRef]
2. Fergen, I.; Seemann, K.; von der Weth, A.; Shüppen, A. Soft ferromagnetic thin films for high frequency applications. *J. Magn. Magn. Mater.* **2002**, *242–245*, 146–151. [CrossRef]
3. Bekker, V.; Seemann, K.; Leiste, H. Development and optimisation of thin soft ferromagnetic Fe–Co–Ta–N and Fe–Co–Al–N films with in-plane uniaxial anisotropy for HF applications. *J. Magn. Magn. Mater.* **2006**, *296*, 37–45. [CrossRef]
4. Seemann, K.; Leiste, H.; Bekker, V. A new generation of CMOS-compatible high frequency microinductors with ferromagnetic cores: Theory, fabrication and characterization. *J. Magn. Magn. Mater.* **2006**, *302*, 321–326. [CrossRef]
5. Rahman, B.M.F.; Divan, R.; Rosenmann, D.; Wang, T.; Peng, Y.; Wang, G. Application of sub-micrometer patterned permalloy thin film in tunable radio frequency inductors. *J. Appl. Phys.* **2015**, *117*, 17C121. [CrossRef]
6. Wang, T.; Peng, Y.; Jiang, W.; Huang, Y.M.; Rahman, B.M.F.; Divan, R.; Rosenmann, D.; Wang, G. Integrating nanopatterned ferromagnetic and ferroelectric thin films for electrically tunable RF applications. *IEEE Trans. Microw. Theory Tech.* **2017**, *65*, 504–512. [CrossRef]
7. Karilainen, A.O.; Ikonen, P.M.T.; Simovski, C.R.; Tretyakov, S.A. Choosing dielectric or magnetic material to optimize the bandwidth of miniaturized resonant antennas. *IEEE Trans. Antennas Propagat.* **2011**, *59*, 3991–3998. [CrossRef]
8. Huitema, L.; Reveyrand, T.; Mattei, J.-L.; Arnaud, E.; Decroze, C.; Monediere, T. Frequency tunable antenna using a magneto-dielectric material for DVB-H application. *IEEE Trans. Antennas Propagat.* **2013**, *61*, 4456–4466. [CrossRef]
9. Ludwig, A.; Tewes, M.; Glasmachers, S.; Löhndorf, M.; Quandt, E. High-frequency magnetoelastic materials for remote-interrogated stress sensors. *J. Magn. Magn. Mater.* **2002**, *242*, 1126–1131. [CrossRef]
10. Sohn, J.; Han, S.H.; Yamaguchi, M.; Limm, S.H. Tunable electromagnetic noise suppressor integrated with a magnetic thin film. *Appl. Phys. Lett.* **2006**, *89*, 103501. [CrossRef]
11. Park, J.; Lee, J.W.; Choi, H.J.; Jang, W.G.; Kim, T.S.; Suh, D.S.; Jeong, H.Y.; Chang, S.Y.; Roh, J.C.; Yoo, C.S.; et al. Electromagnetic interference shielding effectiveness of sputtered NiFe/Cu multi-layer thin film at high frequencies. *Thin Solid Film.* **2019**, *677*, 130–136. [CrossRef]
12. Snoek, J.L. Dispersion and absorption of magnetic ferrites at frequencies above 1 MHz. *Physica* **1948**, *14*, 207–217. [CrossRef]
13. Perrin, G.; Acher, O.; Peuzin, J.C.; Vucadinovic, N. Sum rules for gyromagnetic permeability of ferromagnetic thin films: Theoretical and experimental results. *J. Magn. Magn. Mater.* **1996**, *157/158*, 289–290. [CrossRef]
14. Acher, O.; Queste, S.; Ledieu, M.; Barholz, K.-U.; Mattheis, R. Hysteretic behavior of the dynamic permeability on a Ni-Fe thin film. *Phys. Rev. B* **2003**, *68*, 184414. [CrossRef]
15. Lagarkov, A.N.; Rozanov, K.N.; Simonov, N.A.; Starostenko, S.N. Microwave permeability of magnetic films. In *Handbook of Advanced Magnetic Materials*; Liu, Y., Sellmyer, D.J., Shindo, D., Eds.; Springer: Boston, MA, USA, 2006; pp. 1742–1773.
16. Rozanov, K.N.; Koledintseva, M.Y. Application of generalized Snoek's law over a finite frequency range: A case study. *J. Appl. Phys.* **2016**, *119*, 073901. [CrossRef]
17. Buznikov, N.A.; Rozanov, K.N. The effect of stripe domain structure on dynamic permeability of thin ferromagnetic films with out-of-plane uniaxial anisotropy. *J. Magn. Magn. Mater.* **2005**, *285*, 314–326. [CrossRef]
18. Zezyulina, P.A.; Iakubov, I.T.; Lagarkov, A.N.; Maklakov, S.A.; Maklakov, S.S.; Naboko, A.S.; Osipov, A.V.; Petrov, D.A.; Rozanov, K.N.; Ryzhikov, I.A. The effect of the perpendicular anisotropy and eddy currents on the microwave performance of single-layer and multi-layer permalloy films. *IEEE Magn. Lett.* **2016**, *7*, 3705804. [CrossRef]
19. Bobrovskii, S.Y.; Iakubov, I.T.; Lagarkov, A.N.; Maklakov, S.A.; Maklakov, S.S.; Osipov, A.V.; Rozanov, K.N.; Ryzhikov, I.A.; Petrov, D.A.; Zezyulina, P.A. Variation of microwave magnetic properties for thin films of ferromagnetic metals with the film thickness. *J. Magn. Magn. Mater.* **2018**, *459*, 20–25. [CrossRef]
20. Ben Youssef, J.; Vukadinovic, N.; Billet, D.; Labrune, M. Thickness-dependent magnetic excitations in permalloy films with nonuniform magnetization. *Phys. Rev. B* **2004**, *69*, 174402. [CrossRef]
21. Dastagir, T.; Xu, W.; Sinha, S.; Wu, H.; Cao, Y.; Yu, H.B. Tuning the permeability of permalloy films for on-chip inductor applications. *Appl. Phys. Lett.* **2010**, *97*, 162506. [CrossRef]
22. Izotov, A.V.; Belyaev, B.A.; Boev, N.M.; Burmitskikh, A.V.; Leksikov, A.A.; Skomorokhov, G.V.; Solovev, P.N. Tailoring the microwave properties of thin Permalloy films using a periodically grooved substrate. *Physica B* **2022**, *629*, 413654. [CrossRef]
23. Li, T.; Wang, Y.; Shi, H.; Xi, L.; Xue, D. Impact of skin effect on permeability of Permalloy films. *J. Magn. Magn. Mater.* **2022**, *545*, 168750. [CrossRef]
24. Jiang, X.; Zhang, J.; Song, X.; Wang, H.; Zhang, K.; He, Z.; Wu, C.; Yu, Z.; Lan, Z.; Sun, K. Tunable resonance frequency of NiFe thin films by oblique deposition. *J. Magn. Magn. Mater.* **2022**, *547*, 168946. [CrossRef]

25. Goldman, S.; Celinski, Z. Magnetic properties of $(Ni_{0.81}Fe_{0.19}/SiO_2)_n$ multilayers for high frequency on-wafer inductor applications. *J. Magn. Magn. Mater.* **2023**, *569*, 170440. [CrossRef]
26. Kurlyandskaya, G.V.; Fernández, E.; Svalov, A.; Burgoa Beitia, A.; García-Arribas, A.; Larrañaga, A. Flexible thin film magnetoimpedance sensors. *J. Magn. Magn. Mater.* **2016**, *415*, 91–96. [CrossRef]
27. Agra, K.; Mori, T.J.A.; Dorneles, L.S.; Escobar, V.M.; Silva, U.C.; Chesman, C.; Bohn, F.; Corrêa, M.A. Dynamic Magnetic behavior in non-magnetostrictive multilayered films grown on glass and flexible substrates. *J. Magn. Magn. Mater.* **2014**, *355*, 136–141. [CrossRef]
28. Blyakhman, F.A.; Buznikov, N.A.; Sklyar, T.F.; Safronov, A.P.; Golubeva, E.V.; Svalov, A.V.; Sokolov, S.Y.; Melnikov, G.Y.; Orue, I.; Kurlyandskaya, G.V. Mechanical, electrical and magnetic properties of ferrogels with embedded iron oxide nanoparticles obtained by laser target evaporation: Focus on multifunctional biosensor applications. *Sensors* **2018**, *18*, 872. [CrossRef] [PubMed]
29. Melnikov, G.Y.; Lepalovskij, V.N.; Svalov, A.V.; Safronov, A.P.; Kurlyandskaya, G.V. Magnetoimpedance thin film sensor for detecting of stray fields of magnetic particles in blood vessel. *Sensors* **2021**, *21*, 3621. [CrossRef]
30. Sugita, Y.; Fujiwara, H.; Sato, T. Critical thickness and perpendicular anisotropy of evaporated permalloy films with stripe domains. *Appl. Phys. Lett.* **1967**, *10*, 229–231. [CrossRef]
31. Amos, N.; Fernandez, R.; Ikkawi, R.; Lee, B.; Lavrenov, A.; Krichevsky, A.; Litvinov, D.; Khizroev, S. Magnetic force microscopy study of magnetic stripe domains in sputter deposited Permalloy thin films. *J. Appl. Phys.* **2008**, *103*, 07E732. [CrossRef]
32. Svalov, A.V.; Aseguinolaza, I.R.; Garcia-Arribas, A.; Barandiaran, J.M.; Alonso, J.; Fernández-Gubieda, M.L.; Orue, I.; Kurlyandskaya, G.V. Structure and magnetic properties of thin Permalloy films near the transcritical state. *IEEE Trans. Magn.* **2010**, *46*, 333–336. [CrossRef]
33. Cao, D.; Wang, Z.; Feng, E.; Wei, J.; Wang, J.; Liu, Q. Magnetic properties and microstructure investigation of electrodeposited FeNi/ITO films with different thickness. *J. Alloys Compd.* **2013**, *581*, 66–70. [CrossRef]
34. Silva, E.F.; Corrêa, M.A.; Della Pace, R.D.; Plá Cid, C.C.; Kern, P.R.; Carara, M.; Chesman, C.; Alves Santos, O.; Rodríguez-Suárez, R.L.; Azevedo, A.; et al. Thickness dependence of the magnetic anisotropy and dynamic magnetic response of ferromagnetic NiFe films. *J. Phys. D Appl. Phys.* **2017**, *50*, 185001. [CrossRef]
35. De Melo, A.S.; Bohn, F.; Ferreira, A.; Vaz, F.; Correa, M.A. High-frequency magnetoimpedance effect in meander-line trilayered films. *J. Magn. Magn. Mater.* **2020**, *515*, 167166. [CrossRef]
36. Raj, R.; Kuila, M.; Gupta, M.; Reddy, V.R. 57Fe Mössbauer and magneto-optical Kerr effect (MOKE) study of transcritical state in permalloy (FexNi100-x) thin films. *Hyperfine Interact.* **2021**, *242*, 30. [CrossRef]
37. Komogortsev, S.V.; Vazhenina, I.G.; Kleshnina, S.A.; Iskhakov, R.S.; Lepalovskij, V.N.; Pasynkova, A.A.; Svalov, A.V. Advanced characterization of FeNi-based films for the development of magnetic field sensors with tailored functional parameters. *Sensors* **2022**, *22*, 3324. [CrossRef] [PubMed]
38. Cotón, N.; Andrés, J.P.; Molina, E.; Jaafar, M.; Ranchal, R. Stripe domains in electrodeposited $Ni_{90}Fe_{10}$ thin films. *J. Magn. Magn. Mater.* **2023**, *565*, 170246. [CrossRef]
39. Coïsson, M.; Vinai, F.; Tiberto, P.; Celegato, F. Magnetic properties of FeSiB thin films displaying stripe domains. *J. Magn. Magn. Mater.* **2009**, *321*, 806–809. [CrossRef]
40. Marty, A.; Samson, Y.; Gilles, B.; Belakhovsky, M.; Dudzik, E.; Dürr, H.; Dhesi, S.S.; van der Laan, G.; Goedkoop, J.B. Weak-stripe magnetic domain evolution with an in-plane field in epitaxial FePd thin films: Model versus experimental results. *J. Appl. Phys.* **2000**, *87*, 5472–5474. [CrossRef]
41. Vukadinovic, N.; Vacus, O.; Labrune, M.; Acher, O.; Pain, D. Magnetic excitations in a weak-stripe-domain structure: A 2D dynamic micromagnetic approach. *Phys. Rev. Lett.* **2000**, *85*, 2817–2820. [CrossRef]
42. Sallica Leva, E.; Valente, R.; Martínez Tabares, F.; Vásquez Mansilla, M.; Roshdestwensky, S.; Butera, A. Magnetic domain crossover in FePt thin films. *Phys. Rev. B* **2010**, *82*, 144410. [CrossRef]
43. Kaczer, J.; Murtinova, L. On the demagnetizing energy of periodic magnetic distributions. *Phys. Stat. Solidi A* **1974**, *23*, 79–86. [CrossRef]
44. Ramesh, M.; Ren, E.W.; Artman, J.O.; Kryder, M.H. Domain mode ferromagnetic resonance studies in bismuth-substituted magnetic garnet films. *J. Appl. Phys.* **1988**, *64*, 5483–5485. [CrossRef]
45. Vukadinovic, N.; Ben Youssef, J.; Le Gall, H. Influence of magnetic parameters on microwave absorption of domain mode ferromagnetic resonance. *J. Magn. Magn. Mater.* **1995**, *150*, 213–222. [CrossRef]
46. Rozanov, K.N.; Simonov, N.A.; Osipov, A.V. Microwave measurements of the magnetic film permeability. *J. Commun. Technol. Electron.* **2002**, *47*, 210–216.
47. Iakubov, I.T.; Kashurkin, O.Y.; Lagarkov, A.N.; Maklakov, S.A.; Osipov, A.V.; Rozanov, K.N.; Ryzhikov, I.A.; Starostenko, S.N. A contribution from the magnetoelastic effect to measured microwave permeability of thin ferromagnetic films. *J. Magn. Magn. Mater.* **2012**, *324*, 3385–3388. [CrossRef]
48. Hristoforou, E.; Ktena, A.; Angelopoulos, S. Magnetostrictive materials for sensing applications. In *Encyclopedia of Smart Materials*; Olab, A.-G., Ed.; Springer: Amsterdam, The Netherlands, 2022; Volume 15, pp. 355–365.
49. Munakata, M.; Namikawa, M.; Motoyama, M.; Yagi, M.; Shimada, Y.; Yamaguchi, M.; Arai, K.I. Magnetic properties and frequency characteristics of $(CoFeB)_x$–$(Si0_{1.9})_{1-x}$ and CoFeB films for RF applications. *Trans. Magn. Soc. Japan* **2002**, *2*, 388–393. [CrossRef]

50. Wang, G.; Dong, C.; Wang, W.; Wang, Z.; Chai, G.; Jiang, C.; Xue, D. Observation of rotatable stripe domain in permalloy films with oblique sputtering. *J. Appl. Phys.* **2012**, *112*, 093907. [CrossRef]
51. Liu, M.; Li, Q.; Song, C.; Feng, H.; Son, Y.; Zhong, L.; Pan, L.; Zhao, C.; Li, Q.; Xu, J.; et al. Microwave excitations and hysteretic magnetization dynamics of stripe domain films. *J. Magn. Magn. Mater.* **2022**, *547*, 168939. [CrossRef]

sensors

MDPI

Article

A Remote Two-Point Magnetic Localization Method Based on SQUID Magnetometers and Magnetic Gradient Tensor Invariants

Yingzi Zhang [1], Gaigai Liu [1], Chen Wang [1], Longqing Qiu [2], Hongliang Wang [1] and Wenyi Liu [1,*]

[1] State Key Laboratory of Dynamic Measurement Technology, North University of China, Taiyuan 030051, China; zhangyingzi@nuc.edu.cn (Y.Z.); b200610@st.nuc.edu.cn (G.L.); b20220620@st.nuc.edu.cn (C.W.); wanghongliang@nuc.edu.cn (H.W.)
[2] Shanghai Institute of Microsystem and Information Technology, Chinese Academy of Sciences, Shanghai 200050, China; lq.qiu@mail.sim.ac.cn
* Correspondence: liuwenyi@nuc.edu.cn; Tel.: +86-139-3469-7107

Abstract: In practical application, existing two-point magnetic gradient tensor (MGT) localization methods have a maximum detection distance of only 2.5 m, and the magnetic moment vectors of measured targets are all unknown. In order to realize remote, real-time localization, a new two-point magnetic localization method based on self-developed, ultra-sensitive superconducting quantum interference device (SQUID) magnetometers and MGT invariants is proposed. Both the magnetic moment vector and the relative position vector can be directly calculated based on the linear positioning model, and a quasi-Newton optimization algorithm is adopted to further improve the interference suppression capability. The simulation results show that the detection distance of the proposed method can reach 500 m when the superconducting MGT measurement system is used. Compared with Nara's single-point tensor (NSPT) method and Xu's two-point tensor (XTPT) method, the proposed method produces the smallest relative localization error (i.e., significantly less than 1% in the non-positioning blind area) without sacrificing real-time characteristics. The causes of and solutions to the positioning blind area are also analyzed. The equivalent experiments, which were conducted with a detection distance of 10 m, validate the effectiveness of the localization method, yielding a minimum relative localization error of 4.5229%.

Keywords: magnetic localization; magnetic gradient tensor; superconducting quantum interference device; magnetic anomaly detection

Citation: Zhang, Y.; Liu, G.; Wang, C.; Qiu, L.; Wang, H.; Liu, W. A Remote Two-Point Magnetic Localization Method Based on SQUID Magnetometers and Magnetic Gradient Tensor Invariants. *Sensors* **2024**, *24*, 5917. https://doi.org/10.3390/s24185917

Academic Editors: Galina V. Kurlyandskaya and Arcady Zhukov

Received: 28 July 2024
Revised: 20 August 2024
Accepted: 9 September 2024
Published: 12 September 2024

1. Introduction

Since the magnetic characteristics of underwater targets are difficult to camouflage, magnetic anomaly detection (MAD) technology based on magnetic sensors has received a great deal of attention in the field of underwater target detection and localization. Compared with the traditional magnetic field scalar measurement and magnetic field vector measurement, magnetic gradient tensor (MGT) measurement can not only avoid the influence of the geomagnetic background field and the magnetization direction of the magnetic target, but it can also directly locate the magnetic dipole target through the MGT inversion algorithm [1–3]. As a result, MGT measurement and inversion has become a research hotspot in geophysical exploration, archaeology, and security applications [4–7].

The development of an ultra-sensitive MGT measurement system is essential for detecting the magnetic anomaly changes generated by long-distance underwater targets. MGT measurement systems can be constructed by fluxgates or superconducting quantum interference devices (SQUIDs) [8,9], with the detection distance is related to their sensitivity. As we all know, the sensitivity of SQUID magnetometers is 2 to 3 orders higher than that of fluxgate magnetometers [10]. Meanwhile, the length of the baseline of SQUID gradiometers is much smaller than that of fluxgate gradiometers under the same gradient sensitivity, which is conducive to the miniaturization of the design of measurement systems [11,12]. In addition, the bandwidth of SQUID magnetometers is much higher than that of fluxgate magnetometers, and it is more suitable for measurement occasions where the signal to be measured changes dramatically [13]. However, the current MGT measurement system, which is designed to realize the real-time localization of a single magnetic dipole target, is mainly constructed by fluxgate magnetometers, and the maximum detection distance is only 5 m. Therefore, an MGT measurement system based on SQUID magnetometers was designed in this paper to improve detection distance.

The relative position vector from the observation point to the magnetic dipole object can be calculated using the measured MGT and its invariants, a process called MGT inversion. According to the number of observation points, the current magnetic localization methods based on MGT inversion algorithms are divided into two categories: single-point tensor (SPT) methods and two-point tensor (TPT) methods. SPT methods can be categorized into three main groups: eigenvalue-based localization methods, Nara's single-point tensor (NSPT) methods, and scalar triangulation and ranging (STAR) methods. Eigenvalue-based localization methods were developed rapidly after Wynn [14,15] and Frahm [16] proposed and promoted the innovative application of MGT for point-to-point positionings of magnetic dipoles [17,18]. However, the solution of this method has an inherent fourfold ambiguity; it requires adding a measuring point or additional information to eliminate 'ghost' solutions [19]. The NSPT method, which only needs to measure the MGT and the magnetic field vector at one observation point [20], has attracted a great deal of attention because of its simplicity and high real-time performance. In order to solve the localization dead-zone problem of this method, the Moore–Penrose generalized inverse matrix [21], truncated singular-value decomposition [22], and eigenvector constraint-based method [23] were studied to eliminate the MGT singularity at certain measuring points. However, the above-mentioned NSPT-based methods all involve the measurement of the magnetic field vector of a magnetic dipole target, which is difficult to separate from the geomagnetic field in the actual measurement process, and a small geomagnetic field measurement error leads to a considerable localization error. Certain single-point, high-order MGT localization methods based on NSPT have been researched [24,25], but they are more susceptible to instrument measurement errors. The STAR method, based on tensor invariants, which has also been widely studied, was proposed in order to eliminate the influence of the geomagnetic background field [26–30]. Additionally, it has a unique advantage in highly dynamic magnetic detection applications. However, the inherent aspherical error and greater number of instrument measurement errors introduced by the eight triaxial fluxgate magnetometers contained in the probe structure reduce the positioning accuracy of this method. As a result, the TPT method was gradually developed; it is not affected by the geomagnetic field and does not have aspherical errors. The TPT methods proposed earlier either required prior information [3] or adopted optimization algorithms, such as particle swarm optimization (PSO) [31]. These are complex in computation and relatively poor in real-time performance. Compared to them, Xu's two-point tensor (XTPT) method [2] can provide an analytical solution of the relative position vector that is simple and provided in real time. However, an approximation error is introduced if the distance between the two observation points is not sufficiently small. Therefore, our research group proposed a new two-point tensor (NTPT) method [32] with higher localization accuracy, which is not only unaffected by the geomagnetic field, but also has no approximate errors. In this method, a new simulation model was established that takes into account the influence of different z-coordinates on localization. However, the NTPT method only considers cases where the relative position vectors and the magnetic moment vector are not coplanar and the noise suppression ability needs to be enhanced in the actual experiment process.

In this paper, a two-point magnetic gradient tensor (TMGT) localization method for remote, single-magnetic-dipole targets based on SQUID magnetometers and MGT invariants is proposed. In order to improve the detection distance, an MGT probe composed of eight self-developed SQUID magnetometers was designed. A linear localization model based on the spatial position relationship between a magnetic moment vector and relative position vectors—which can not only realize the high-precision localization of magnetic target, but also calculate its magnetic moment vector—is also proposed. At the same time, an objective function based on the relationship between the measured MGTs and the relative position vectors was constructed to optimize the inversion results, and the magnetic interference suppression ability was further improved. A long-distance spherical trajectory simulation model was established to analyze the positioning accuracy and positioning blind area of this method. Both simulations and equivalent experiments helped to demonstrate that the TMGT method has a better localization performance for remote magnetic targets.

2. Methods

2.1. Magnetic Gradient Tensor and Tensor Invariants

The magnetization of a magnetic target by the geomagnetic field causes a local distortion in the geomagnetic field, thus forming a magnetic anomaly. The aim behind the detection of magnetic anomalies is to locate the target by detecting these distortions. The magnetic target can be equivalent to a magnetic dipole when the detection distance is greater than 2.5 times the maximum size of the magnetic target. According to the Biot–Savart law, the magnetic induction intensity vector $\boldsymbol{B} = (B_x, B_y, B_z)$ generated by a magnetic dipole at a certain observation point can be expressed as follows:

$$\boldsymbol{B} = \frac{\mu}{4\pi}\frac{3(\boldsymbol{M}\cdot\boldsymbol{r})\boldsymbol{r} - \boldsymbol{M}r^2}{r^5} = \frac{\mu}{4\pi r^5}\begin{bmatrix} 3r_x^2 - r^2 & 3r_x r_y & 3r_x r_z \\ 3r_x r_y & 3r_y^2 - r^2 & 3r_y r_z \\ 3r_x r_z & 3r_y r_z & 3r_z^2 - r^2 \end{bmatrix}\begin{bmatrix} M_x \\ M_y \\ M_z \end{bmatrix}, \tag{1}$$

where $\boldsymbol{r} = (r_x, r_y, r_z)$ is the relative position vector from the magnetic dipole to the observation point, r is the modulus of $\boldsymbol{r}$, $\boldsymbol{M} = (M_x, M_y, M_z)$ is the magnetic moment vector of the magnetic dipole, and $\boldsymbol{\mu}$ is the permeability of the medium at the observation point. Since the observation point is usually in the air, then $\mu \approx \mu_0 = 4\pi \times 10^{-7} N \cdot A^{-2}$, where μ_0 is the vacuum permeability.

The magnetic gradient tensor $\boldsymbol{G}$ is the spatial variation rate of the three components of the magnetic induction intensity vector $\boldsymbol{B}$ in the orthogonal direction, and it has nine elements in total. Among them, the curl and divergence of the magnetic field vector at a certain point in the passive static magnetic field are zero [32]. Hence, the magnetic gradient tensor $\boldsymbol{G}$ has only five independent elements and can be denoted as follows:

$$\boldsymbol{G} = \begin{bmatrix} B_x \\ B_y \\ B_z \end{bmatrix}\begin{bmatrix} \frac{\partial}{\partial x} & \frac{\partial}{\partial y} & \frac{\partial}{\partial z} \end{bmatrix} = \begin{bmatrix} B_{xx} & B_{xy} & B_{xz} \\ B_{xy} & B_{yy} & B_{yz} \\ B_{xz} & B_{yz} & -B_{xx} - B_{yy} \end{bmatrix}. \tag{2}$$

The five independent elements of $\boldsymbol{G}$ are expressed as follows:

$$\begin{bmatrix} B_{xx} \\ B_{xy} \\ B_{xz} \\ B_{yy} \\ B_{zz} \end{bmatrix} = \frac{\mu}{4\pi r^7}\begin{bmatrix} (9r^2 - 15r_x^2)r_x & (3r^2 - 15r_x^2)r_y & (3r^2 - 15r_x^2)r_z \\ (3r^2 - 15r_x^2)r_y & \left(3r^2 - 15r_y^2\right)r_x & -15r_x r_y r_z \\ (3r^2 - 15r_x^2)r_z & -15r_x r_y r_z & (3r^2 - 15r_z^2)r_x \\ \left(3r^2 - 15r_y^2\right)r_x & \left(3r^2 - 15r_y^2\right)r_y & \left(3r^2 - 15r_y^2\right)r_z \\ -15r_x r_y r_z & \left(3r^2 - 15r_y^2\right)r_z & (3r^2 - 15r_z^2)r_y \end{bmatrix}\begin{bmatrix} M_x \\ M_y \\ M_z \end{bmatrix}. \tag{3}$$

The eigenvalues of the characteristic equation of $\boldsymbol{G}$ can be derived as follows [18]:

$$\begin{cases} \lambda_1 = \dfrac{3\mu_0 M}{8\pi r^4}\left(-\cos\theta + \sqrt{5(\cos\theta)^2 + 4}\right) \\ \lambda_2 = \dfrac{3\mu_0 M}{4\pi r^4}\cos\theta \\ \lambda_1 = \dfrac{3\mu_0 M}{8\pi r^4}\left(-\cos\theta - \sqrt{5(\cos\theta)^2 + 4}\right) \end{cases}, \tag{4}$$

where $M = \|\boldsymbol{M}\|$ is the modulus of $\boldsymbol{M}$, and θ is the angle between $\boldsymbol{r}$ and $\boldsymbol{M}$. From Equation (3), $\cos\theta$ can be deduced and denoted as follows:

$$\cos\theta = \frac{\boldsymbol{M} \cdot \boldsymbol{r}}{\|\boldsymbol{M}\| \cdot \|\boldsymbol{r}\|} = \frac{\lambda_2}{\sqrt{-\lambda_2^2 - \lambda_1\lambda_3}}, \tag{5}$$

and the relationships between the three eigenvalues above are

$$\begin{cases} \lambda_1 + \lambda_2 + \lambda_3 = 0 \\ \lambda_1 \geq \lambda_2 \geq \lambda_3 \\ |\lambda_1| \geq |\lambda_2| \\ |\lambda_3| \geq |\lambda_2| \end{cases}. \tag{6}$$

These eigenvalues of $\boldsymbol{G}$, and any combination of them, which are called tensor invariants, are independent of the coordinate system choice and are kept unchanged when the attitude of the magnetic gradient tensor measurement system changes. Therefore, tensor invariants are resistant to motion noise.

As the magnetic anomalies caused by magnetic dipoles have positive and negative peaks, the normalized source strength (NSS), a tensor invariant, is often used to locate the magnetic target because it is isotropic around the magnetic dipole and is not affected by the direction of magnetization. The NSS can be expressed as [18]

$$NSS = \sqrt{-\lambda_2^2 - \lambda_1\lambda_3} = \frac{3\mu_0 M}{4\pi r^4}. \tag{7}$$

The three eigenvectors corresponding to the three real eigenvalues are expressed as follows:

$$\boldsymbol{u}_i = \begin{bmatrix} \alpha_i \\ \beta_i \\ \gamma_i \end{bmatrix} = \begin{bmatrix} B_{yz}B_{xy} + (\lambda_i - B_{yy})B_{xz} \\ B_{xz}B_{xy} + (\lambda_i - B_{xx})B_{yz} \\ (B_{xx} - \lambda_i)(B_{yy} - \lambda_i) - B_{xy}^2 \end{bmatrix}, \ i = 1,2,3, \tag{8}$$

which are orthogonal to each other. In the above equation, $\boldsymbol{n}_i = \boldsymbol{u}_i/|\boldsymbol{u}_i|$ is the corresponding unit eigenvectors, and $\boldsymbol{n}_2$ is the unit normal vector unit of the plane defined by $\boldsymbol{M}$ and $\boldsymbol{r}$, corresponding to the eigenvalue λ_2 with the smallest absolute value.

As a result, only five independent elements of $\boldsymbol{G}$ need to be measured, and the magnetic target can be located with the inversion algorithm based on the magnetic gradient tensor invariant.

2.2. Superconducting MGT Measurement System

The superconducting MGT measurement system is mainly composed of an MGT probe, a non-magnetic liquid helium dewar, SQUID readout circuits, a data acquisition system, and a control software. The structure diagram of the probe is shown in Figure 1, and it is mainly composed of eight uniaxial SQUID magnetometers and one stainless steel probe rod. These SQUID magnetometers are divided into three groups, corresponding to the three axes of the Cartesian coordinate system, and the magnetometers are designed in the "face to face" form in order to construct an MGT measurement probe. When obtaining the MGT, the difference calculation is often used to replace the differential calculation in the MGT. As a result, the output of this system is expressed as follows:

$$G = \begin{bmatrix} \frac{-(B_{y1}-B_{y2})}{d_{23}} - \frac{(B_{z1}-B_{z2})}{d_{18}} & \frac{B_{x1}-B_{x3}}{d_{67}} & \frac{B_{x1}-B_{x2}}{d_{47}} \\ \frac{B_{x1}-B_{x3}}{d_{67}} & \frac{B_{y1}-B_{y2}}{d_{23}} & \frac{B_{y1}-B_{y3}}{d_{25}} \\ \frac{B_{x1}-B_{x2}}{d_{47}} & \frac{B_{y1}-B_{y3}}{d_{25}} & \frac{B_{z1}-B_{z2}}{d_{18}} \end{bmatrix}, \quad (9)$$

where B_{ij} represents the output of the SQUID magnetometers, i indicates the direction of the magnetometers ($i = x, y, z$), j indicates the number of magnetometers in a certain direction, and d_{mn} indicates the distance between any two magnetometers in a probe.

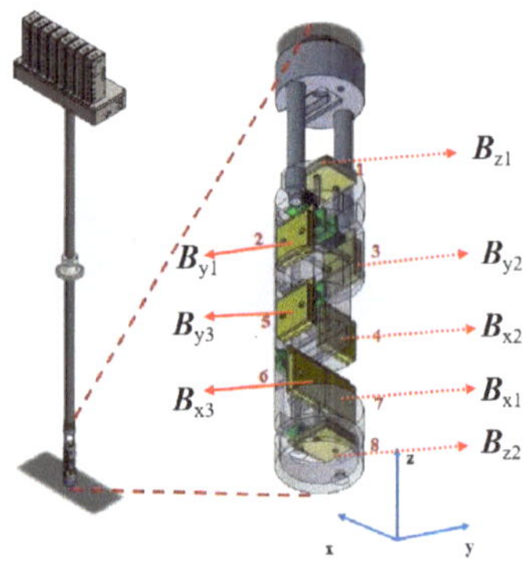

Figure 1. A structure diagram of the probe.

2.3. Inversion Algorithm Based on MGT Invariants

A schematic diagram of the inversion algorithm based on the two-point MGT proposed in this paper is shown in Figure 2, where T represents the magnetic target; A and B represent the two observation points; $\boldsymbol{r}_A$ and $\boldsymbol{r}_B$ are the relative position vectors of the magnetic target to observation point A and observation point B, respectively; φ is the angle between $\boldsymbol{r}_A$ and $\boldsymbol{r}_B$; $\boldsymbol{M}$ is the magnetic moment vector of the magnetic target; θ_A is the angle between $\boldsymbol{r}_A$ and $\boldsymbol{M}$; θ_B is the angle between $\boldsymbol{r}_B$ and $\boldsymbol{M}$; $\boldsymbol{n}_{A2}$ is the unit normal vector of the plane ATP formed by $\boldsymbol{M}$ and $\boldsymbol{r}_A$; and $\boldsymbol{n}_{B2}$ is the unit normal vector of the plane BTP formed by $\boldsymbol{M}$ and $\boldsymbol{r}_B$.

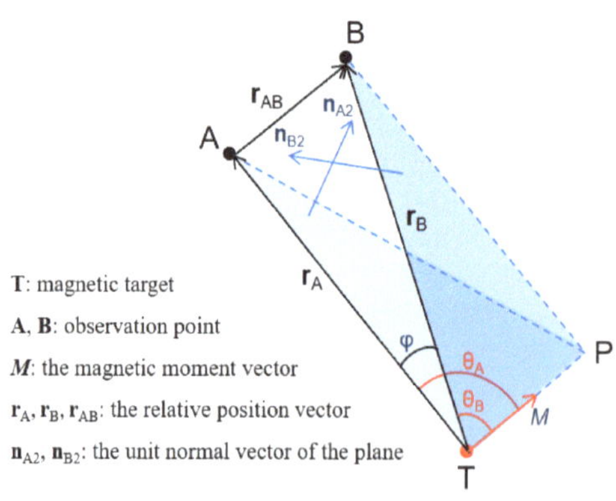

Figure 2. A schematic diagram of the non-coplanar inversion algorithm.

As can be seen from Figure 2, the relative position vector $\boldsymbol{r}_{AB}$ from observation point A to observation point B can be expressed as follows:

$$\boldsymbol{r}_{AB} = \boldsymbol{r}_B - \boldsymbol{r}_A. \tag{10}$$

In the actual measurement process, since the positions of observation points A and B were known, the position of the magnetic target T could be obtained as long as $\boldsymbol{r}_A$ or $\boldsymbol{r}_B$ was solved. According to the law of cosine, the relationship between $\boldsymbol{r}_A$ and $\boldsymbol{r}_B$ can be denoted as follows:

$$r_{AB}^2 = r_A^2 + r_B^2 - 2r_A r_B \cos\varphi, \tag{11}$$

where r_A, r_B, and r_{AB} are the moduli of $\boldsymbol{r}_A$, $\boldsymbol{r}_B$, and $\boldsymbol{r}_{AB}$, respectively. According to Equation (6), the relation between $\boldsymbol{r}_A$ and $\boldsymbol{r}_B$ can also be deduced as follows:

$$\frac{r_A^4}{r_B^4} = \frac{NSS_B}{NSS_A}. \tag{12}$$

As the magnetic gradient tensor $\boldsymbol{G}$ at the observation point can be obtained via the superconducting MGT measurement system, the NSS of the observation point can be calculated through the eigenvalues of G. Therefore, as long as $\cos\varphi$ is solved, r_A can be obtained. The procedure for solving $\cos\varphi$ is as follows:

(a) When $\boldsymbol{n}_{A2} \times \boldsymbol{n}_{B2} \neq 0$, which means $\boldsymbol{M}$ is not coplanar with $\boldsymbol{r}_A$ and $\boldsymbol{r}_B$, the three vectors can form a triangular pyramid, as shown in Figure 2. Since the distance between observation points A and B is particularly small compared to the distance between the observation point and the magnetic target, φ and the dihedral angle α (the angle between the plane ATP and BTP) must be less than 90 degrees. According to the relation between the angle of the lines and the angle of the planes in the triangular pyramid, $\cos\varphi$ can be deduced as follows:

$$\cos\varphi = \cos\theta_A \cos\theta_B + \sin\theta_A \sin\theta_B \cos\alpha, \tag{13}$$

where $\cos\theta_A$ and $\cos\theta_B$ can be calculated from Equation (4). The value of $\sin\theta_A \sin\theta_B$ can be uniquely determined as both θ_A and θ_B are less than 180 degrees. Based on the relation between the dihedral angle α and the unit normal vector of the two planes, as well as considering the fact that α is an acute angle, $\cos\alpha$ can be expressed as follows:

$$\cos\alpha = |\boldsymbol{n}_{A2} \cdot \boldsymbol{n}_{B2}|. \tag{14}$$

(b) When $\boldsymbol{n}_{A2} \times \boldsymbol{n}_{B2} = 0$, $\boldsymbol{M}$ is coplanar with $\boldsymbol{r}_A$ and $\boldsymbol{r}_B$, as shown in Figure 3, then $\boldsymbol{M}_i$ is the magnetic moment vector of the magnetic dipole, where i represents different regions. If $\boldsymbol{n}_{A2} \cdot \boldsymbol{n}_{B2} < 0$, which means that $\boldsymbol{M}$ is located at the acute angle area formed by the relative position vectors $\boldsymbol{r}_A$ and $\boldsymbol{r}_B$, then $\cos\varphi$ can be solved by

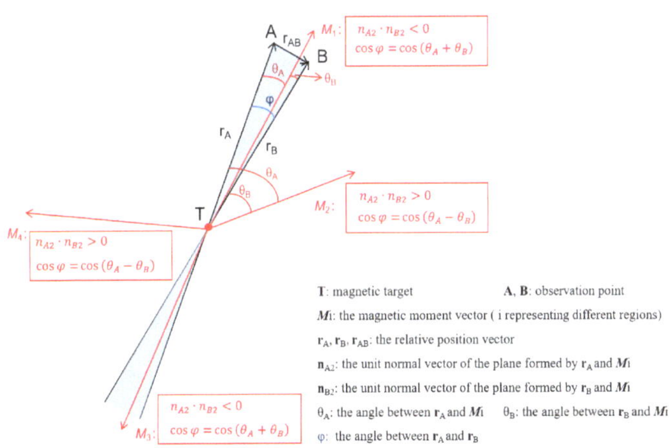

Figure 3. A schematic diagram of the coplanar localization algorithm.

$$\cos\varphi = \cos(\theta_A + \theta_B) = \cos\theta_A \cos\theta_B - \sin\theta_A \sin\theta_B. \quad (15)$$

If $\boldsymbol{n}_{A2} \cdot \boldsymbol{n}_{B2} > 0$, which means the relative position vectors $\boldsymbol{r}_A$ and $\boldsymbol{r}_B$ are on the same side of $\boldsymbol{M}$, then $\cos\varphi$ can be solved by

$$\cos\varphi = \cos(\theta_A - \theta_B) = \cos\theta_A \cos\theta_B + \sin\theta_A \sin\theta_B, \quad (16)$$

where $\boldsymbol{r}_A$ can be calculated through Equations (11)–(13). Then, M, the modulus of the magnetic moment, can be solved by substituting $\boldsymbol{r}_A$ into Equation (7).

The cosine of the angle between $\boldsymbol{r}_A$ and $\boldsymbol{r}_B$ can also be expressed as follows:

$$\cos\varphi = \frac{\boldsymbol{r}_A \cdot \boldsymbol{r}_B}{r_A r_B} = \frac{\boldsymbol{r}_A \cdot (\boldsymbol{r}_A + \boldsymbol{r}_{AB})}{r_A r_B} = \frac{r_A^2 + r_A \boldsymbol{n}_A \cdot \boldsymbol{r}_{AB}}{r_A r_B}, \quad (17)$$

where $\boldsymbol{n}_A$ is the unit direction vectors of $\boldsymbol{r}_A$. According to the properties of the plane unit normal vector, the following equations can be obtained:

$$\boldsymbol{n}_A \cdot \boldsymbol{n}_{A2} = 0, \quad (18)$$

$$\boldsymbol{n}_B \cdot \boldsymbol{n}_{B2} = (\boldsymbol{n}_A + \boldsymbol{n}_{AB}) \cdot \boldsymbol{n}_{B2} = 0, \quad (19)$$

$$\boldsymbol{M} \cdot \boldsymbol{n}_{A2} = M\boldsymbol{n}_M \cdot \boldsymbol{n}_{A2} = 0, \quad (20)$$

$$\boldsymbol{M} \cdot \boldsymbol{n}_{B2} = M\boldsymbol{n}_M \cdot \boldsymbol{n}_{B2} = 0, \quad (21)$$

where $\boldsymbol{n}_B$, $\boldsymbol{n}_{AB}$, and $\boldsymbol{n}_M$ are the unit direction vectors of $\boldsymbol{r}_B$, $\boldsymbol{r}_{AB}$, and $\boldsymbol{M}$. By combining Equations (5) and (17), $\boldsymbol{n}_A$, $\boldsymbol{n}_B$, and $\boldsymbol{n}_M$ can be solved. As a result, $\boldsymbol{r}_A$, $\boldsymbol{r}_B$, and $\boldsymbol{M}$ can be obtained via the following:

$$\boldsymbol{r}_A = r_A \boldsymbol{n}_A, \quad (22)$$

$$\boldsymbol{r}_B = r_B \boldsymbol{n}_B, \quad (23)$$

$$\boldsymbol{M} = M\boldsymbol{n}_M. \quad (24)$$

However, in the actual detection process, the $\boldsymbol{r}_A$ calculated by this algorithm is accurate only when the signal-to-noise ratio is particularly high. Therefore, due to the presence of interference, the solution of $\boldsymbol{r}_A$ is transformed into an optimization problem. In order to improve the anti-interference ability of this localization method, the objective function can be constructed as follows [2]:

$$f = \min\|(\boldsymbol{G_A} - \boldsymbol{G_B})\boldsymbol{n_A} - (3\boldsymbol{G_A} + \boldsymbol{G_B})\boldsymbol{r_{AB}}\|_2, \quad (25)$$

where $\boldsymbol{G_A}$ and $\boldsymbol{G_B}$ are the magnetic gradient tensor of observation points A and B, respectively. The relative position vector $\boldsymbol{r}_A$ obtained from Equation (22) is taken as the initial solution, and a more accurate $\boldsymbol{r}_A$ can be solved via the quasi-Newton optimization algorithm. Naturally, the magnetic moment of the magnetic dipole $\boldsymbol{M}$ can be obtained via the final $\boldsymbol{r}_A$.

As a result, in the whole inversion process, only the positions of two observation points and the corresponding magnetic gradient tensors are needed in the TMGT algorithm, which is not affected by the geomagnetic field and has strong anti-interference ability.

3. Simulations

As shown in Figure 4, a spherical trajectory model was established to analyze the performance of the algorithms under various orientations of the magnetic target in order to verify the feasibility of the TMGT localization algorithm. In this model, observation point A is at the origin of the coordinate system and observation point B is 10 m away from observation point A on the x-axis. The magnetic target T moves along the surface of the spherical trajectory model, where the polar angle β varies from 0 to 180 degrees and the azimuthal angle γ varies from 0 to 360 degrees (both with an interval of 1 degree). The radius r_A of the model is 500 m, and the magnetic moment M of the magnetic target T is

$(50, 50, 70.7) \times 10^6$ A·m^2. In order to quantitatively evaluate the localization performance of the inversion algorithm, the relative localization error ε is used and defined as

$$\varepsilon = \sqrt{(-x_t + x_{t0})^2 + (-y_t + y_{t0})^2 + (-z_t + z_{t0})^2} / r_A \times 100\%, \tag{26}$$

where (x_{t0}, y_{t0}, z_{t0}) is the real coordinate of the magnetic target, and (x_t, y_t, z_t) is the estimated coordinate of the inversion algorithm. During the remote magnetic target localization process, if ε exceeds 10%, then the localization is considered a failure.

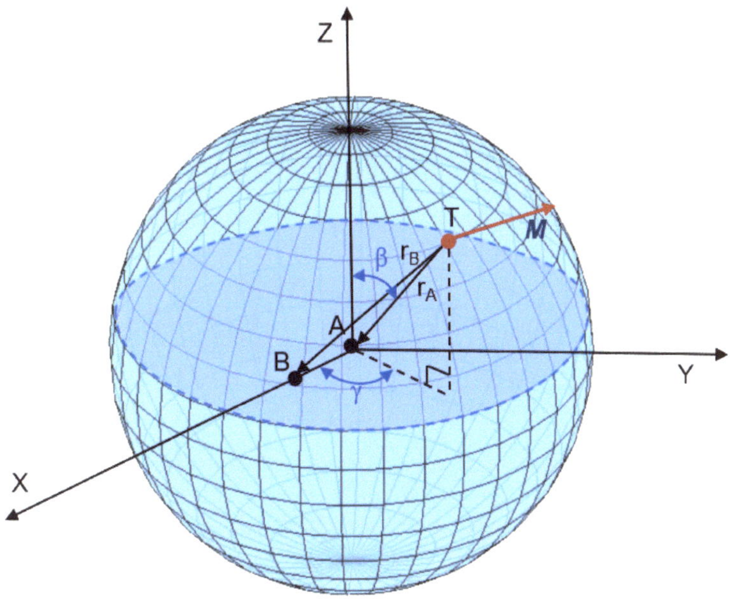

Figure 4. Spherical trajectory model.

3.1. Without the Influence of Noise

In order to compare the localization performance of the NSPT, XTPT, and TMGT methods, a set of simulations without noise were carried out, and the simulation results are shown in Figure 5. It can be seen that the relative localization errors were particularly large when the magnetic target was at certain points, which can be called positioning blind spots. Without the influence of noise, the location and number of the positioning blind spots in each method were kept unchanged after repeated simulations. Among these three methods, XTPT had the largest number of positioning blind spots. TMGT had three positioning blind spots, and NSPT had two positioning blind spots. In the non-positioning blind area, the relative localization error of the TMGT method was much smaller than those of the NSPT and XTPT methods, which were at about a magnitude of 10^{-12} or below. Meanwhile, the relative localization error of the NSPT method was three orders of magnitude smaller than that of the XTPT method. In summary, under ideal conditions, the TMGT method proposed in this paper had the best localization performance, followed by the NSPT and XTPT methods.

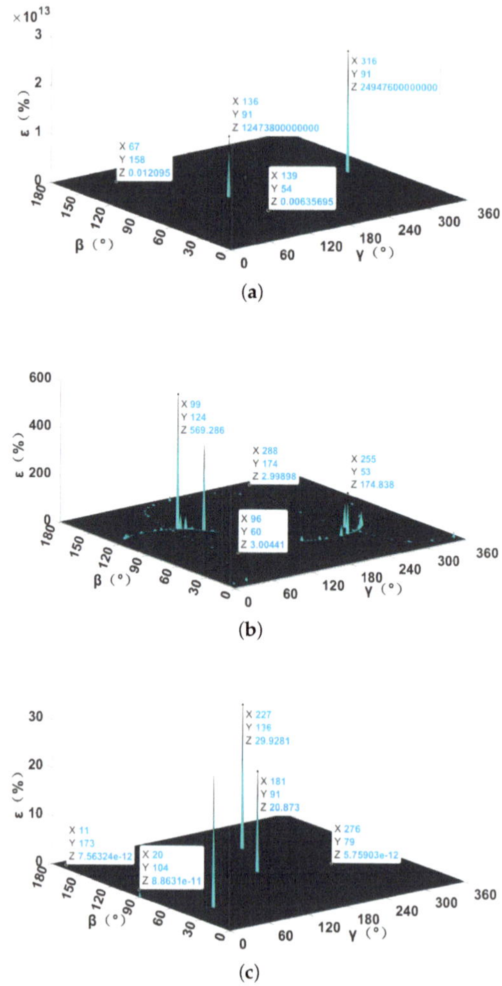

Figure 5. Relative localization error without noise: (**a**) NSPT method; (**b**) XTPT method (**c**) TMGT method.

3.2. With the Influence of Noise

Magnetic field measurement noise, which mainly includes geomagnetic background measurement noise and instrument measurement noise, are unavoidable in practical localization applications. Since the slightest geomagnetic activity could cause magnetic field fluctuations of ±20 nT, a Gaussian distribution geomagnetic noise with a mean of 0 and a standard deviation of 20 nT was added to the three components of the calculated magnetic induction vector $\boldsymbol{B}$. At the same time, instrument measurement noise with an average of 0 and a standard deviation of 10^{-13} T was added to each element of the calculated magnetic gradient tensor $\boldsymbol{G}$ since the sensitivity of the superconducting MGT measurement system, which is determined by the bottom line of magnetic noise, was generally in the order of 10^{-13} T. Then, a set of simulations involving magnetic field measurement noise was carried out.

As illustrated in Figure 6, under the same simulation conditions, each method had its own positioning blind area, and the location of the area was kept unchanged in repeated simulations. In these three methods, the positioning blind areas of the TMGT method and

NSPT method were distributed like a sine wave curve, while that of the XTPT method was distributed along two curves and had the largest number of positioning blind spots. In the non-positioning blind area, the NSPT method needs to measure the magnetic induction intensity vector B at the observation point caused by the magnetic target, which is inversely proportional to the third power of the detection distance. When the detection distance was 500 m, the magnetic induction intensity vector signal generated by the magnetic target was submerged in the background noise of the geomagnetic field, and the relative localization errors were greater than 10%, resulting in a failure of positioning. As the TMGT and XTPT methods only need to measure the MGT at the two observation points, they were mainly affected by the instrument measurement noise. With the same instrument measurement noise, the relative localization error of the XTPT method was about 3% in the non-positioning blind area with or without of noise, and the relative localization error of the TMGT method was two to three orders smaller than that of the XTPT method. Compared with the other two methods, the TMGT method has better noise resistance and localization accuracy.

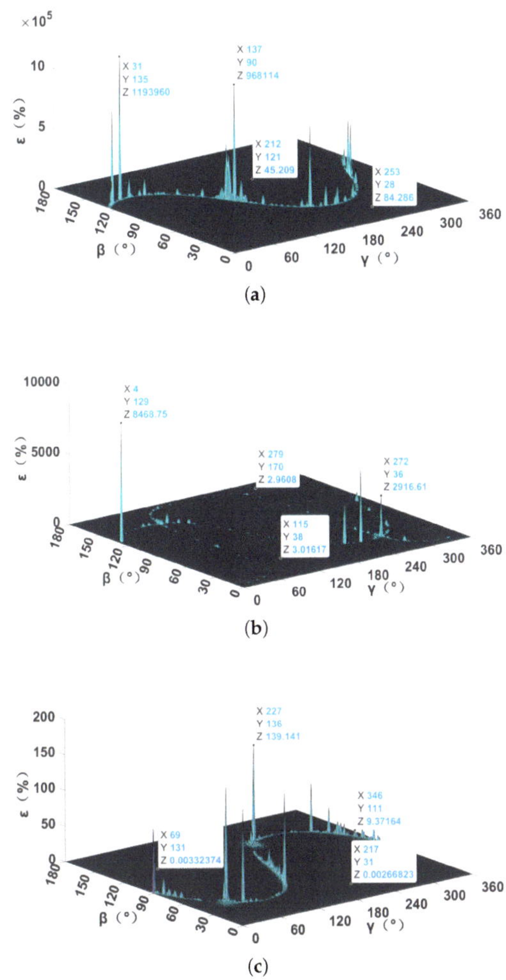

Figure 6. The relative localization error with noise: (**a**) NSPT method; (**b**) XTPT method; and (**c**) TMGT method.

The influence of different instrument noise on the localization error of the TMGT method was clear when the detection distance and magnetic moment were the same, as shown in Figure 7. Evidently, the greater the instrument noise, the greater the localization error and vice versa. It can be inferred that under the same magnetic moment and localization error, the smaller the instrument noise, the farther the detection distance. Therefore, the probe structure and readout circuits of the measurement system must be optimized to reduce the instrument noise of the whole system in order to improve the positioning accuracy or detection distance of the measuring system.

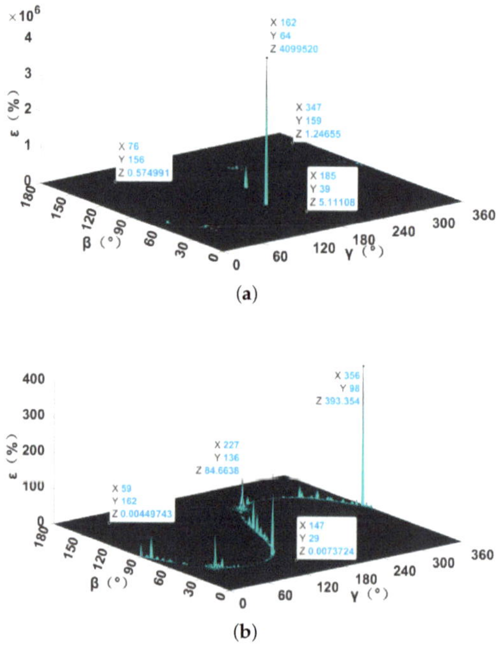

Figure 7. The relative localization error with different instrument noise: (**a**) standard deviation of 10^{-12} T; (**b**) standard deviation of 10^{-14} T.

3.3. Positioning Blind-Area Analysis

When the magnetic target is located at some points, the MGT matrix at the observation points is singular, and the area where these points converge is called the positioning blind area. Although the TMGT method does not calculate the inverse of the MGT, it may influence the solving of the spatial positioning model. In the previous simulation, it was found that different localization methods have their own fixed positioning blind area under the fixed magnetic moment vectors of a magnetic target. In order to further verify this conjecture, a simulation that only changed the magnetic moment vector was carried out, the results of which are shown in Figure 8. Different magnetic moment vectors were found to only change the relative localization error in the positioning blind area. The location of the positioning blind area was kept unchanged, and the localization performance in the non-positioning blind area was not affected. In order to eliminate the effect of the irreversible MGT matrix, the Moore–Penrose generalized inverse matrix [21], truncated singular value decomposition [22], or eigenvector constraint-based method [23] can be adopted. In practical detection applications, multiple localization methods or measurement systems can also be combined to eliminate the positioning blind area.

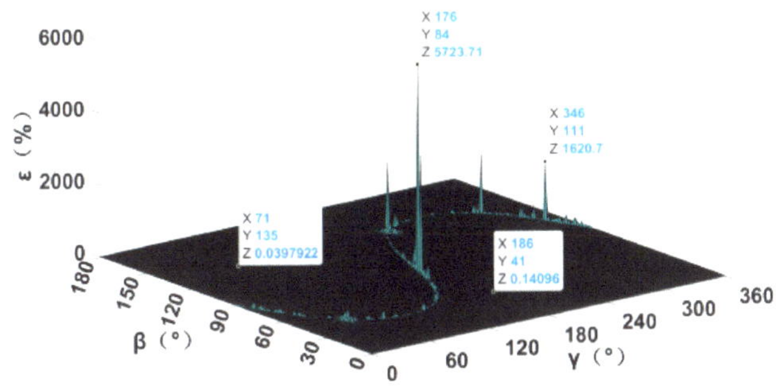

(a)

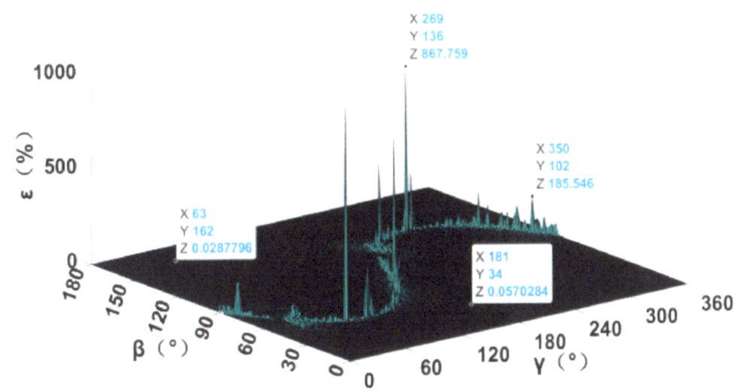

(b)

Figure 8. Relative localization error with different magnetic moment vectors: (**a**) M = (50, 70.7, 50) $\times$ 10^6 A·m^2; (**b**) M = (70.7, 50, 50) $\times$ 10^6 A·m^2.

3.4. With and without the Optimization Algorithm

The quasi-Newton optimization algorithm was added to the TMGT method in order to further improve the noise resistance, and the NTPT method was selected as a comparison in order to analyze the effect of the optimization algorithm on the localization performance. As shown in Figure 9, the relative localization error of the TMGT method was less than 10% at most points in the positioning blind area. Conversely, the relative localization errors of the NTPT method in the blind area were almost all greater than 10%. At the same time, the relative localization error of the TMGT method was about two orders of magnitude smaller than that of the NTPT method in the non-positioning blind area.

Since the introduction of a non-linear optimization algorithm will generally increase the localization time, the average localization time of the TMGT method and NTPT method were also compared and analyzed, and the maximum number of iterations of the TMGT method was set to 1000. The simulation code was performed on a laptop equipped with an Intel ® Core ™ Ultra 7 155H processor running at a nominal clock speed of 1.4 GHz. In each simulation cycle, the magnetic targets that moved along 65,341 points on the spherical trajectory model were localized. The total time required by the TMGT method was 1987.299689 s, and the average time for locating each point was about 0.030414 s. The total time of the NTPT method was 1999.893442 s, and the average time for locating each point was about 0.030607 s. It can be seen that the TMGT method proposed in this paper does not deteriorate in real time due to the introduction of an optimization algorithm, and it is also demonstrated that it can meet real-time localization requirements.

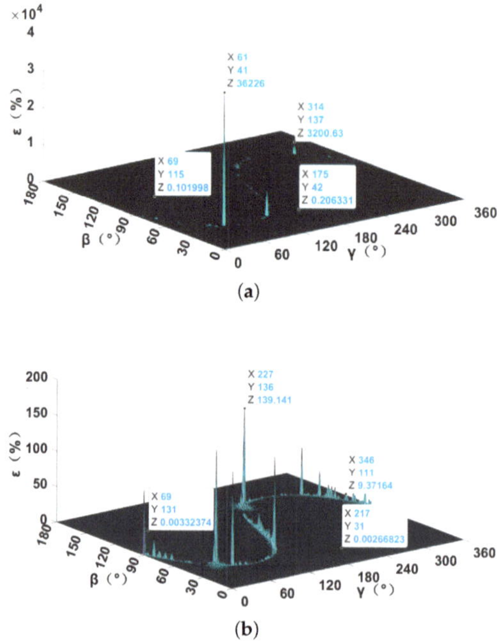

Figure 9. Relative localization error of the (**a**) NTPT method and (**b**) TMGT method.

4. Experiments and Results Analysis

4.1. Equivalent Experimental Setup

Since it is difficult to find a site that can meet the requirement of a detection distance of 500 m with few magnetic interference factors, an equivalent experiment was adopted to verify the effectiveness of the TMGT method. In Section 3, the magnetic induction intensity generated by the magnetic dipole target at observation point A was on the order of 10^{-10} T to 10^{-13} T. Therefore, as seen in Figure 10b, a one-dimensional coil was designed and made as the magnetic dipole target in the equivalence experiment, and it was able to produce a magnetic induction intensity of about 10^{-10} T at a distance of 10 meters. The design parameters of the equivalent magnetic dipole object are shown in Table 1.

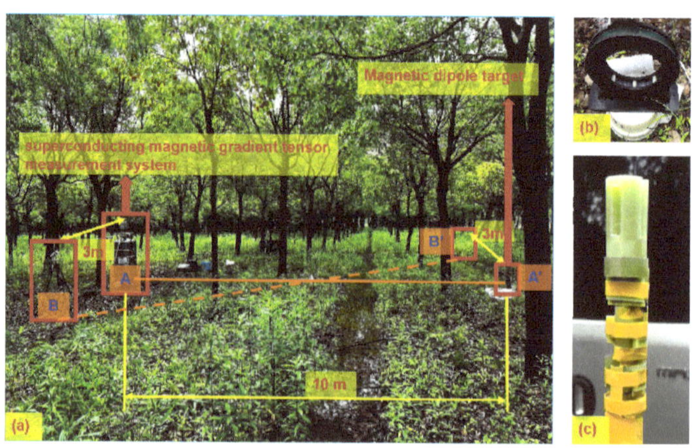

Figure 10. The equivalence experiment: (**a**) the experimental site; (**b**) the magnetic dipole target; and (**c**) the superconducting magnetic gradient tensor probe.

Table 1. The design parameters of the equivalent magnetic dipole object.

Name	Coil Specification	Coil Turns	Effective Diameter	Total Coil Inductance	Total Coil Resistance (25 °C)
One-dimensional coil	1.5 mm	150	200 mm	6.823 mH	0.93 Ω

The SQUID magnetometers, the SQUID readout circuits, and the control software of the superconducting MGT system used in the experiment were all independently developed by the Shanghai Institute of Microsystem and Information Technology. The sensitivity of the SQUID magnetometers was 5 fT/Hz^2. The pendulum rate of the superconducting magnetic gradient tensor measurement system was 2 mT/s, and the bandwidth affected by the metal dewar was about 500 Hz. The experimental data measured by the superconducting MGT probe were stored in the data acquisition system (NI 9235) through the SQUID readout circuits. The probe is shown in Figure 10c, and it is described in detail in Section 2.2. The baseline distance between the SQUID sensors was determined, and the output of the probe was as follows:

$$G = 100\begin{bmatrix} -\frac{1}{2}(B_{1y} - B_{2y}) - \frac{1}{11}(B_{1z} - B_{2z}) & \frac{10}{17}(B_{1x} - B_{3x}) & \frac{5}{14}(B_{1x} - B_{2x}) \\ \frac{10}{17}(B_{1x} - B_{3x}) & \frac{1}{2}(B_{1y} - B_{2y}) & \frac{5}{16}(B_{1y} - B_{3y}) \\ \frac{5}{14}(B_{1x} - B_{2x}) & \frac{5}{16}(B_{1y} - B_{3y}) & \frac{1}{11}(B_{1z} - B_{2z}) \end{bmatrix}. \quad (27)$$

As shown in Figure 10, the site of this experiment was Hengsha Island in Shanghai, where the background magnetic field is relatively clean. Before the experiment, in order to further reduce the magnetic interference caused by vibration and other factors, we buried the superconducting MGT measurement system underground. Observation point A was set as the origin of the Cartesian coordinate system, and the positive direction of the x-axis was set as south on the compass. The coordinates of the magnetic target position A′ were about (−10, 0, 1) m, the coordinates of observation point B were about (0, −3, 0) m, and the coordinates of the magnetic target position B′ were about (−10, 3, 0) m. The TMGT method needs to measure the magnetic gradient tensor at two observation points, but there was only one superconducting magnetic gradient tensor measurement system. According to the relativity of measurement, the MGT at observation point B generated by the magnetic dipole target T at position A′ is the same as the MGT at observation point A that is generated by the magnetic dipole target T at position B′. Therefore, in order to facilitate the experiment—after measuring the magnetic gradient tensor $\boldsymbol{G}_{\mathrm{A}}$ at observation point A, which is generated by the magnetic dipole target T at the position A′—we moved the magnetic dipole target T to position B′, and we then measured the magnetic gradient tensor $\boldsymbol{G}_{\mathrm{B}}$ at observation point A. Finally, the TMGT method and XTPT method were used to locate the magnetic target at the position A′ based on the measured $\boldsymbol{G}_{\mathrm{A}}$ and $\boldsymbol{G}_{\mathrm{B}}$.

4.2. Analysis of Experimental Results

Compared with the other localization methods mentioned above, the TMGT method not only has higher localization accuracy but can also calculate the magnetic moment vector of the measured target. The higher the localization accuracy, the closer the magnetic moment vector is to the true value. Eight different currents were successively added to the one-dimensional coil to simulate the magnetic dipole target with different magnetic moments, and eight sets of measurement data were obtained. Due to the excessive, uncontrolled magnetic interference in the field experiment, the localization errors of the other two-point methods were all much greater than 10% in the eight experiments; as such, only the experimental results of the TMGT method are listed in Table 2. Although the superconducting MGT measurement system is highly susceptible to magnetic interference due to its ultra-high sensitivity, the TMGT method can still achieve a relative good localization effect when the current is greater than 1.242 A in the field experiment. The minimum relative localization error was 4.5229%, and the corresponding magnetic moment amplitude was 3.3940×10^{10} A·m^2. The magnetic gradient tensor generated by the magnetic dipole target at the observation point was overwhelmed by the magnetic noise when the current was too low. Hence, the magnetic noise mechanism and compensation algorithm of the superconducting MGT measurement system will be studied in order to further reduce the influence of magnetic noise and improve the detection distance. At the same time, the localization method proposed in this paper was mainly designed for a single magnetic dipole target; thus, it is not applicable when there are multiple magnetic dipole targets or if the measurement distance is too small. Therefore, the positioning method will be further optimized for different magnetic source models [33–36] in the future.

Table 2. The experimental results of the TMGT method.

Sets	Current (A)	Relative Localization Error (%)	Magnetic Moment (A·m^2)
1	0.828	338.5185	9.2624×10^{12}
2	0.966	104.9234	3.8083×10^{11}
3	1.104	42.2661	1.1361×10^{11}
4	1.242	10.1966	2.4411×10^{10}
5	1.380	4.5229	3.3940×10^{10}
6	2.760	5.0574	2.8727×10^{10}
7	4.140	6.0444	3.2539×10^{10}
8	5.520	10.3482	3.0586×10^{10}

5. Conclusions

At present, the MGT inversion algorithms are mainly applied to the MGT detection system based on fluxgate magnetometers. In this case, the detection distance is limited and the magnetic moment vector of the magnetic target cannot be calculated. In order to improve the detection range and obtain more information about the magnetic target, a remote magnetic localization method based on ultra-sensitive SQUID magnetometers and MGTs at two observation points was proposed. A spherical trajectory simulation model with a detection distance of 500 m was established in order to analyze the localization performance of the TMGT method for magnetic targets in any azimuth. The simulation results show the following: (a) the relative positioning error of the proposed TMGT method is much smaller than that of the NSPT method and the XTPT method regardless of noise; (b) the smaller the instrument measurement noise, the higher the localization accuracy and the farther the detection distance; (c) the positioning blind area of the TMGT method can be eliminated; and (d) the TMGT method, when using a quasi-Newton optimization algorithm, has better interference suppression ability and can meet the requirements of real-time detection. The equivalent test results show that the TMGT method not only has better localization performance in practical applications, but it can also calculate the magnetic moment vector of a magnetic dipole target. When the detection distance is 10 m, the minimum relative localization error of the TMGT method is 4.5229%. However, when the magnetic moment of the magnetic target is too small, the localization method cannot work. In the future, we will optimize the probe structure of the superconducting MGT system to reduce the instrument measurement noise so as to further improve the detection distance for the magnetic target.

Author Contributions: Conceptualization, Y.Z. and W.L.; methodology, Y.Z. and G.L.; software, Y.Z., G.L. and H.W.; validation, Y.Z., C.W. and L.Q.; formal analysis, Y.Z. and C.W.; investigation, Y.Z., C.W. and G.L.; resources, Y.Z. and W.L.; data curation, Y.Z., C.W. and L.Q.; writing—original draft preparation, Y.Z.; writing—review and editing, Y.Z., H.W. and W.L.; visualization, Y.Z. and H.W.; supervision, Y.Z. and W.L.; project administration, Y.Z. and H.W.; funding acquisition, Y.Z. All authors have read and agreed to the published version of the manuscript.

Funding: This research was supported by the Foundation of State Key Laboratory of Dynamic Measurement Technology (2023-SYSJJ-05), the National Natural Science Foundation of China (62201523), and the Fundamental Research Program of Shanxi Province (20210302124329).

Data Availability Statement: Dataset available on request from the authors (The raw data supporting the conclusions of this article will be made available by the authors on request.).

Conflicts of Interest: The authors declare no conflicts of interest.

References

1. Gao, X.; Yan, S.; Li, B. A Novel Method of Localization for Moving Objects with an Alternating Magnetic Field. *Sensors* **2017**, *17*, 923. [CrossRef] [PubMed]
2. Xu, L.; Gu, H.; Chang, M.; Fang, L.; Lin, P.; Lin, C. Magnetic Target Linear Location Method Using Two-Point Gradient Full Tensor. *IEEE Trans. Instrum. Meas.* **2021**, *70*, 6007808. [CrossRef]
3. Liu, H.; Wang, X.; Zhao, C.; Ge, J.; Dong, H.; Liu, Z. Magnetic Dipole Two-Point Tensor Positioning Based on Magnetic Moment Constraints. *IEEE Trans. Instrum. Meas.* **2021**, *70*, 9700410. [CrossRef]
4. Stolz, R.; Zakosarenko, V.; Schulz, M.; Chwala, A.; Fritzsch, L.; Meyer, H.G.; Kostlin, E. Magnetic full-tensor SQUID gradiometer system for geophysical applications. *Lead. Edge* **2006**, *25*, 178–180. [CrossRef]
5. von der Osten-Woldenburg, H.; Chaume, B.; Reinhard, W. New archaeological discoveries through magnetic gradiometry: The early Celtic settlement on Mont Lassois, France. *Lead. Edge* **2006**, *25*, 46–48. [CrossRef]
6. Billings, S.D. Superconducting Magnetic Tensor Gradiometer System for Detection of Underwater Military Munitions. Ph.D. Thesis, Black Tusk Geophysics Inc., Vancouver, BC, Canada, 2012.
7. Chwala, A.; Schmelz, M.; Zakosarenko, V.; Schiffler, M.; Schneider, M.; Thürk, M.; Bräuer, S.; Bauer, F.; Schulz, M.; Krüger, A.; et al. Underwater operation of a full tensor SQUID gradiometer system. *Supercond. Sci. Technol.* **2019**, *32*, 024003. [CrossRef]
8. Liu, G.; Zhang, Y.; Liu, W. Structural Design and Parameter Optimization of Magnetic Gradient Tensor Measurement System. *Sensors* **2024**, *24*, 4083. [CrossRef]
9. Young, J.; Keenan, S.; Clark, D.; Sullivan, P.; Billings, S. Development of a high temperature superconducting magnetic tensor gradiometer for underwater UXO detection. In Proceedings of the OCEANS'10 IEEE SYDNEY, Sydney, NSW, Australia, 24–27 May 2010; pp. 1–4. [CrossRef]
10. Stolz, R.; Schmelz, M.; Zakosarenko, V.; Foley, C.; Tanabe, K.; Xie, X.; Fagaly, R.L. Superconducting sensors and methods in geophysical applications. *Supercond. Sci. Technol.* **2021**, *34*, 033001. [CrossRef]
11. Li, Q.; Li, Z.; Zhang, Y.; Fan, H.; Yin, G. Integrated Compensation and Rotation Alignment for Three-Axis Magnetic Sensors Array. *IEEE Trans. Magn.* **2018**, *54*, 4001011. [CrossRef]
12. Stolz, R.; Schmelz, M.; Anders, S.; Kunert, J.; Franke, D.; Zakosarenko, V. Long baseline LTS SQUID gradiometers with sub-μm sized Josephson junctions. *Supercond. Sci. Technol.* **2020**, *33*, 055002. [CrossRef]
13. Stolz, R.; Schiffler, M.; Becken, M.; Thiede, A.; Schneider, M.; Chubak, G.; Marsden, P.; Bergshjorth, A.B.; Schaefer, M.; Terblanche, O. SQUIDs for magnetic and electromagnetic methods in mineral exploration. *Miner. Econ.* **2022**, *35*, 467–494. [CrossRef]
14. Wynn, W. *Dipole Tracking with a Gradiometer*; Technical Report 3493; Informal Report NSRDL/PC; Naval Ship Research and Development Laboratory: Bethesda, MD, USA, 1972.
15. Wynn, W.; Frahm, C.; Carroll, P.; Clark, R.; Wellhoner, J.; Wynn, M. Advanced superconducting gradiometer/Magnetometer arrays and a novel signal processing technique. *IEEE Trans. Magn.* **1975**, *11*, 701–707. [CrossRef]
16. Frahm, C.P. *Inversion of the Magnetic Field Gradient Equation for a Magnetic Dipole Field*; Technical report; Informal Report; NCSL: New Orleans, LA, USA, 1972.
17. Wilson, H. Analysis of the Magnetic Gradient Tensor. In *Technical Memorandum, Defence Research Establishment Pacific*; Defence Research Establishment Pacific: Research and Development Branch, Department of National Defence Canada: Victoria, BC, Canada, 1985.
18. Clark, D. New methods for interpretation of magnetic vector and gradient tensor data I: Eigenvector analysis and the normalised source strength. *Explor. Geophys.* **2012**, *43*, 267–282. [CrossRef]
19. Yin, G.; Zhang, L.; Jiang, H.; Wei, Z.; Xie, Y. A closed-form formula for magnetic dipole localization by measurement of its magnetic field vector and magnetic gradient tensor. *J. Magn. Magn. Mater.* **2020**, *499*, 166274. [CrossRef]
20. Nara, T.; Suzuki, S.; Ando, S. A Closed-Form Formula for Magnetic Dipole Localization by Measurement of Its Magnetic Field and Spatial Gradients. *IEEE Trans. Magn.* **2006**, *42*, 3291–3293. [CrossRef]
21. Nara, T.; Ito, W. Moore–Penrose generalized inverse of the gradient tensor in Euler's equation for locating a magnetic dipole. *J. Appl. Phys.* **2014**, *115*, 17E504. [CrossRef]
22. Higuchi, Y.; Nara, T.; Ando, S. A Truncated Singular Value Decomposition Approach for Locating a Magnetic Dipole with Euler's Equation. *Int. J. Appl. Electromagn. Mech.* **2016**, *52*, 67–72. [CrossRef]
23. Tang, W.; Huang, G.; Li, G.; Yang, G. Eigenvector Constraint-Based Method for Eliminating Dead Zone in Magnetic Target Localization. *Remote Sens.* **2023**, *15*, 4959. [CrossRef]
24. Sui, Y.; Leslie, K.; Clark, D. Multiple-Order Magnetic Gradient Tensors for Localization of a Magnetic Dipole. *IEEE Magn. Lett.* **2017**, *8*, 6506605. [CrossRef]

25. Wang, B.; Ren, G.; Li, Z.; Li, Q. A third-order magnetic gradient tensor optimization algorithm based on the second-orde improved central difference method. *AIP Adv.* **2021**, *11*, 065302. [CrossRef]
26. Wiegert, R.; Lee, K.; Oeschger, J. Improved magnetic STAR methods for real-time, point-by-point localization of unexplode ordnance and buried mines. In Proceedings of the OCEANS 2008, Quebec City, QC, Canada, 15–18 September 2008; pp. 1– [CrossRef]
27. Wiegert, R.F. Magnetic STAR technology for real-time localization and classification of unexploded ordnance and buried mine In Proceedings of the Defense + Commercial Sensing, Orlando, FL, USA, 1 May 2009.
28. Li, X.; Yan, S.; Liu, J.; Yan, Y. Novel Magnetic Localization Methods for Minimizing the Ellipse Error Based on Tensor Invariant *IEEE Magn. Lett.* **2022**, *13*, 8105205. [CrossRef]
29. Lin, S.; Pan, D.; Wang, B.; Liu, Z.; Liu, G.; Wang, L.; Li, L. Improvement and Omnidirectional Analysis of Magnetic Gradien Tensor Invariants Method. *IEEE Trans. Ind. Electr.* **2021**, *68*, 7603–7612. [CrossRef]
30. Wang, C.; Zhang, X.; Qu, X.; Pan, X.; Fang, G.; Chen, L. A Modified Magnetic Gradient Contraction Based Method fo Ferromagnetic Target Localization. *Sensors* **2016**, *16*, 2168. [CrossRef] [PubMed]
31. Liu, J.; Li, X.; Zeng, X. Online magnetic target location method based on the magnetic gradient tensor of two points. *Chin. Geophys.* **2017**, *60*, 3995–4004. (In Chinese) [CrossRef]
32. Liu, G.; Zhang, Y.; Wang, C.; Li, Q.; Li, F.; Liu, W. A New Magnetic Target Localization Method Based on Two-Point Magneti Gradient Tensor. *Remote Sens.* **2022**, *14*, 6088. [CrossRef]
33. Beiki, M.; Clark, D.; Austin, J.; Foss, C. Estimating source location using normalized magnetic source strength calculated fron magnetic gradient tensor data. *Geophysics* **2012**, *77*, J23–J37. [CrossRef]
34. Spantideas, S.T.; Kapsalis, N.C.; Kakarakis, S.D.J.; Capsalis, C.N. A Method of Predicting Composite Magnetic Sources Employin Particle Swarm Optimization. *Prog. Electromagn. Res. M* **2014**, *39*, 161–170. [CrossRef]
35. Trenkel, C.; Engelke, S.; Bubeck, K.; Lange, J.; Tenacci, Z.; Junge, A. Reliable Distance Scaling of AC Magnetic Fields for Spac Mission Verification Campaigns. In Proceedings of the 2019 ESA Workshop on Aerospace EMC (Aerospace EMC), Budapes Hungary, 20–22 May 2019; pp. 1–9. [CrossRef]
36. Spantideas, S.T.; Giannopoulos, A.E.; Kapsalis, N.C.; Capsalis, C.N. A Deep Learning Method for Modeling the Magneti Signature of Spacecraft Equipment Using Multiple Magnetic Dipoles. *IEEE Magn. Lett.* **2021**, *12*, 2100905. [CrossRef]

sensors

Article

Microfluidic Detection of SPIONs and Co-Ferrite Ferrofluid Using Amorphous Wire Magneto-Impedance Sensor

Gabriele Barrera [1,*], Federica Celegato [1], Marta Vassallo [1], Daniele Martella [2,3], Marco Coïsson [1], Elena S. Olivetti [1], Luca Martino [1], Hüseyin Sözeri [4], Alessandra Manzin [1] and Paola Tiberto [1]

1 Department of Advanced Materials Metrology and Life Science, Istituto Nazionale di Ricerca Metrologica (INRiM), Strada delle Cacce, 91, 10135 Turin, Italy; f.celegato@inrim.it (F.C.); marta.vassallo95@gmail.com (M.V.); m.coisson@inrim.it (M.C.); e.olivetti@inrim.it (E.S.O.); l.martino@inrim.it (L.M.); a.manzin@inrim.it (A.M.); p.tiberto@inrim.it (P.T.)
2 European Laboratory for Non Linear Spectroscopy (LENS), via N. Carrara, 1, 50019 Florence, Italy; martella@lens.unifi.it
3 Department of Chemistry "Ugo Schiff", University of Florence, Via della Lastruccia 3-13, 50019 Florence, Italy
4 Magnetics Laboratory, TÜBITAK Ulusal Metroloji Enstitüsü (UME), Gebze Yerleşkesi, 41470 Kocaeli, Turkey; huseyin.sozeri@tubitak.gov.tr
* Correspondence: g.barrera@inrim.it

Abstract: The detection of magnetic nanoparticles in a liquid medium and the quantification of their concentration have the potential to improve the efficiency of several relevant applications in different fields, including medicine, environmental remediation, and mechanical engineering. To this end, sensors based on the magneto-impedance effect have attracted much attention due to their high sensitivity to the stray magnetic field generated by magnetic nanoparticles, their simple fabrication process, and their relatively low cost. To improve the sensitivity of these sensors, a multidisciplinary approach is required to study a wide range of soft magnetic materials as sensing elements and to customize the magnetic properties of nanoparticles. The combination of magneto-impedance sensors with ad hoc microfluidic systems favors the design of integrated portable devices with high specificity towards magnetic ferrofluids, allowing the use of very small sample volumes and making measurements faster and more reliable. In this work, a magneto-impedance sensor based on an amorphous $Fe_{73.5}Nb_3Cu_1Si_{13.5}B_9$ wire as the sensing element is integrated into a customized millifluidic chip. The sensor detects the presence of magnetic nanoparticles in the ferrofluid and distinguishes the different stray fields generated by single-domain superparamagnetic iron oxide nanoparticles or magnetically blocked Co-ferrite nanoparticles.

Keywords: magneto-impedance sensor; magnetic nanoparticles; magnetic wire; SPIONs; Co-ferrite; microfluidic system

Citation: Barrera, G.; Celegato, F.; Vassallo, M.; Martella, D.; Coïsson, M.; Olivetti, E.S.; Martino, L.; Sözeri, H.; Manzin, A.; Tiberto, P. Microfluidic Detection of SPIONs and Co-Ferrite Ferrofluid Using Amorphous Wire Magneto-Impedance Sensor. *Sensors* **2024**, *24*, 4902. https://doi.org/10.3390/s24154902

Academic Editor: Grigoris Kaltsas

Received: 4 June 2024
Revised: 17 July 2024
Accepted: 25 July 2024
Published: 28 July 2024

1. Introduction

Magnetic nanoparticles (MNPs) suspended in aqueous or non-aqueous liquids (ferrofluids) have the potential to improve the efficiency of relevant applications in key areas of society, including medicine, environmental treatment, and mechanical engineering [1–4].

In the field of water purification, MNPs have emerged as a promising material due to their large surface area and fast response to magnetic fields [5–9]. The former enables the effective trapping of large amounts of contaminants through adsorption or catalytic degradation, while the latter facilitates their separation from the treated water. However, MNPs themselves can act as contaminants and are potentially harmful to the environment and human health [6,10]. Therefore, a reliable assessment to determine the complete removal of MNPs from water resources (after their use for contaminant removal) is a significant undertaking that can be addressed with the development of dedicated sensors.

Contextually, in the field of biomedicine, the detection of the stray magnetic field generated by MNPs with selected surface functionalization promotes the development of a range of diagnostic sensors and lab-on-chip devices with increasing sensitivity, biocompatibility, reliability, safety, and energy efficiency [2,11–16]. MNP-based biosensors make biomedical diagnostics fast, simple, robust, and high-throughput, improving the detection, separation, and transport of various bioanalytes [14,17], such as cancer biomarkers, which are present at very low concentrations in the early stages of the disease [18].

In both aforementioned fields, the crucial goal of detecting MNPs and quantifying their concentration in a liquid medium is leading to the development of a large number of magnetic sensors [19–22], among which those based on the magneto-impedance effect are of great interest due to their exceptional sensitivity to the small magnetic fields (up to 10^{-12} T [23]), simple fabrication process, and relatively low cost [24–28].

In particular, the magneto-impedance (MI) effect is defined as a significant change in the electrical impedance of a soft ferromagnetic material when an alternating current flows through it and an external magnetic field is present [29–33].

Improving the sensitivity of MI sensors to the stray magnetic field generated by the MNPs is an interesting ongoing multidisciplinary research topic [16,25,26,28]. This goal is being pursued by exploring a wide range of soft magnetic materials as sensing elements with different compositions, microstructures, and shapes, including microwires, films, and ribbons [34–38]. Furthermore, with equal importance, attention is being paid to customizing the magnetic properties of MNPs by varying their composition, shape, size, functionalization, and concentration in the ferrofluid [25,35–38].

The combined application of MI sensors for the detection of MNPs suspended in a liquid medium with a microfluidic system offers promising opportunities for the design of relatively inexpensive miniaturized integrated portable devices with high specificity and sensitivity [16,17,39]. In particular, the milli- and microfluidic chips enable the use of very small volumes of liquid, reducing the amount of sample required and the amount of waste produced, as well as making measurements faster and more reliable [24,40].

This study presents a magneto-impedance (MI) sensor designed to detect the stray magnetic field generated by a ferrofluid. The sensor utilizes amorphous $Fe_{73.5}Nb_3Cu_1Si_{13.5}B_9$ microwire as the MI sensing element, positioned within a custom millifluidic chip that facilitates the flow of the ferrofluid around the wire.

The response of the MI sensor is tested using two ferrofluids with different magnetic behaviors. The first (i.e., the commercial Synomag ferrofluid) contains single-domain superparamagnetic iron-oxide nanoparticles (SPIONs), which are characterized by anhysteretic behavior and zero magnetic moment in the absence of an applied external magnetic field; the second contains magnetically blocked Co-ferrite nanoparticles, which exhibit a net magnetic moment at zero applied magnetic field. The Co-ferrite nanoparticles are synthesized by a co-precipitation method, and before incorporating them into the liquid medium, their structural and morphological properties as well as the magnetic behavior are studied in detail.

2. Materials and Methods

2.1. Materials

Co-ferrite nanoparticles were synthesized using the conventional co-precipitation method described in [41]. In brief, an aqueous solution of cobalt(II) nitrate hexahydrate and iron(III) chloride hexahydrate with a 1:2 molar ratio was prepared. The pH of the solution was raised to 13 by adding 10 M aqueous NaOH, and then the solution was heated to 100 °C under nitrogen flow and stirring. After 30 min, iron(III) nitrate nonahydrate was added to the solution to achieve a Co^{2+}/Fe^{3+} molar ratio of 1:3, followed by cooling to room temperature. The precipitate was decanted using a magnet and washed multiple times with deionized water until reaching a neutral pH. The nanoparticles are dispersed in aqueous medium, resulting in a stable ferrofluid with selected concentrations c equal to 0.8, 4.0, 8.1, 16.2, and 28.3 mg/mL.

Ferrofluid containing superparamagnetic iron-oxide nanoparticles (SPIONs) was purchased from Micromod Partikeltechnologie GmbH, Rostock, Germany [42]. The selected concentrations of SPIONs in the liquid medium are c = 0.76, 1.4, 7.6 mg/mL. The SPIONs are composed of iron oxide (mainly γ-Fe_2O_3) and result in a multicore structure with a nanoflower shape covered by a dextran shell. The magnetic core size is 9.0 $\pm$ 1.7 nm, while the particle size is $\approx$66 nm [42].

The amorphous magnetic wire of nominal composition $Fe_{73.5}Nb_3Cu_1Si_{13.5}B_9$ (FINEMET) was produced by the melt spinning in water technique [43,44]; it is called Fe-based wire in the following text.

2.2. Characterization Techniques

The structure and crystallinity of Co-ferrite NPs were investigated by X-ray diffraction (XRD) in a Panalytical X'Pert PRO MPD diffractometer with Cu K_α radiation. The dried nanopowders were placed in the cavity of a silicon zero-background sample holder and analyzed in Bragg–Brentano configuration. The atomic ratio between Fe and Co in Co-ferrite NPs was estimated by standardless semi-quantitative analysis through energy dispersive spectrometry coupled to a scanning electron microscope (SEM-EDS) after deposition of the nanoparticles on the carbon adhesive tape used for electron microscopy.

Morphology of the Co-ferrite NPs was determined by transmission electron microscopy (TEM, JEM-2100, JEOL, Tokyo, Japan). TEM images were analyzed by the open-source software ImageJ [45] to estimate the NP size distribution.

Magnetic characterization of all samples was performed at room temperature with a highly sensitive vibrating sample magnetometer (VSM, LakeShore, Carson, CA, USA) operating in the magnetic field range of $\pm$17 kOe. Hysteresis loops of the magnetic wire were evaluated by applying the magnetic field along both the longitudinal (i.e., parallel, PA) and transverse (i.e., perpendicular, PE) directions to its major geometric axis. In contrast, SPIONs and Co-ferrite nanoparticles in powder form were randomly dispersed in a suitable sample holder whose diamagnetic signal was duly taken into account and properly subtracted.

Moreover, isothermal residual magnetization (IRM) and continuous demagnetization remanence (DCD) curves were also measured [46] for Co-ferrite nanoparticles. The IRM curve reports the remanence values obtained from a magnetic field progressively increasing towards positive values applied to the initially demagnetized magnetic sample. Conversely, the DCD curve reports the remanence values obtained from a magnetic field progressively increasing towards negative values applied to the magnetic sample initially at positive remanence [46].

A microfluidic system was assembled with the objective of facilitating the flow of the ferrofluid around the Fe-based wire. A schematic representation of this system is provided in Figure 1.

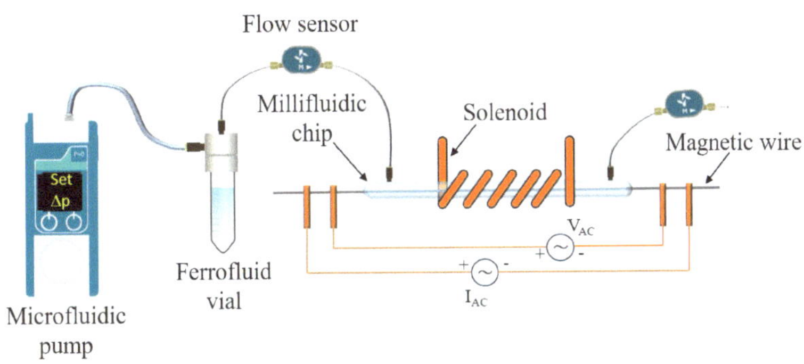

Figure 1. Scheme of the microfluidic system.

The millifluidic chip consists of a commercial quartz channel with a length of 13 cm and an inner/outer diameter of 1/1.5 mm. The quartz channel ends were inserted into plastic nozzles, which were threaded by lathe to connect the chip to a vial containing the magnetic ferrofluid via external fluorinated ethylene-propylene (FEP) tubes with an inner/outer diameter of 0.5/1.58 mm.

The magnetic wire was positioned within the quartz tube, with the two ends passing through the chip to create a dry area available for the electrical contacts used to measure the magneto-impedance effect. The application of glue ensured that the system was perfectly watertight.

The ferrofluid was made to flow into the chip via an advanced pressure-based flow control system (LineUp Flow EZ, Fluident). In the microfluidic circuit, two flow sensors were positioned at the ends of the chip to measure the flow rate and thus facilitate a reliable estimate of the velocity of the ferrofluid inside the quartz tube as a function of the pressure set on the flow control. When the microfluidic system was turned on, the chip was filled with ferrofluid, which completely submerged the magnetic wire. The entire chip was placed inside a solenoid, which was used to apply the DC magnetic field along the longitudinal axis of the magnetic wire. Some photos of the experimental microfluidic setup, millifluidic chip, and section of the quartz tube are available in the Supplementary Materials; in this paper, the term microfluidic refers to the entire experimental system used, as the pump is suitable for microliter volumes and the FEP tubes are in the micrometer range, while the term millifluidic refers only to the chip used due to the size of its inner diameter.

Magneto-impedance measurements were performed on the Fe-based wire using a conventional four-contact volt-ampere technique in the presence of an external DC field.

The four electrical contacts were made using conductive silver paste on the wire ends outside the millifluidic chip. The two external contacts were used to drive the sinusoidal AC drive current (I_{AC}) through the wire, while the two internal ones were used to measure the voltage signal (V_{AC}) across the wire, see Figure 1.

A signal generator (SD6022X Siglent, Helmond,the Netherlands) was exploited to generate I_{AC}, the frequency of which was set in the range 0.5–1.5 MHz, while the intensity was estimated to be 20 mA peak-to-peak by measuring the voltage across a precisely known resistor placed in series with the magnetic wire. A digital multimeter (3478A Hewlett Packard, Palo Alto, CA, USA) was used to measure the voltage signal across the Fe-based magnetic wire (whose resistance was 26 Ω). The impedance of the wire is given by $Z = V_{AC}/I_{AC}$.

The impedance measurements were performed as a function of the external DC magnetic field (generated by a solenoid powered by a Hewlett Packard 6654A DC power supply) with an amplitude in the range of $H = \pm 150$ Oe and directed along the main axis of the magnetic wire. The modulation of the external DC field affected the configuration of the magnetic domains in the wire, leading to a large variation in the measured voltage (ΔV_{AC}) and consequently in the impedance (ΔZ). The magneto-impedance ratio (MI) is defined as

$$\frac{\Delta Z}{Z}(\%) = \frac{Z(H) - Z(H_{max})}{Z(H_{max})} \times 100 \tag{1}$$

where $Z(H_{max})$ is the impedance measured at the maximum amplitude of the applied longitudinal DC field.

3. Results and Discussion

3.1. Structural and Morphological Characterization

The XRD pattern of the Co-ferrite NPs is shown in Figure 2a: the distinctive peaks reveal that the NPs are crystalline materials and all the reflections are compatible with the ones of cubic iron-spinel phases like magnetite and its substituted relatives such as $CoFe_2O_4$, where all the Fe^{2+} ions have been replaced by Co^{2+} ions. Since these two phases have the same cubic structure and very similar lattice parameters, their XRD pattern is almost identical; consequently, distinguishing cobalt ferrite from magnetite on the basis of

X-ray diffraction is very difficult. For this reason, the actual presence and concentration of Co ions in the synthesized NPs was checked and confirmed by SEM-EDS elemental semi-quantitative analysis: the atomic ratio between Co and Fe turned out to be 1:3, corresponding to the one used in the synthesis recipe. Thus, the actual composition of the Co-ferrite NPs can be written as $Co_{0.75}Fe_{2.25}O_4$. The crystallite size of the Co-ferrite NPs, calculated from XRD by means of Scherrer's formula after subtraction of the instrumental contribution to peak broadening, is around 11 nm.

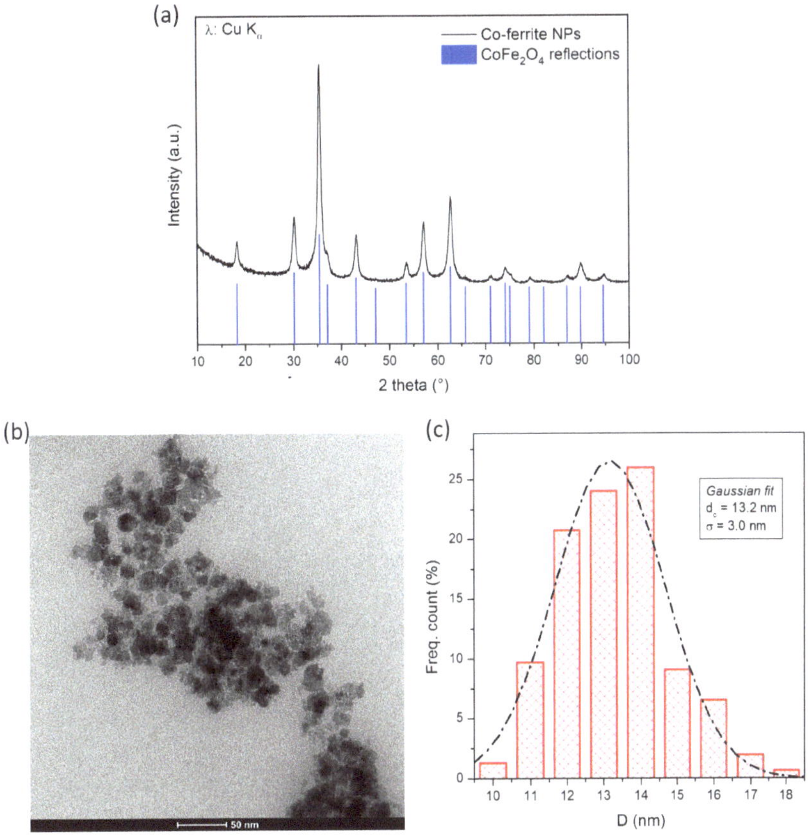

Figure 2. Structural and morphological properties of Co-ferrite NPs: (**a**) XRD pattern, together with the reference lines of $CoFe_2O_4$; (**b**) representative TEM image; (**c**) NP size distribution obtained by the statistical analysis of TEM images and fitted by a Gaussian function (dashed line).

A representative TEM image of the Co-ferrite NPs is shown in Figure 2b. The NPs appear well defined, with an almost spherical shape. Statistical analysis of some TEM images yields the NP size distribution shown in Figure 2c, which is well fitted by a Gaussian function with a mean value of 13.2 nm and a standard deviation of 3.0 nm, in reasonable agreement with the value obtained by XRD analysis.

The morphology of the Fe-based wire is investigated by the SEM images shown in Figure 3. The top-view image (panel a) reveals that the diameter of the wire is not perfectly constant but is characterized by fluctuations of a few tens of microns around an average value of 150 µm. Furthermore, the cross-sectional image (panel b) shows a homogeneous morphology of the wire throughout its thickness with no visible cracks, crystals, or inclusions.

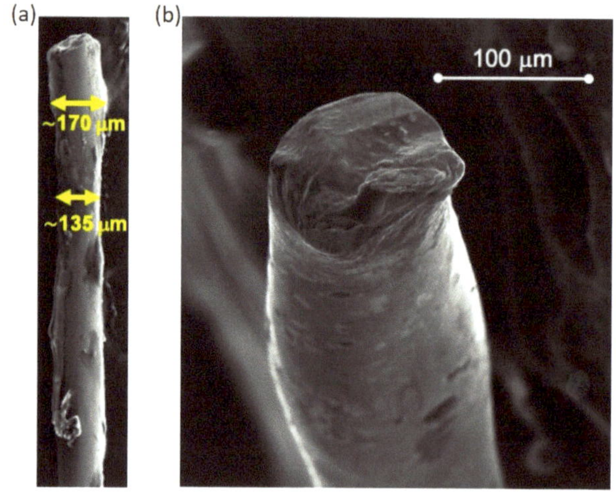

Figure 3. SEM images of the Fe-based wire: (**a**) top view and (**b**) cross section.

3.2. Magnetic Characterization of the MNPs

Room-temperature hysteresis loops for SPIONs and Co-NP samples in dry powder form are reported in Figure 4a.

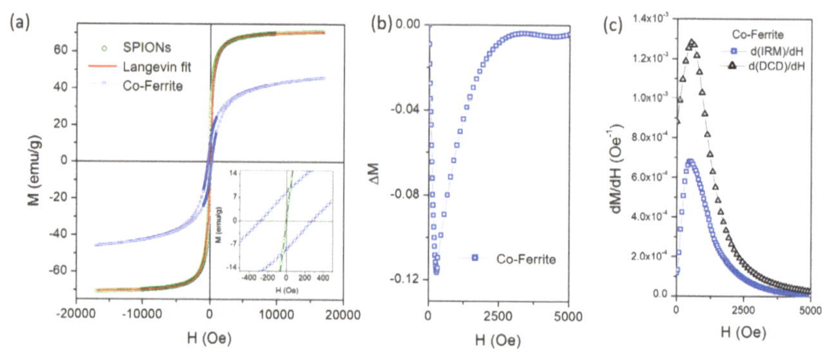

Figure 4. (**a**) Room-temperature hysteresis loop for Co-NPs (blue curve) and SPIONs (green curve) fitted with a Langevin curve (red line); (**b**) $\Delta M(H)$ curve obtained combining linearly the IRM and DCD measurements for the Co-NPs (see text for details); (**c**) field derivative for IRM and DCD curves for the Co-NPs.

The magnetization curve for SPIONs shows an anhysteretic behavior with a sigmoidal trend with a non-saturating behavior at high fields. This curve is well fitted by the superimposition of two Langevin curves (red line), proving the genuine superparamagnetic state of these nanoparticles [47,48].

The value of the saturation magnetization (M_S) is estimated from the high-field extrapolation of the Langevin fit curve, which gives $M_S \approx 71$ emu/g, a value lower than the saturation magnetization of bulk magnetite (92 emu/g) [49] due to the non-negligible role of magnetic disorder induced by surface magnetic anisotropy. Moreover, Langevin's fit leads to estimate the average size of SPIONs at around 12 nm, in good agreement with the size values obtained from TEM and XRD analyses.

Conversely, the $M(H)$ curve for the Co-NP sample shows a hysteretic behavior with a sigmoidal trend, characterized by an unsaturated behavior at the maximum applied field. The M_S values are estimated by fitting the high-field portion of the $M(H)$ curve with the well-known expression [50] $M(H) = M_S(1 - \delta/H - \gamma/H^2) + \chi H$, where δ and γ are free parameters, while χ is set to zero to neglect any paramagnetic contribution. This procedure gives $M_S \approx 50$ emu/g; such a value is fully compatible with Co-ferrite nanoparticles [51,52], although it is lower than that of bulk Co-ferrite (80 emu/g) [49]. This reduction can be attributed to the effects of a non-equilibrium distribution of Co^{2+} and $Fe^{2+/3+}$ cations in the spinel ferrite structure [51] and spin canting/disorder on the nanoparticles surface [51]. Additionally, the $M(H)$ curve is characterized by a coercive field (H_C) value of about 280 Oe and a remanence magnetization (M_R) of about 8.5 emu/g (see the inset of Figure 4a).

The remanence curves, i.e., isothermal remanence magnetization (IRM) and dc-demagnetization remanence (DCD), for the hysteretic Co-NP sample are measured by means of VSM and are reported in the Supplementary Materials.

These curves represent non-equilibrium magnetic states measured after the application and removal of a positive DC field with increasing amplitude on the sample in different initial magnetic configurations: demagnetized state and saturation remanence for IRM and DCD measurements, respectively [46].

The parameters of isothermal remanence coercivity ($H_{C_{IRM}}$) and demagnetization remanence coercivity ($H_{C_{DCD}}$) indicate the field at which the IRM curve is equal to 0.5 and the field at which the DCD curve crosses zero, respectively; the estimated values for Co-NPs are $H_{C_{IRM}}$ = 927 Oe and $H_{C_{DCD}}$ = 850 Oe.

The evidence of $H_{C_{IRM}} > H_{C_{DCD}}$ indicates that the Co-NP sample turns out to be more difficult to magnetize than to demagnetize; consequently, it can be inferred that interactions among Co-ferrite NPs occur and play a non-negligible role in the magnetization process [46,53,54]. This assumption is also supported by the shape of the $\Delta M(H)$ curve (see Figure 4b) obtained by combining linearly the IRM and DCD measures: $\Delta M = DCD(H) - [1 - 2IRM(H)]$. In particular, the well-defined negative dip confirms the existence of dipole–dipole interparticle interactions, which tends to demagnetize the entire sample [55–57].

Moreover, the ratios $H_{C_{DCD}}/H_C$ = 3 and M_R/M_S = 0.17 collocate the studied Co-NPs in the pseudo-single-domain (PSD) region of the Day's plot [58].

The field derivative for both normalized remanence curves is reported in Figure 4c to highlight the switching field distribution required to magnetize the Co-NP sample. Both curves show a rapid increase from a low value to a well-defined maximum. The presence of a single peak indicates that a single reversal mechanism in the magnetization process takes place [55]. After that, the curves very slowly decrease, approaching zero, to a field value of approximately H = 5 kOe, which indicates the minimum amplitude of the magnetic field required to activate all irreversible mechanisms of the magnetization process. At this value, the hysteresis curve shows the overlap of the two branches (see Figure 4a) but has not yet reached its maximum value, which will only be achieved by increasing the applied magnetic field by a rotational and reversible mechanism of magnetization.

3.3. Magnetic Characterization of Magnetic Fe-Based Wire

Room-temperature hysteresis loops for the Fe-based wire obtained by applying a magnetic field along the directions parallel (PA) and perpendicular (PE) to its geometric axis are reported in Figure 5a; both curves are normalized to the value of the magnetic moment at H = 10 kOe.

The $M(H)$ curve along PA (black symbols) shows a magnetization reversal in a narrow magnetic field range by a sharp and irreversible jump of the magnetic moment, defining a high magnetic susceptibility at the coercive field (χ_{Hc} = 1.64 × 10^{-2} Oe^{-1}) and magnetic remanence (M_r/M_s = 0.75), as well as an extremely low coercive field (below the VSM resolution). Conversely, the magnetization reversal along the PE direction (red symbols) occurs by a continuous rotation of the magnetic vector over a wide field range (H = 6.5 kOe) with a perfectly linear behavior, resulting in a low magnetic susceptibility (χ_{H_c} = 1.85 × 10^{-4} Oe^{-1}) and a magnetic remanence and coercive field close to zero. No irreversible processes characterized by magnetization jumps are visible.

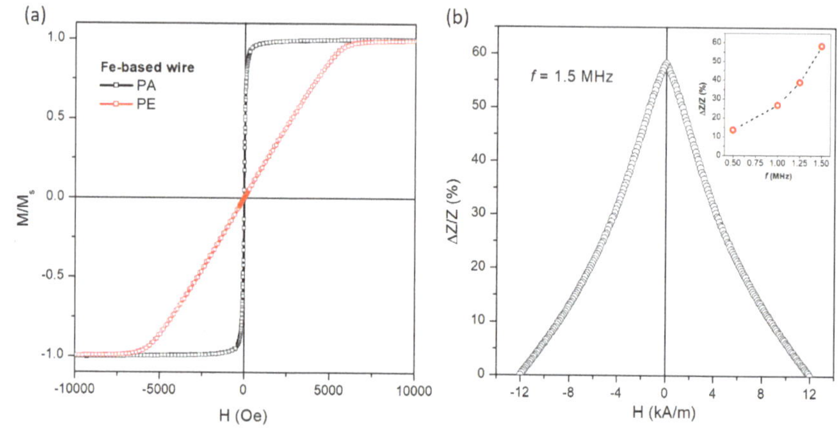

Figure 5. (**a**) Room-temperature hysteresis loops of the Fe-based wire evaluated along the directions parallel (PA) and perpendicular (PE) to its geometric axis; (**b**) field dependence of $\Delta Z/ZV$ at f = 1.5 MHz for the Fe-based wire and the frequency dependence of the $\Delta Z/ZV$ at zero applied field.

These soft magnetic properties of the Fe-based wire are consistent with an ideal single domain structure characterized by an extremely low crystalline anisotropy with the easy and hard axes aligned along the PA and PE directions, respectively. Such effective magnetic anisotropy is mainly dominated by the contribution of the shape anisotropy resulting from the high aspect ratio of the wire [49].

An example of the magneto-impedance (MI) effect on the magnetic Fe-based wire is shown in Figure 5b, where the $\Delta Z/Z$ ratio is reported as a function of the DC magnetic field applied along the PA direction of the wire. The frequency of the AC current applied to the wire is 1.5 MHz.

The $\Delta Z/ZV$ curve shows a single peak at $H \approx 0$ Oe and a monotonous and symmetric decrease as the magnetic field amplitude increases.

Such behavior of the MI curve indicates that the easy magnetization direction for magnetic anisotropy is nearly parallel to the wire axis and the transversal magnetization is always dominated by the rotational process, in perfect agreement with the evidence obtained from the hysteresis loops (see Figure 5a) [29,33,59].

The dependence of the maximum of $\Delta Z/Z$ (H = 0) as a function of the AC current frequency is shown in the inset of Figure 5b. The observed increase is due to the relationship between the current frequency and the skin penetration depth. In fact, as the frequency increases, the skin penetration depth decreases, and thus, the circular permeability increases, leading to an increase in Z [60]. The highest value results at 1.5 MHz, which is the value selected as the working frequency for the detection of MNPs using an amorphous wire magneto-impedance sensor in a microfluidic setup.

3.4. *Microfluidic Detection of MNPs Using Amorphous Wire Magneto-Impedance Sensor*

A magneto-impedance signal on Fe-based wire can be exploited to detect the presence of a magnetic ferrofluid surrounding it, allowing the wire to act as a sensing element. The ferrofluid is pumped by the microfluidic system up to surround the Fe-based wire, after which it is stopped before measuring the magnetic impedance.

Figure 6a shows that the $\Delta Z/Z$ (%) curve of the bare wire (here displayed as a dashed line and already shown previously in Figure 5b) is significantly modified by the presence of SPION and Co-ferrite NPs dispersed in a liquid medium at selected concentrations; each curve shown in Figure 6a is the average of six repeated measurements under the same experimental conditions.

In particular, the magnitude of the MI effect is adjusted according to the type and concentration of the NPs; conversely, the shape of the $\Delta Z/Z$ (%) curve maintains its maximum at $H = 0$ Oe with a monotonous trend decreasing as the external magnetic field increases. This evidence shows that the ferrofluid magnetically interacts with the wire but does not affect its uniaxial magnetic anisotropy. In fact, the magnetic NPs generate a stray field that affects the MI behavior by superimposing on the external longitudinal DC field and the transverse AC magnetic field generated by the AC excitation current.

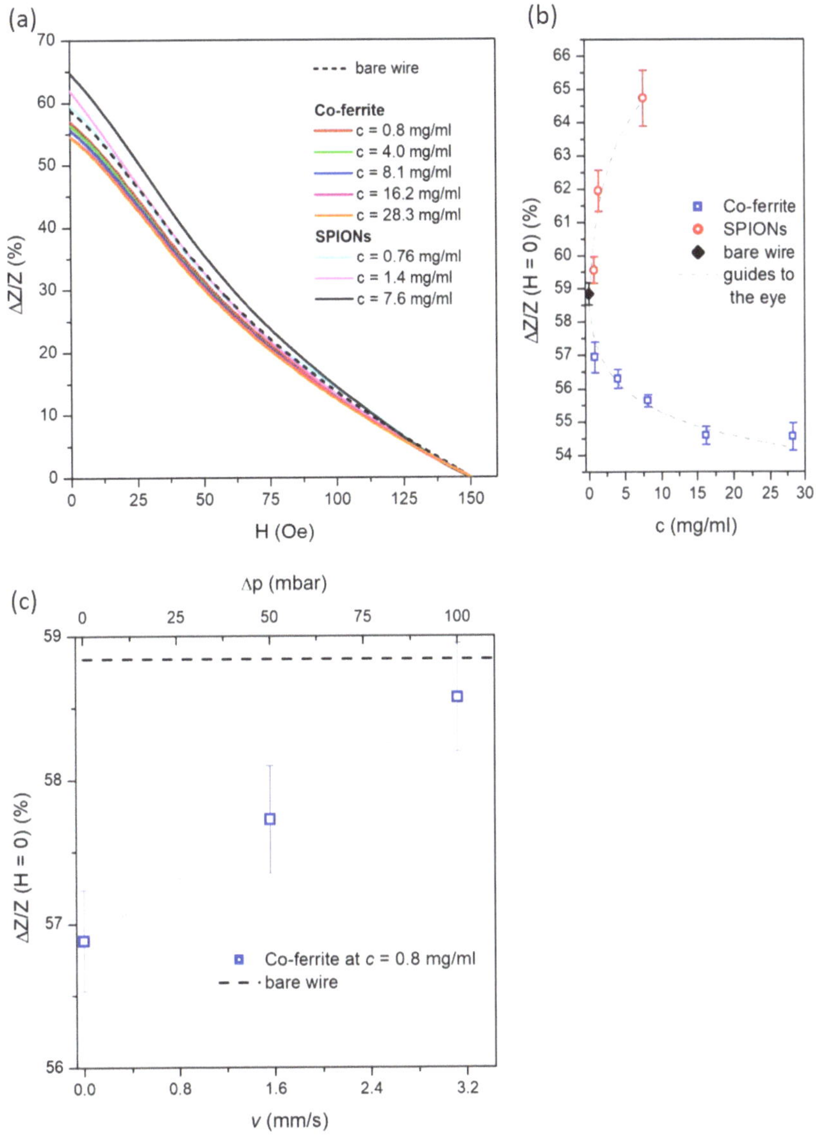

Figure 6. (**a**) Field dependence of $\Delta Z/ZV$ at $f = 1.5$ MHz for the Fe-based wire immersed on the static ferrofluid at different concentrations of Co-ferrite NPs or SPIONs; (**b**) $\Delta Z/ZV$ values at zero DC field as a function of the static ferrofluid concentration for Co-ferrite NPs and SPIONs; (**c**) $\Delta Z/ZV(H = 0)$ ratio as a function of the pressure difference (Δp) set to the flow of the Co-ferrite ferrofluid.

The $\Delta Z/Z$ values at $H = 0$, corresponding to the maximum MI variation, are shown as a function of nanoparticle concentration in Figure 6b. The SPIONs and Co-ferrite NPs affect the MI signal in opposite ways with increasing concentration. The former leads to an increase in the MI amplitude, while the latter induces a reduction. The explanation for this opposite behavior should be sought in the different magnetic states of the two types of NPs [35–38,61].

As a matter of fact, the magnetic behavior of the SPIONs, which refers to a single domain magnetic state with extremely low magnetic anisotropy energy and negligible magnetic interaction (see Figure 4a and Section 3.2 for more details), allows the SPIONs themselves to instantaneously generate a magnetic moment in response to the transverse AC and longitudinal DC magnetic fields applied during magneto-impedance measurements. Therefore, under the action of the AC transverse magnetic field alone (i.e., $H = 0$) whose magnitude is related to the current intensity and whose maximum is near the wire surface, the magnetic moment of SPIONs quickly reacts, promoting the domain wall mobility of the Fe-based wire, thus enhancing the MI effect as a function of SPION concentration as shown in Figure 6b [37,38]. The superimposition of the longitudinal DC magnetic field progressively overcomes the effect of the transverse field and (depending on its strength gradually orients the magnetic moments of the SPIONs along the major axis of the wire. This results in a reduction in the transverse magnetic permeability of the wire, resulting in a monotonic reduction of the $\Delta Z/Z(\%)$ curve as a function of H, as shown in Figure 6a.

In contrast, the hysteretic magnetic behavior of the Co-ferrite nanoparticles, characterized by a significant magnetic anisotropy energy even in the absence of an applied magnetic field (see Figure 4a and Section 3.2 for more details), hinders the action of the transverse AC magnetic field, which is no longer sufficient to orient the magnetic moments of the Co-ferrite nanoparticles in a transverse direction.

Furthermore, the intrinsic magnetic moments of these nanoparticles favor dipole magnetic interactions, which end up with the formation of nanoparticle arrangements such as clusters or chains. The resulting magnetic stray field tends to hinder the domain wall motion in the Fe-based wire, leading to a progressive reduction of the MI effect as a function of the Co-NP concentrations, as shown in Figure 6b [35,36,61]. Again, the longitudinal DC magnetic field gradually orients the magnetic moments of Co-ferrite NPs or clusters along the major axis of the wire; thus, a monotonic reduction of the $\Delta Z/Z(\%)$ curve as a function of H is measured, as shown in Figure 6a.

The evolution of $\Delta Z/Z$ ($H = 0$) values as a function of c was fitted to determine the calibration curve of the amorphous wire MI sensors for both types of nanoparticles.

In particular, the same logarithmic function, represented by the equation $\Delta Z/Z$ ($H = 0$) $= P_1 + P_2 \times \ln(c + P_3)$, was able to fit well the experimental data obtained from both nanoparticle systems with suitable coefficients (P_1, P_2, and P_3), as shown by the dashed gray lines in Figure 6b.

Moreover, these calibration curves enable the determination of the limit of detection (LOD) of the studied MI sensor, which is a pivotal parameter for characterizing it [62]. The LOD is defined as the smallest concentration of nanoparticles that produces a statistically significantly greater MI signal than that obtained from the repeated measurements of the bare wire [62,63]. This quantity is given by LOD = 3.3 σ_b/m, where σ_b is the standard deviation of the MI value for the bare wire and m is the slope of the calibration curve for c $\to$ 0 [63]. The LOD estimates are $\approx$0.3 and 0.7 mg/mL for SPIONs and Co-ferrite, respectively.

In addition, the detection range of the MI sensor is evaluated using Co-ferrite ferrofluid, taking advantage of the availability of more concentrated samples. It is approximately between 0 and 16.2 mg/ml; in fact, at higher concentrations, the calibration curve becomes so flat that the measured MI values tend to be statistically indistinguishable. It is evident that the experimental $\Delta Z/Z$ ($H = 0$) value 54.55 $\pm$ 0.41 % obtained for c = 28.3 mg/mL overlaps with the value 54.59 $\pm$ 0.27 % obtained for c = 16.2 mg/mL. A similar detection range would also be expected for the SPION ferrofluid if the calibration curve were extended to higher concentrations.

Magneto-impedance measurements are also capable of detecting magnetic nanoparticles under continuous flow of the ferrofluid.

The trend of $\Delta Z/Z$ ($H = 0$) as a function of the velocity (v) of the Co-ferrite ferrofluid at $c = 0.8$ mg/ml through the millifluidic chip is shown in Figure 6c. The flow velocity is related to the pressure difference (Δp) set in the pressure-based flow control system and is determined by averaging the flow rate measurements through two flow sensors placed in the microfluidic setup (see Figure 1).

In the range Δp = 0–100 mbar, the ferrofluid flows inside the millifluidic chip with a v = 0–3.125 mm/s dominated by a laminar flow as confirmed by the Reynolds number (Re) of 3.125 for maximum speed. The Reynolds number is calculated by $Re = \rho v s.L/\mu$, where v is the flow velocity, $L = 1$ mm is the inner diameter of the quartz channel, and $\rho = 1000$ kg/m^3 and $\mu = 0.001$ Pa·s are the density and viscosity of the ferrofluid, respectively, which are reasonably approximated to those of pure water (at room temperature) due to the very low concentration of nanoparticles.

As a matter of fact, a linear approach of the $\Delta Z/Z$ ($H = 0$) ratio from the value measured in static flow conditions ($v = 0$) towards the value of the bare wire is observed with increasing v.

This behavior can be reasonably explained by considering the competing Brownian and laminar forces that drive the displacement of the Co-ferrite NPs within the millifluidic chip. The balance of these hydrodynamic forces determines the time taken for the NPs to move in the multichannel during MI measurements and thus the efficiency of their magnetic interaction with the Fe-based wire. MI measurements take approximately 2 min due to the time required to bring the DC field to its maximum amplitude.

In the static flow condition ($v = 0$), Brownian forces are dominant, resulting in the random displacement of each NP within the liquid. This results in a high probability of the NPs remaining within the millifluidic chip throughout the duration of the MI measurement. Consequently, the NPs, subjected to AC and DC fields, maximize the interaction with the domain walls of the Fe-based magnetic wire, thereby inducing a substantial variation in the MI signal.

In the continuous flow condition, the hydrodynamic forces typical of the laminar state transport the NPs into the flow. Therefore, the travel time of the Co-ferrite NPs in the millichannel is progressively reduced as the ferrofluid velocity increases. Consequently, during MI measurements, the magnetic interaction of the NPs with the walls of the magnetic wires becomes progressively less effective and the variation in the MI signal becomes less significant.

4. Conclusions

A magneto-impedance sensor aimed at detecting the stray magnetic field generated by magnetic nanoparticles dispersed in a ferrofluid was successfully obtained by exploiting an amorphous $Fe_{73.5}Nb_3Cu_1Si_{13.5}B_9$ magnetic wire (average diameter of 150 μm), characterized by soft magnetic properties and anisotropy that is easy-axis-aligned along its main geometrical axis.

The detection of the magnetic nanoparticles takes place in a custom millifluidic chip to facilitate the flow of the ferrofluid around the Fe-based wire and to reduce the required sample and the amount of waste produced. In addition, this combined application of MI sensors with a microfluidic system promotes the development of miniaturized and portable devices, making measurements faster and more reliable.

In particular, it is observed that the magnitude of the MI signal is significantly influenced by the intrinsic magnetic properties of the NPs and their concentration in the ferrofluid. The superparamagnetic state of the SPIONs enables them to respond rapidly to the AC transverse magnetic field with a magnetic moment that generates a stray field capable of increasing the transverse magnetic permeability of the wire with a consequent enhancement of the MI effect. Conversely, the magnetically blocked state of the Co-ferrite NPs is characterized by an intrinsic magnetic moment that favors dipole magnetic interaction, resulting in a magnetic stray field that tends to hinder the domain wall motion in the Fe-based microwire. This leads to a reduction in the MI effect.

Under static flow conditions, the MI sensor demonstrates sensitivity to ferrofluid concentration with a detection limit of approximately 0.3 and 0.7 mg/ml for SPION and Co-ferrite, respectively. In dynamic flow conditions, the travel time of the Co-ferrite NPs decreases as a function of fluid velocity, and therefore, the magnetic interaction of the NPs with the wires becomes progressively less effective and the variation of the MI signal is observed to be less significant.

In conclusion, a simple and robust amorphous wire magneto-impedance sensor was proposed that is characterized by a simple fabrication process and relatively low cost. Its sensitivity to the stray magnetic field generated by MNPs in the ferrofluid makes it useful for reliably assessing the complete removal of NPs from water resources (after their use for contaminant removal), as well as for promoting the development of diagnostic sensors and lab-on-chip devices based on the detection of functionalized NPs.

Supplementary Materials: The following supporting information can be downloaded at: https://www.mdpi.com/article/10.3390/s24154902/s1, - Photo of experimental microfluidic detection setup; - photo of millifluidic chip equipped with nozzles and magnetic wire; - photo of the section of quartz tube; - IRM and DCD curves for hysteretic Co-ferrite NP sample.

Author Contributions: Conceptualization, G.B., F.C., M.C. and D.M.; formal analysis, G.B. and M.C.; investigation, G.B., F.C., M.V., E.S.O., L.M., M.C. and H.S.; data curation, G.B., E.S.O. and D.M.; writing—original draft preparation, G.B. and E.S.O.; visualization, A.M. and P.T.; supervision, A.M. and P.T. All authors have read and agreed to the published version of the manuscript.

Funding: Financial support by the Italian Ministry of University and Research (MUR) under the PNR program with DM 737/2021 is acknowledged here.

Data Availability Statement: The data are available from the corresponding author upon reasonable request.

Conflicts of Interest: The authors declare no conflicts of interest.

References

1. Socoliuc, V.; Avdeev, M.; Kuncser, V.; Turcu, R.; Tombácz, E.; Vékás, L. Ferrofluids and bio-ferrofluids: Looking back and stepping forward. *Nanoscale* **2022**, *14*, 4786–4886. [CrossRef] [PubMed]
2. Imran, M.; Affandi, A.M.; Alam, M.M.; Khan, A.; Khan, A.I. Advanced biomedical applications of iron oxide nanostructures based ferrofluids. *Nanotechnology* **2021**, *32*, 422001. [CrossRef] [PubMed]
3. Philip, J. Magnetic nanofluids (Ferrofluids): Recent advances, applications, challenges, and future directions. *Adv. Colloid Interface Sci.* **2023**, *311*, 102810. [CrossRef] [PubMed]
4. Oehlsen, O.; Cervantes-Ramirez, S.I.; Cervantes-Aviles, P.; Medina-Velo, I.A. Approaches on ferrofluid synthesis and applications: Current status and future perspectives. *ACS Omega* **2022**, *7*, 3134–3150. [CrossRef] [PubMed]
5. Dalla Vecchia, E.; Coisson, M.; Appino, C.; Vinai, F.; Sethi, R. Magnetic characterization and interaction modeling of zerovalent iron nanoparticles for the remediation of contaminated aquifers. *J. Nanosci. Nanotechnol.* **2009**, *9*, 3210–3218. [CrossRef] [PubMed]
6. Govan, J. Recent advances in magnetic nanoparticles and nanocomposites for the remediation of water resources. *Magnetochemistry* **2020**, *6*, 49. [CrossRef]
7. Liosis, C.; Papadopoulou, A.; Karvelas, E.; Karakasidis, T.E.; Sarris, I.E. Heavy metal adsorption using magnetic nanoparticles for water purification: A critical review. *Materials* **2021**, *14*, 7500. [CrossRef] [PubMed]
8. Shukla, S.; Khan, R.; Daverey, A. Synthesis and characterization of magnetic nanoparticles, and their applications in wastewater treatment: A review. *Environ. Technol. Innov.* **2021**, *24*, 101924. [CrossRef]
9. Punia, P.; Bharti, M.K.; Chalia, S.; Dhar, R.; Ravelo, B.; Thakur, P.; Thakur, A. Recent advances in synthesis, characterization, and applications of nanoparticles for contaminated water treatment-A review. *Ceram. Int.* **2021**, *47*, 1526–1550. [CrossRef]
10. Buchman, J.T.; Hudson-Smith, N.V.; Landy, K.M.; Haynes, C.L. Understanding nanoparticle toxicity mechanisms to inform redesign strategies to reduce environmental impact. *Accounts Chem. Res.* **2019**, *52*, 1632–1642. [CrossRef] [PubMed]
11. Mitu, S.A.; Ahmed, K.; Bui, F.M. Magnetic biosensors. In *Biosensors Nanotechnology*; Scrivener Publishing LLC: Beverly, MA, USA 2023; pp. 201–213.
12. Li, Y.; Cheng, H.; Alhalili, Z.; Xu, G.; Gao, G. The progress of magnetic sensor applied in biomedicine: A review of non-invasive techniques and sensors. *J. Chin. Chem. Soc.* **2021**, *68*, 216–227. [CrossRef]
13. Jimenez, V.O.; Hwang, K.Y.; Nguyen, D.; Rahman, Y.; Albrecht, C.; Senator, B.; Thiabgoh, O.; Devkota, J.; Bui, V.D.A.; Lam, D.S.; et al. Magnetoimpedance biosensors and real-time healthcare monitors: progress, opportunities, and challenges. *Biosensors* **2022**, *12*, 517. [CrossRef] [PubMed]

14. Tripathy, A.; Nine, M.J.; Silva, F.S. Biosensing platform on ferrite magnetic nanoparticles: Synthesis, functionalization, mechanism and applications. *Adv. Colloid Interface Sci.* **2021**, *290*, 102380. [CrossRef]
15. Jiao, J.; Zhang, H.; Zheng, J. Ferrofluids transport in bioinspired nanochannels: Application to electrochemical biosensing with magnetic-controlled detection. *Biosens. Bioelectron.* **2022**, *201*, 113963. [CrossRef] [PubMed]
16. Melnikov, G.Y.; Lepalovskij, V.N.; Svalov, A.V.; Safronov, A.P.; Kurlyandskaya, G.V. Magnetoimpedance thin film sensor for detecting of stray fields of magnetic particles in blood vessel. *Sensors* **2021**, *21*, 3621. [CrossRef] [PubMed]
17. Khizar, S.; Ben Halima, H.; Ahmad, N.M.; Zine, N.; Errachid, A.; Elaissari, A. Magnetic nanoparticles in microfluidic and sensing: From transport to detection. *Electrophoresis* **2020**, *41*, 1206–1224. [CrossRef] [PubMed]
18. Narlawar, S.; Coudhury, S.; Gandhi, S. Magnetic properties-based biosensors for early detection of cancer. In *Biosensor Based Advanced Cancer Diagnostics*; Elsevier: Amsterdam, The Netherlands, 2022; pp. 165–178.
19. Khan, M.A.; Sun, J.; Li, B.; Przybysz, A.; Kosel, J. Magnetic sensors-A review and recent technologies. *Eng. Res. Express* **2021**, *3*, 022005. [CrossRef]
20. Rocha-Santos, T.A. Sensors and biosensors based on magnetic nanoparticles. *TrAC Trends Anal. Chem.* **2014**, *62*, 28–36. [CrossRef]
21. Jin, Z.; Koo, T.M.; Kim, M.S.; Al-Mahdawi, M.; Oogane, M.; Ando, Y.; Kim, Y.K. Highly-sensitive magnetic sensor for detecting magnetic nanoparticles based on magnetic tunnel junctions at a low static field. *AIP Adv.* **2021**, *11*, 015046. [CrossRef]
22. Gloag, L.; Mehdipour, M.; Chen, D.; Tilley, R.D.; Gooding, J.J. Advances in the application of magnetic nanoparticles for sensing. *Adv. Mater.* **2019**, *31*, 1904385. [CrossRef]
23. Uchiyama, T.; Ma, J. Development of pico tesla resolution amorphous wire magneto-impedance sensor for bio-magnetic field measurements. *J. Magn. Magn. Mater.* **2020**, *514*, 167148. [CrossRef]
24. Yang, Z.; Liu, Y.; Lei, C.; Sun, X.c.; Zhou, Y. Ultrasensitive detection and quantification of *E. coli* O157: H7 using a giant magnetoimpedance sensor in an open-surface microfluidic cavity covered with an antibody-modified gold surface. *Microchim. Acta* **2016**, *183*, 1831–1837. [CrossRef]
25. Sayad, A.; Skafidas, E.; Kwan, P. Magneto-impedance biosensor sensitivity: Effect and enhancement. *Sensors* **2020**, *20*, 5213. [CrossRef] [PubMed]
26. Garcia-Arribas, A. The performance of the magneto-impedance effect for the detection of superparamagnetic particles. *Sensors* **2020**, *20*, 1961. [CrossRef] [PubMed]
27. Beato-López, J.; Pérez-Landazábal, J.; Gómez-Polo, C. Enhanced magnetic nanoparticle detection sensitivity in non-linear magnetoimpedance-based sensor. *IEEE Sensors J.* **2018**, *18*, 8701–8708. [CrossRef]
28. Wang, T.; Chen, Y.; Wang, B.; He, Y.; Li, H.; Liu, M.; Rao, J.; Wu, Z.; Xie, S.; Luo, J. A giant magnetoimpedance-based separable-type method for supersensitive detection of 10 magnetic beads at high frequency. *Sensors Actuators A Phys.* **2019**, *300*, 111656. [CrossRef]
29. Vázquez, M.; Sinnecker, J.P.; Kurlyandskaya, G.V. Hysteretic Behavior and Anisotropy Fields in the Magneto-Impedance Effect. *Mater. Sci. Forum* **1999**, *302–303*, 209–218. [CrossRef]
30. Vázquez, M. Giant magneto-impedance in soft magnetic Wires. *J. Magn. Magn. Mater.* **2001**, *226–230*, 693–699. [CrossRef]
31. Bukreev, D.A.; Derevyanko, M.S.; Moiseev, A.A.; Svalov, A.V.; Semirov, A.V. The study of the distribution of electrical and magnetic properties over the conductor cross-section using magnetoimpedance tomography: Modeling and experiment. *Sensors* **2022**, *22*, 9512. [CrossRef]
32. Buznikov, N.; Safronov, A.; Orue, I.; Golubeva, E.; Lepalovskij, V.; Svalov, A.; Chlenova, A.; Kurlyandskaya, G. Modelling of magnetoimpedance response of thin film sensitive element in the presence of ferrogel: Next step toward development of biosensor for in-tissue embedded magnetic nanoparticles detection. *Biosens. Bioelectron.* **2018**, *117*, 366–372. [CrossRef]
33. Knobel, M.; Pirota, K.R. Giant magnetoimpedance: Concepts and recent progress. *J. Magn. Magn. Mater.* **2002**, *242–245*, 33–40. [CrossRef]
34. Yang, Z.; Lei, C.; Zhou, Y.; Liu, Y.; Sun, X.c. A GMI biochip platform based on Co-based amorphous ribbon for the detection of magnetic Dynabeads. *Anal. Methods* **2015**, *7*, 6883–6889. [CrossRef]
35. Kurlyandskaya, G.; Jantaratana, P.; Bebenin, N.; Vas' kovskiy, V. Magnetic properties and magnetoimpedance of electroplated wires. *Solid State Phenom.* **2012**, *190*, 581–584. [CrossRef]
36. Yuvchenko, A.; Lepalovskii, V.; Vas' kovskii, V.; Safronov, A.; Volchkov, S.; Kurlyandskaya, G. Magnetic impedance of structured film meanders in the presence of magnetic micro-and nanoparticles. *Tech. Phys.* **2014**, *59*, 230–236. [CrossRef]
37. Kurlyandskaya, G.; Sanchez, M.; Hernando, B.; Prida, V.; Gorria, P.; Tejedor, M. Giant-magnetoimpedance-based sensitive element as a model for biosensors. *Appl. Phys. Lett.* **2003**, *82*, 3053–3055. [CrossRef]
38. Amirabadizadeh, A.; Lotfollahi, Z.; Zelati, A. Giant magnetoimpedance effect of Co68. 15 Fe4. 35 Si12. 5 B15 amorphous wire in the presence of magnetite ferrofluid. *J. Magn. Magn. Mater.* **2016**, *415*, 102–105. [CrossRef]
39. Rezvani Jalal, N.; Mehrbod, P.; Shojaei, S.; Labouta, H.I.; Mokarram, P.; Afkhami, A.; Madrakian, T.; Los, M.J.; Schaafsma, D.; Giersig, M.; et al. Magnetic nanomaterials in microfluidic sensors for virus detection: A review. *ACS Appl. Nano Mater.* **2021**, *4*, 4307–4328. [CrossRef]
40. Zhong, Z.; He, J.; Li, G.; Xia, L. Recent Advances in Magnetic Nanoparticles-Assisted Microfluidic Bioanalysis. *Chemosensors* **2023**, *11*, 173. [CrossRef]
41. Vassallo, M.; Vicentini, M.; Salzano De Luna, M.; Barrera, G.; Tiberto, P.; Manzin, A.; Martella, D. Magnetic Hyperthermia to Promote Acrylamide Radical Polymerizations. *ACS Appl. Polym. Mater.* **2024**, *6*, 4696–4707. [CrossRef]

42. Riahi, K.; van de Loosdrecht, M.M.; Alic, L.; ten Haken, B. Assessment of differential magnetic susceptibility in nanoparticles: Effects of changes in viscosity and immobilisation. *J. Magn. Magn. Mater.* **2020**, *514*, 167238. [CrossRef]
43. Waseda, Y.; Ueno, S.; Hagiwara, M.; Aust, K.T. Formation and mechanical properties of Fe- and Co-base amorphous alloy wires produced by in-rotating-water spinning method. *Prog. Mater. Sci.* **1990**, *34*, 149–260. [CrossRef]
44. Ogasawara, I.; Ueno, S. Preparation and Properties of Amorphous Wires. *IEEE Trans. Magn.* **1995**, *31*, 1219–1223. [CrossRef]
45. Available online: https://imagej.net/ij/ (accessed on 3 June 2024).
46. Hillion, A.; Tamion, A.; Tournus, F.; Gaier, O.; Bonet, E.; Albin, C.; Dupuis, V. Advanced magnetic anisotropy determination through isothermal remanent magnetization of nanoparticles. *Phys. Rev. B* **2013**, *88*, 094419. [CrossRef]
47. Elfimova, E.A.; Ivanov, A.O.; Camp, P.J. Static magnetization of immobilized, weakly interacting, superparamagnetic nanoparticles. *Nanoscale* **2019**, *11*, 21834–21846. [CrossRef]
48. Gubin, S.P. *Magnetic Nanoparticles*; John Wiley & Sons: Hoboken, NJ, USA, 2009.
49. Cullity, B.D.; Graham, C.D. *Introduction to Magnetic Materials*; John Wiley & Sons: Hoboken, NJ, USA, 2011.
50. Zhang, H.; Zeng, D.; Liu, Z. The law of approach to saturation in ferromagnets originating from the magnetocrystalline anisotropy. *J. Magn. Magn. Mater.* **2010**, *322*, 2375–2380. [CrossRef]
51. Barrera, G.; Coisson, M.; Celegato, F.; Raghuvanshi, S.; Mazaleyrat, F.; Kane, S.; Tiberto, P. Cation distribution effect on static and dynamic magnetic properties of Co1-xZnxFe2O4 ferrite powders. *J. Magn. Magn. Mater.* **2018**, *456*, 372–380. [CrossRef]
52. Karaagac, O.; Yildiz, B.B.; Köçkar, H. The influence of synthesis parameters on one-step synthesized superparamagnetic cobalt ferrite nanoparticles with high saturation magnetization. *J. Magn. Magn. Mater.* **2019**, *473*, 262–267. [CrossRef]
53. Fearon, M.; Chantrell, R.; Wohlfarth, E. A theoretical study of interaction effects on the remanence curves of particulate dispersions. *J. Magn. Magn. Mater.* **1990**, *86*, 197–206. [CrossRef]
54. Del Bianco, L.; Spizzo, F.; Sgarbossa, P.; Sieni, E.; Barucca, G.; Ruggiero, M.R.; Geninatti Crich, S. Dipolar magnetic interactions in Mn-doped magnetite nanoparticles loaded into PLGA nanocapsules for nanomedicine applications. *J. Phys. Chem. C* **2019**, *123*, 30007–30020. [CrossRef]
55. Ovejero, J.G.; Spizzo, F.; Morales, M.P.; Del Bianco, L. Mixing iron oxide nanoparticles with different shape and size for tunable magneto-heating performance. *Nanoscale* **2021**, *13*, 5714–5729. [CrossRef]
56. Garcia-Acevedo, P.; Gonzalez-Gomez, M.A.; Arnosa-Prieto, A.; de Castro-Alves, L.; Pineiro, Y.; Rivas, J. Role of dipolar interactions on the determination of the effective magnetic anisotropy in iron oxide nanoparticles. *Adv. Sci.* **2023**, *10*, 2203397. [CrossRef] [PubMed]
57. Varvaro, G.; Agostinelli, E.; Laureti, S.; Testa, A.; Garcia-Martin, J.M.; Briones, F.; Fiorani, D. Magnetic anisotropy and intergrain interactions in L10 CoPt (1 1 1)/Pt (1 1 1)/MgO (1 0 0) PLD granular films with tilted easy axes. *J. Phys. D Appl. Phys.* **2008**, *41*, 134017. [CrossRef]
58. Day, R.; Fuller, M.; Schmidt, V. Hysteresis properties of titanomagnetites: Grain-size and compositional dependence. *Phys. Earth Planet. Inter.* **1977**, *13*, 260–267. [CrossRef]
59. Tandon, P.; Sahu, R.; Mishra, A.C. Giant magnetoimpedance effect in electrodeposited CoNiFe/Cu composite wire: Experimental study and analytical modelling. *Phys. B Condens. Matter* **2022**, *642*, 414131. [CrossRef]
60. Pal, S.; Manik, N.; Mitra, A. Dependence of frequency and amplitude of the ac current on the GMI properties of Co based amorphous wires. *Mater. Sci. Eng. A* **2006**, *415*, 195–201. [CrossRef]
61. Chiriac, H.; Herea, D.D.; Corodeanu, S. Microwire array for giant magneto-impedance detection of magnetic particles for biosensor prototype. *J. Magn. Magn. Mater.* **2007**, *311*, 425–428. [CrossRef]
62. Kruse, J.; Wörner, J.; Schneider, J.; Dörksen, H.; Pein-Hackelbusch, M. Methods for Estimating the Detection and Quantification Limits of Key Substances in Beer Maturation with Electronic Noses. *Sensors* **2024**, *24*, 3520. [CrossRef]
63. Hayashi, Y.; Matsuda, R.; Ito, K.; Nishimura, W.; Imai, K.; Maeda, M. Detection limit estimated from slope of calibration curve: An application to competitive ELISA. *Anal. Sci.* **2005**, *21*, 167–169. [CrossRef]

sensors

MDPI

Article

Inkjet Printing Magnetostrictive Materials for Structural Health Monitoring of Carbon Fibre-Reinforced Polymer Composite

Nisar Ahmed [1,2,*], Patrick J. Smith [3] and Nicola A. Morley [2]

1 Centre for Additive Manufacturing, Faculty of Engineering, University of Nottingham, Nottingham NG7 2RD, UK
2 Department of Materials Science and Engineering, University of Sheffield, Sheffield S1 3JD, UK; n.a.morley@sheffield.ac.uk
3 Department of Mechanical Engineering, University of Sheffield, Sheffield S1 3JD, UK; patrick.smith@sheffield.ac.uk
* Correspondence: nisar.ahmed2@notingham.ac.uk

Abstract: Inkjet printing of magnetic materials has increased in recent years, as it has the potential to improve research in smart, functional materials. Magnetostriction is an inherent property of magnetic materials which allows strain or magnetic fields to be detected. This makes it very attractive for sensors in the area of structural health monitoring by detecting internal strains in carbon fibre-reinforced polymer (CFRP) composite. Inkjet printing offers design flexibility for these sensors to influence the magnetic response to the strain. This allows the sensor to be tailored to suit the location of defects in the CFRP. This research has looked into the viability of printable soft magnetic materials for structural health monitoring (SHM) of CFRP. Magnetite and nickel ink dispersions were selected to print using the JetLab 4 drop-on-demand technique. The printability of both inks was tested by selecting substrate, viscosity and solvent evaporation. Clogging was found to be an issue for both ink dispersions. Sonicating and adjusting the jetting parameters helped in distributing the nanoparticles. We found that magnetite nanoparticles were ideal as a sensor as there is more than double increase in saturation magnetisation by 49 Am^2/kg and more than quadruple reduction of coercive field of 5.34 kA/m than nickel. The coil design was found to be the most sensitive to the field as a function of strain, where the gradient was around 80% higher than other sensor designs. Additive layering of 10, 20 and 30 layers of a magnetite square patch was investigated, and it was found that the 20-layered magnetite print had an improved field response to strain while maintaining excellent print resolution. SHM of CFRP was performed by inducing a strain via bending and it was found that the magnetite coil detected a change in field as the strain was applied.

Keywords: magnetostriction; magnetite; nickel; ink dispersion; inkjet printing; JetLab 4

Citation: Ahmed, N.; Smith, P.J.; Morley, N.A. Inkjet Printing Magnetostrictive Materials for Structural Health Monitoring of Carbon Fibre-Reinforced Polymer Composite. *Sensors* **2024**, *24*, 4657. https://doi.org/10.3390/s24144657

Academic Editor: Arcady Zhukov

Received: 22 May 2024
Revised: 5 July 2024
Accepted: 11 July 2024
Published: 18 July 2024

1. Introduction

Printed electronics via inkjet printing onto substrates such as on printed circuit boards (PCB) have been around for decades. This method has gained an interest in printing smart magnetic materials for sensors and actuators [1]. One area of research is printing magnetostrictive sensors for structural health monitoring (SHM) of carbon fibre-reinforced polymer (CFRP) composite. The current state of the art in SHM shown in [2–5] conclude that fibre Bragg grating (FBG) and acoustic methods are suitable for SHM due to reliability in sensing damages in CFRP, whereas magnetostrictive sensors are limited in manufacturing methods and cost.

Therefore, this paper exploits the gap in reducing the manufacturing process time and cost of materials used for SHM devices via inkjet printing soft magnetic materials. Printing magnetostrictive materials has been difficult due to the materials' availability and compatibility, which so far has been restricted to three-dimensional (3D) extrusion or powdered metal additive manufacturing [6,7]. Inkjet printing is desirable in applications where the weight and cost of material are important. The objective of this paper is to demonstrate the printability of magnetostrictive sensors and be able to monitor damage in aircraft CFRP material. The aim is to develop and print magnetostrictive via an inkjet printing technique; this allows additional freedom in designing various patterns and layers which could improve the performance of the sensor. Therefore, this paper will print characterise and develop a tension to test the sensor's magnetic field response to strain.

In the literature, ref. [8] have shown that magnetostrictive materials can detect damage present in CFRP under bending and impact forces. One of the main advantages of using magnetostrictive materials is that magnetostriction is an inherent property of magnetic materials and does not degrade the CFRP. Ref. [9] have designed and developed an SHM device consisting of a magnetostrictive ribbon embedded in the CFRP while detecting the magnetic field using a coil inductor. Wireless monitoring can also be implemented and will be useful in aerospace applications. Another method shown by [10], made a printable copper inductor coil, which was used for detecting small changes in the magnetic field from the magnetostrictive sensor.

The two main forms of magnetoelastic effects [11] can be described by Equation (1): Joule magnetostriction, where the material changes in length under a magnetic field, and Equation (2): Inverse-Joule or Villari Effect, where the magnetic flux density changes direction under an applied strain.

$$\varepsilon = \sigma / E_y^H + d_{33} H \quad (1)$$

$$B = d_{33}\sigma + \mu^{\sigma} H \quad (2)$$

where in Equation (1), ε is the strain, σ is stress, E_y^H is the compliance coefficient at constant field strength, $d_{33} = d\varepsilon/dH$ (strain/field), and H is the magnetic field. Where in Equation (2), B is the magnetic flux density in Tesla (T), $d_{33} = dB/d\sigma$ (induction/stress), which is the magnetostrictive constant, and $\mu\sigma$ is the permeability at constant mechanical stress.

Fe-based soft magnetic materials are ideal for structural health monitoring [6] as they have low coercivity and high saturation magnetisation. However, inkjet-printed magnetostrictive materials have not yet been tested for SHM of CFRP composite.

Printing magnetic materials by inkjet printing is achievable; however, there are certain factors that could influence the base metal [12,13]:

- Nozzle and droplets;
- Alignment of moments;
- Magnetic and non-magnetic printing.

Magnetic inks can form long liquid bridges from the nozzle, which makes the drop on demand (DOD) system difficult to print with magnetic inks. The inkjet system uses a piezoelectric actuator to push inks from the ink chamber to the nozzle therefore controlling the number of drops to the substrate. However, ref. [12], mentioned that magnetic inks could be pushed by a magnetic field instead of an actuator or any mechanical device. This would mean controlling the magnetic particles further by ensuring each drop contains magnetic metal particles and breaking the bridge with the drop to the nozzle, which could improve densification and further control of the droplets.

Thick film magnetostrictive material has been produced and measured by [14]. They have produced a paste containing Terfenol-D and glass frit, then printed using screen printing. By measuring the magnetostriction using the change in length from lasers, they reported that it has a lower magnetostriction constant than the bulk materials due to void formation in the thick film.

Nickel and magnetite are soft magnetic materials (ferromagnetic) that can be used for printing magnetostrictive sensors. In literature, nickel and magnetite nanoparticles have been used and successfully printed/deposited by [12,15–24]. Although nickel and magnetite have been shown to be printable, there have been issues with ink synthesis, droplet formation, substrate and post-printing.

There are various ways to synthesise magnetite and nickel inks, such as additives, sol–gel and solvents to improve the printing and post-printing process. The substrate onto which the ink is printed plays a key role in retaining the nanoparticles and evaporating the solvent. Ref. [20], have shown that nickel can be printed onto glass slides using a Dimatix Materials Printer inkjet system for electronic applications. The nickel inks printed were preheated to 140 °C on a hotplate and then heated to 180 °C after printing to avoid wetting and to sufficiently evaporate the solvent. However, glass as a substrate material cannot be used as a magnetostrictive sensor as there needs to be greater flexibility in the substrate. Polymer materials can be used to print nickel ink; however, there are limitations; for example, ref. [24], shows that the PET substrate material causes shrinkage and blistering problems when curing nickel inks. Kapton film could be used as the melting point for Kapton is around 400 °C, well above the curing temperature of nickel. Ref. [19], have shown that nickel ink (with ethanol solvent) can be printed on Kapton and cured on a hot plate at 150 °C for 20 min and then flashlight sintered.

Ref. [23] have printed magnetite inks via inkjet printing onto sacrificial paper to produce a turntable actuator and explored the effect of synthesising magnetite solvent to form a photo-curable ink such as oleic acid. Polymeric resin is used to strengthen and form bonds between magnetite nanoparticles. The magnetic properties at room temperature were measured for both magnetite inks in the original and with polymeric resin. It was shown that the polymeric resin reduced the magnetic properties, such as saturation magnetisation (M_s), slightly, which was due to the higher level of non-magnetic material within the print. Ref. [22] explored the effect of printing magnetite inks onto paper by changing the jetting (piezoelectric or thermal), temperature of print and size of magnetite nanoparticles. The magnetic properties remain unchanged for both piezoelectric and thermal print heads. Ref. [18] produced an inkjet-printed magnetite inductor core on paper at room temperature and polyimide (sintered at 300 °C) by JetLab 4 by MicroFab Inc. (Plano, TX, USA). Oleic acid was used to cover the ink and was treated with potassium hydroxide. Aggregation in some areas was formed; however, the print was successful on both paper and polyimide.

2. Methods

The metal dispersion inks that were selected to be studied in this research were 20% Magnetite with Dimethylformamide (DMF) and 2% Nickel with N-Methyl-2-pyrrolidone (NMP) nanoparticles (NPs) inks, manufactured by Nanoshel. Table 1 shows the composition and physical properties of the inks. These include the solvent within the ink, the viscosity of the ink, surface tension, particle size and solvent evaporation temperature. The composition and physical properties are vital for printing NPs using an inkjet system to avoid clogging and printability on a substrate. For example, particle size above 100 nm would be prone to clogging depending on the nozzle size. The ink composition and particles were readily available and tested by the manufacturer. However, for inkjet printing applications, it is critical to know the viscosity of the ink and tailor the composition for the printing system. Equation (3) shows the Z number that determines the printability of the inks [25,26].

$$Z = \frac{\sqrt{\rho d_o \gamma_t}}{\mu_v} \tag{3}$$

where ρ is density, d_o is the orifice diameter, γ_t is the surface tension of the ink, and μ_v is the viscosity of the ink. An ink can be considered printable with a Z number to be between 1 to 10.

A viscometer was used to measure the viscosity of both inks. The viscometer was calibrated by a one-point calibration technique with water at room temperature before measuring the inks. Distilled water was measured at a known temperature and then compared to published values, as shown in Table A1 [27]. Magnetite and nickel ink were placed in a polycarbonate container and placed in a sample holder. The viscosity was measured by placing two vibrating probes and a thermocouple lowered and aligned to the meniscus of the ink.

Table 1. Nickel and magnetite ink properties.

Metal NP	Solvent Composition	Viscosity at 298 K (cP)	Density (g/cm^3)	Surface Tension (mN/m)	Particle Size (nm)	Solvent Evaporation Temperature (K)
Ni (2 wt. %)	NMP and water	7.5–10	2.07	72.8 (water)	80–100	473 (NMP)
Fe_3O_4 (20 wt. %)	Organic Solvent (DMF), IPA Ethanol, Water (ddH_2O)	7.5–10	5.17	72.8 (water)	50–80	426 (DMF)

2.1. Printing Process

The experimental procedure for printing the metal NPs in the JetLab 4 was to first prepare the ink. An ultrasonic bath was used for 30 min before transferring approximately 3 mL of ink to the ink reservoir. The reservoir was then placed into the JetLab 4 printing machine. The jetting parameters were then calibrated. Before printing onto the substrates they were cleaned by air drying to remove any dust or impurities.

The piezoelectric print head was bought from MicroFab Inc. (Plano, TX, USA); it had a nozzle size of 60 μm with controllable voltage output. The printing parameters used were a standard wave at a rise time of 5 μs, dwell time of 5 μs, fall time of 50 μs, echo time of 6 μs, rise time of 10 μs, idle voltage of 0 V, dwell voltage of 60 V and echo voltage of −60 V as seen in Figure 1. There is a slight variation in these parameters depending on the ink mixture. For example, the dwell voltage may change to 65 V, or the dwell time may change to 3 μs to prevent clogging of satellites in each droplet. For example, nickel ink has heavier metal particles, which were not easily mixed into the solution, therefore creating an additional issue when there was a difference in the density at the nozzle. This created a clogging problem where the jetting parameters were not calibrated for the heavier particles of the solution. This results in an irregular print, which cannot be altered during printing.

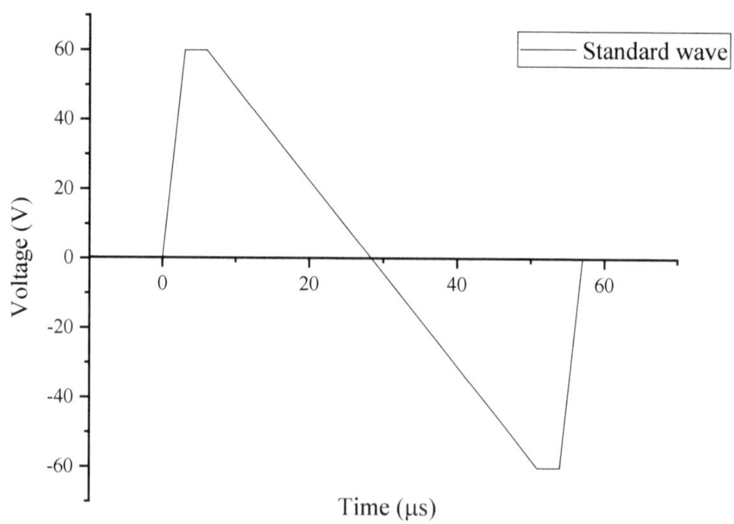

Figure 1. JetLab signal input standard wave for each droplet.

Designs for printing were made in Ansys CAD Space Claim 2019 R2 and converted into bitmap monochrome file. The resolution of the image depends on the number of pixels in the image.

2.2. Post Treatment

After the printing process, the printed design required further treatment in order to be used as a sensor. The first stage involved drying the print after the printing process, which involved heating to evaporate the solvent. The print was left at room temperature, leaving the metal NPs on the substrate.

Coating is another treatment, which serves to protect the inkjet-printed designs from external environments. Two ways of applying a coating were used, these were by applying a layer of silicone via spin coater or spray coating a layer of acrylic polymer. Both coating methods have advantages and disadvantages of applying and protecting the print. Spray coating is an easy way of applying and protecting the print, as it needs no additional machines or equipment. However, the spin coater ensures an even coating, which is useful without compromising thickness and detection of the magnetic field.

An Ossila spin coating machine was used to apply a layer of polydimethylsiloxane (PDMS) to the print under a fume hood. The PDMS (Slygard 184) used was purchased from Merck Life Science (Dorset, UK), which came in pre-packed solutions of resin and hardener. As the PDMS solution was viscous, the spin coater was set at 3000 RPM for 12 s. The coated print was left to dry at room temperature for 24 h. The print with the PDMS layer was then placed in an oven at 60 °C for 1 h to cure. An acrylic conformal coating spray was purchased from RS component. The spray was used under a fume hood and sprayed directly onto the print from 15 cm away.

2.3. Production of Carbon Fibre Composite Sample

The carbon fibre prepreg 4-ply twill weave (VTC401-C200T-HS-3K-42%RW) was used in this project and was supplied by SHD Composites Ltd. (Lincolnshire, UK). The composite laminate was formed using vacuum bagging. The prepreg CFRP were cut to size and layered on top of each other on a glass substrate and sealed in a vacuum bag at −28 Hg. The autoclave was used to apply pressure of 6 bar and heated to 60 °C at a rate of 3 °C/min and held for 60 min. Then the temperature increased to 120 °C at a rate of 3 °C/min for 60 min and then cooled down to room temperature. The cure cycle was recommended for the specific carbon fibre prepreg (VTC-401) from the supplier [28], as demonstrated in [9]. A tile cutter was used to cut the CFRP to size. This included a blade submerged in water to prevent excess dust from cutting the CFRP samples. For the strain bending test, the laminates were made to a size of 25 mm × 50 mm × 0.75 mm

2.4. Villari Effect Magnetostriction

Strain Bending Test

An inductance measurement was used to measure the magnetoelastic performance of the soft magnetic printed designs under strain. These methods were selected from literature as they have proven to detect magnetic fields. The magnetic prints were strained by placing them on a bending rig to apply a bending force. The bending test setup shown in Figure 2a shows the force applied on two ends, which forces the sample around a known curvature radius. The dimensions of the bend rig and radius of curvature were measured and produced in the Ansys CAD design package. The CAD design was converted to an STL file for printing. A resin photo-polymer printer was used to print 3D bend rigs, as shown in Figure 2b. As the print is at the top of the paper, the bend will produce a tensile force and reorientate the magnetic moments, therefore producing a change in the magnetic dipole and field.

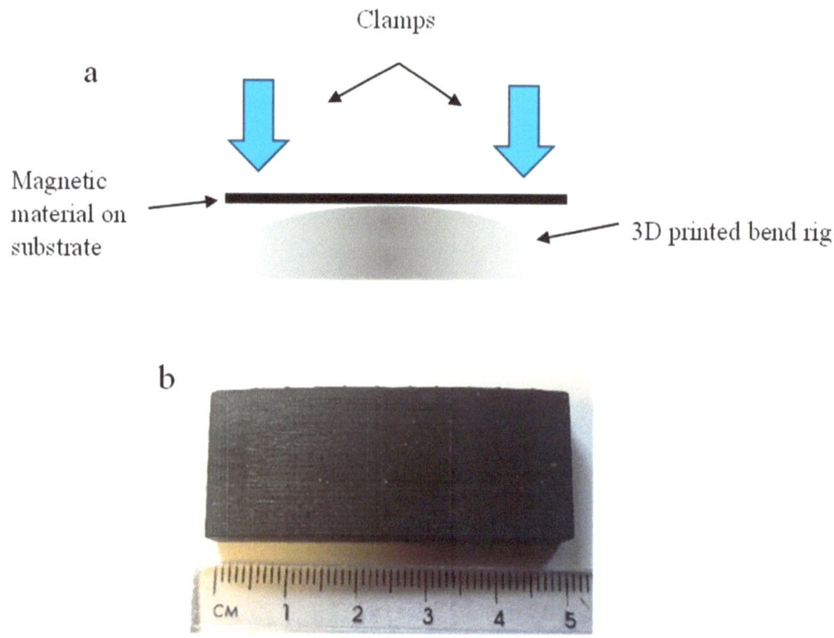

Figure 2. (**a**) Bending test on a known radius of curvature (**b**) 3D printed bend rig.

The radius of curvature is converted to strain by using Equation (4). Where ε is strain, y is the distance from the neutral axis, and R is the radius of curvature. By converting the radius of curvature of R1000, R900, R800, R700, R600, R500, R400, R300, R200 and R100 (where R1000 = 1000 mm radius) to strain values of 0.13, 0.14, 0.16, 0.19, 0.22, 0.26, 0.33, 0.43, 0.65 and 1.3 με respectively for paper substrate. The strain depends on the distance of curvature from the neutral axis; therefore, calculated values for paper and CFRP are presented in Appendix A in Figure A1. The 3D-printed bending rig dimensions were measured again after printing for recalculation of the strain.

$$\varepsilon = \frac{y}{R} \quad (4)$$

$$\mathrm{L} = \Phi \mathrm{I} \quad (5)$$

$$\mathrm{L} = \mu_0 \frac{N^2 A}{l} \quad (6)$$

Measuring a change in inductance is a direct way of measuring the change in the magnetic field of a material. For this measurement, a coil was made to measure the change in inductance when the strain was applied to the print. The inductance is proportional to magnetic flux, as shown in Equation (5), where L is in Henries, Φ is magnetic flux, and I is the current. Therefore, the magnetic flux can be derived from the inductance at a constant current in the coil [29]. This allows flexible measurement to adjust the sensitivity of the inductance. In Equation (6), where L is in Henries, μ_0 is the permeability of free space, N is the number of turns, A is the inner core area in m^2, l is the length of the coil in metres. This shows that in a coil, the number of turns increases the inductance greatly, therefore increasing the area. The dimensions of the 3D-printed coil holder are shown in Figure 3a, where the diameter of the inner air core is 5 mm, and the diameter of the coil holder is 16 mm. The inductor was made from copper wire of 0.1 mm thickness and an air core size radius of 1.5 mm.

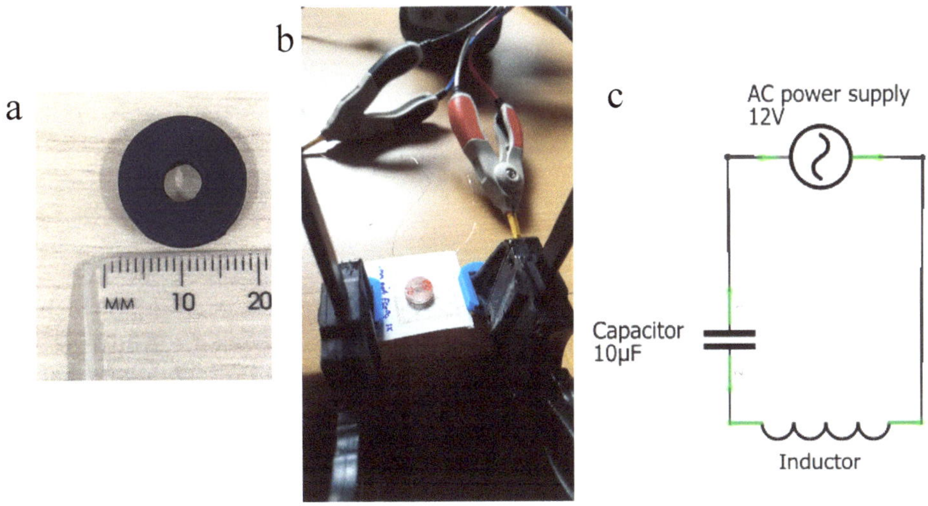

Figure 3. (**a**) Dimension of the air core coil holder used for inductance measurements, (**b**) inductance measurement with coil and clamp on 3d printed bend rig and (**c**) circuit schematic with inductor, capacitor (where 1 and 2 are positive and negative connection) and AC power supply in series.

The experimental setup to measure the inductance of the printed sample consisted of an 891-bench top LCR meter from BK precision, connected to a coil inductor, as seen in Figure 3b,c. A current was applied at a frequency of 1 kHz with a voltage of 1 vrms for the benchtop LCR meter.

3. Results and Discussion

3.1. Magnetic Properties

The magnetic properties were measured by measuring the hysteresis loop using a Quantum Design MPMS magnetometer at room temperature, as shown in Figure 4 and Table 2. From the hysteresis loops, it was observed that the magnetite NP had more than double the saturation magnetisation (M_s) and a lower coercivity (H_c) and remanence (M_r) than the nickel NP, which is preferable for sensing application.

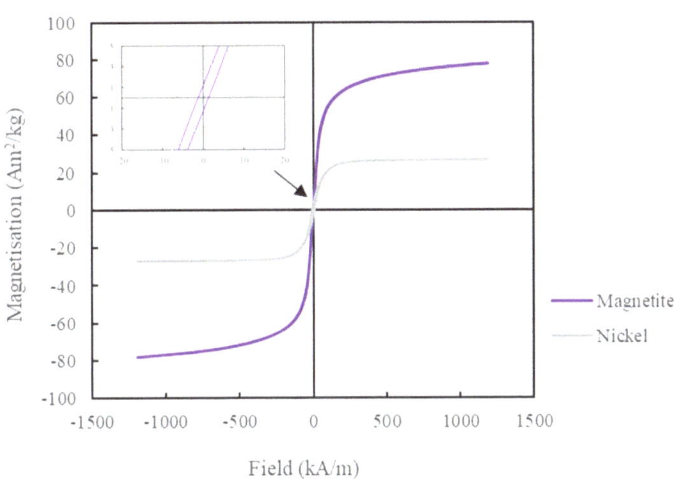

Figure 4. Hysteresis loop of magnetite and nickel NP from −1200 to 1200 kA/m field.

Table 2. Magnetite and nickel saturation magnetisation, remanence and coercivity calculated from the hysteresis loop.

Material	M_s (Am^2/kg)	M_r (Am^2/kg)	H_c (kA/m)
Magnetite	76.5	1.25 ± 0.02	1.22 ± 0.02
Nickel	27.1	2.53 ± 0.01	6.56 ± 0.10

From the literature, the magnetic properties of both magnetite and nickel NP have been measured [15,16]. The M_s of the magnetite was 76 Am^2/kg, which is close to published values at 70–80 Am^2/kg. There are similarities in magnetic properties for all literature shown except for one published paper [15], where H_c and M_r show a 10 kA/m and 10 Am^2/kg difference, respectively, from the current work and other published papers. This may be due to the technique used by [15]; they used an electrode bath (electrolysis) to form magnetite, which may have reduced the particle size, hence increasing the H_c and M_r values considerably. It is reported that bulk magnetite has M_s at 92 Am^2/kg [15], which is similar to the measured magnetite M_s in Table 2, as the difference is due to surface effects such as spin canting.

Whereas the published values and current work for Nickel NPs [30,31] show that there is considerable variation in the magnetic properties. For example, in [30], the data is closest to this work where the M_s is around 30 Am^2/kg; however, in [31], the M_s is around 55 Am^2/kg, which was higher than all published and current work. The gap of 20 Am^2/kg is significant, more than the published values for magnetite. The size and composition of NPs of nickel do affect the M_s value, which could explain the high variation between the published values. For example, a particle size of 20 nm has a single domain, whereas above this size, they are multi-domain. For example, ref. [31], reported that the hysteresis loop for Nickel NPs with a diameter of 22 nm has a similar M_s value of around 30 Am^2/kg from the hysteresis loop. However, this work has a particle diameter of 80 nm, far larger than in [31] This may be down to their process in obtaining Nickel NPs, where thermal fluctuation, solvent and technique could affect their magnetic properties.

3.2. Viscosity Measurement

Table 3 shows the result of magnetite and nickel ink viscosity at room temperature measured using a viscometer SV–1 A from A&D Company Ltd. (Tokyo, Japan). This shows that both inks are suitable for inkjet printing as the value is below 20 cP and above 1 cP However, the measurement was taken at the meniscus, where there is more solvent than heavy NPs. As the heavy NPs tend to drop to the bottom of the polycarbonate container This creates an issue when measuring, as the measured value is not the true viscosity of the ink; rather, it is more of the viscosity of the solvent. Therefore, the process of setting up the measurement was performed quickly to prevent the heavy NPs from dropping to the bottom before the viscosity measurement. The true viscosity is likely to be higher. The viscosity for nickel is slightly higher than magnetite ink even though magnetite contains 20 wt.% NPs and Nickel has 2 wt.% NPs in the ink. The higher viscosity is due to the higher molecular weight and density of the solvent, as NMP in nickel ink has a higher density than DMF in magnetite ink.

Table 3. Viscosity measurement for magnetite and nickel NP.

Metal NP	Solvent	Temperature (K)	Viscosity (cP)
Magnetite	DMF, IPA and water	294	1.92
Nickel	NMP and water	293.6	2.06

The Z number was calculated by using the equation in [26] and found to be around 6–8 cP for nickel ink and 9.7–12.9 cP for magnetite ink. However, the inks can be diluted to improve the viscosity and reduce the Z number. The Z number is less or around the critical value of 10 cP; therefore, this makes it suitable for inkjet printing. In the published paper by [32], the magnetite dispersion from the same company was shown to have a viscosity value of around 1.6–1.7 cP at 297 K. Although the test was not performed at around 297 K, the viscosity in the published values would be slightly lower than in this work. Nevertheless, the inks have shown a good level of viscosity for inkjet printing using the JetLab 4 printing system.

3.3. Print Analysis

Figure 5 shows the (a) image and (b) printed design for (i) magnetite uniaxial patch, (ii) magnetite coil, (iii) magnetite 3 mm grid design and (iiiv) nickel lines on photo paper printed by JetLab 4 with 60 μm print head orifice, respectively. All designs are measured at 25×25 mm^2 in length. The grid designs are adjusted by changing the infill size as track gaps are increased from 3 mm to 5 mm in distance.

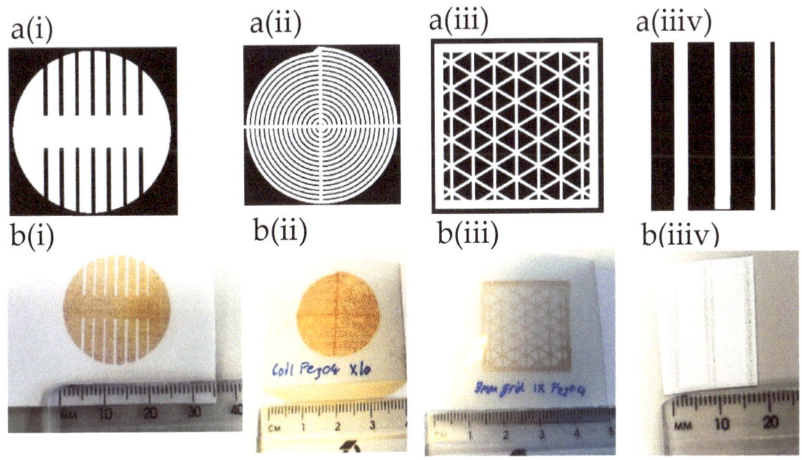

Figure 5. (**a**) image and (**b**) printed design: (**i**) magnetite uniaxial patch, (**ii**) magnetite coil and (**iii**) grid design and (**iiiv**) nickel lines print.

The microscope image of magnetite print on photo paper presented in Figure 6a shows that each droplet has a non-spherical shape, but there is good adhesion between the magnetite nanoparticles on paper. Some droplets, as seen in Figure 6a, have joined, which may have been due to the droplets being too close together during the printing process; this could be improved by adjusting the jetting parameters to reduce satellite droplets, as seen in Figure A2. The radius of droplets (R1, R2 and R3) are all similar to each other with a mean of 78.20 ± 2.5 μm, while the diameter (D1, D2 and D3) have a mean of 159.45 ± 5.3 μm, which is due to the misalignment of droplets. Although the droplets are shown to have a slight variation in diameter and radius, the size of the droplets is relatively good. This would not affect the magnetisation change when strain is applied.

The nickel coil is printed as seen in Figure 6b, which shows the microscope image of nickel ink on photo paper. The microscope image shows that the nickel drops are erratic and do not show good cohesion on the paper. Compared to magnetite, where multiple prints show a relatively good accuracy (print direction on top of each drop), nickel does not print well in additive layering as each drop does not overlap each other. There are areas where the print is inconsistent, as clogging may have disrupted the printing process. The size of the droplets shown in Figure 6b has a mean diameter of 141.6 ± 2.3 μm (D1, D2 and D3) and a spherical mean radius of 70.7 ± 1.5 μm (R1, R2 and R3). In comparison, the nickel drop measured shows that the drops are about 20 μm smaller than magnetite droplets, which is due to the lower concentration of nickel in each droplet and are prone to forming voids and non-uniform NPs due to excess solvent in each drop.

a

b

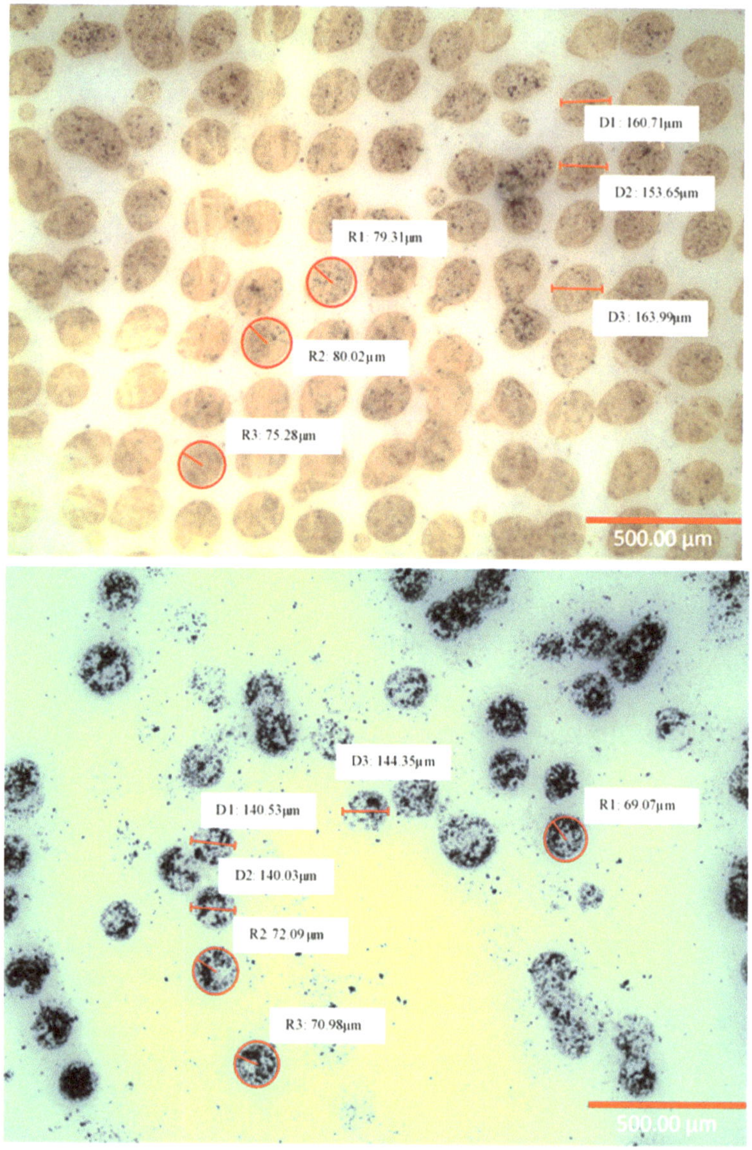

Figure 6. Optical microscope of (**a**) magnetite uniaxial patch design on paper and (**b**) nickel coil design on paper.

The magnetite print has a resolution of around 132 dots per inch (DPI), whereas in the literature [11], the resolution is much higher at 185 DPI. Reducing the droplet size can be achieved by reducing the wettability of the ink on the substrate. Layer-by-layer printing depends on the evaporation of the solvent to prevent the NPs from spreading. Therefore, applying heat during printing which increases solvent evaporation and reduces migration of NPs to the edge increases the density of the NPs in the droplet.

SEM and EDS Analysis

Magnetite and nickel NPs were further analysed by SEM and EDS using AZtec one software (https://nano.oxinst.com/products/aztec/aztecone (accessed on 30 September

2023)). Figure 7 shows the SEM and EDS analysis of magnetite print on paper where (a) and (c) show the magnetite droplet on paper substrate. The distribution of magnetite NPs is not homogeneous across the droplet, as larger amalgamated NPs can be seen clearly in the SEM image. This will affect the roughness and height of the droplet. Achieving homogeneous droplets is essential to reduce porosity and controllable anisotropy. Instead, the droplet will have random alignment if the porosity increases. The EDS image in (b) and spectrum in (d) and (e) shows elements of iron, carbon and oxygen present in the droplet. Iron elements in magnetite droplets can be clearly seen. Carbon and oxygen can be seen all over the substrate and the droplet, as oxygen is present in magnetite and paper, while carbon is present in paper but not in magnetite; however, due to the thinness of the magnetite drops, the electron beam is likely also to be detecting the substrate as well as the droplets.

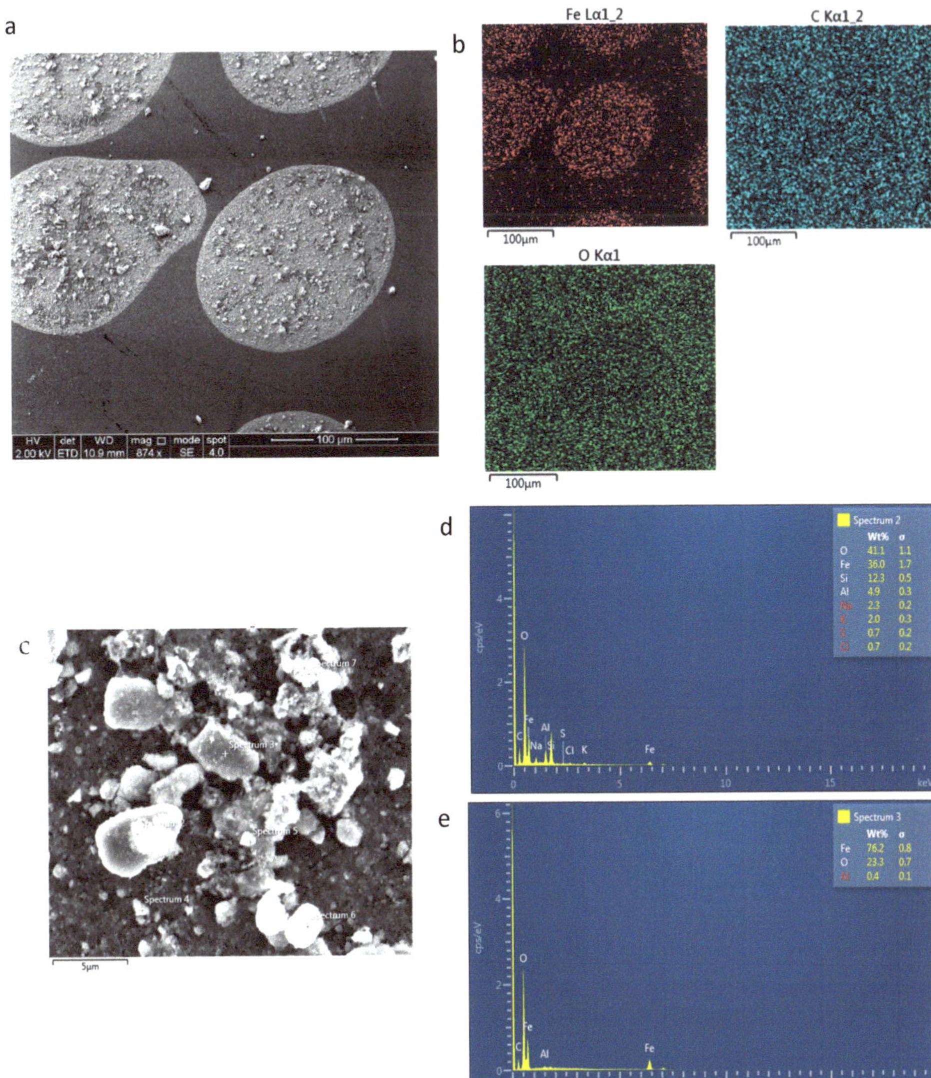

Figure 7. (**a**) SEM image of magnetite on paper at 2 kV, (**b**) EDS element map at 10 kV, (**c**) SEM spectrum label at 2 kV, (**d**) EDS spectrum 2 at 10 kV and (**e**) EDS spectrum 3 at 10 kV.

Nickel droplets on paper SEM and EDS mapping/spectrum can be seen in Figure 8. Compared to magnetite, nickel NPs on paper contain higher levels of porosity, which can be clearly seen in the SEM layered image in (a). This is due to the higher level of solvent than magnetite which is prone to porosity when it is evaporated or absorbed into the paper. The EDS mapping in (b) shows elements of nickel, oxygen, silicon and carbon present. The droplet shows a clear presence of nickel and carbon elements, whereas silicon and oxygen are present in the paper. This is as expected as the nickel NPs contain only nickel and paper showing oxygen. Spectrum (c) and (d) show that the SEM image in (a) contained a majority of nickel and oxygen.

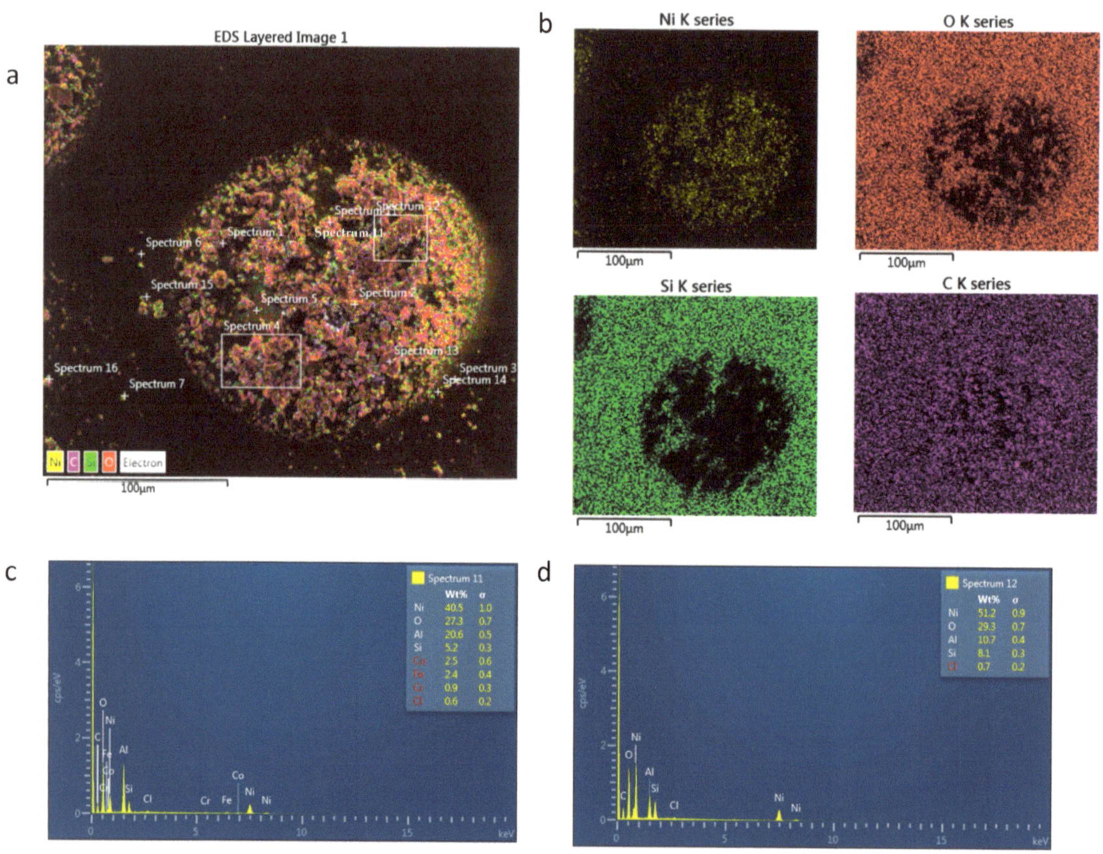

Figure 8. (**a**) Nickel droplet on paper SEM at 2 kV, (**b**) EDS layered map at 10 kV, (**c**) nickel EDS spectrum 11 and (**d**) nickel EDS spectrum 12 at 10 kV.

3.4. Inkjet Printing Sensor Measurement

The graph in Figure 9a shows the mean value of magnetite print inductance, measured on each bend rig, while Figure 9b shows the inductance as a function of strain for nickel print with a deviation of ± 0.005 µH. Most of the results shown in Figure 9 appear to show a negative inductance as the strain is increased. As the strain is increased, the distance between the coil and magnetite print is increased and the tension caused by the bend rig would align the moments toward the stress direction, which would naturally assume a negative trend. The measured magnetic field may be improved by increasing the density or layers of the print. In comparison, nickel has shown a higher inductance than magnetite; for example, the coil design in nickel at 0.22 µε is at 0.025 ΔµH compared to 0.005 ΔµH for magnetite, as shown in Figure 9a,b.

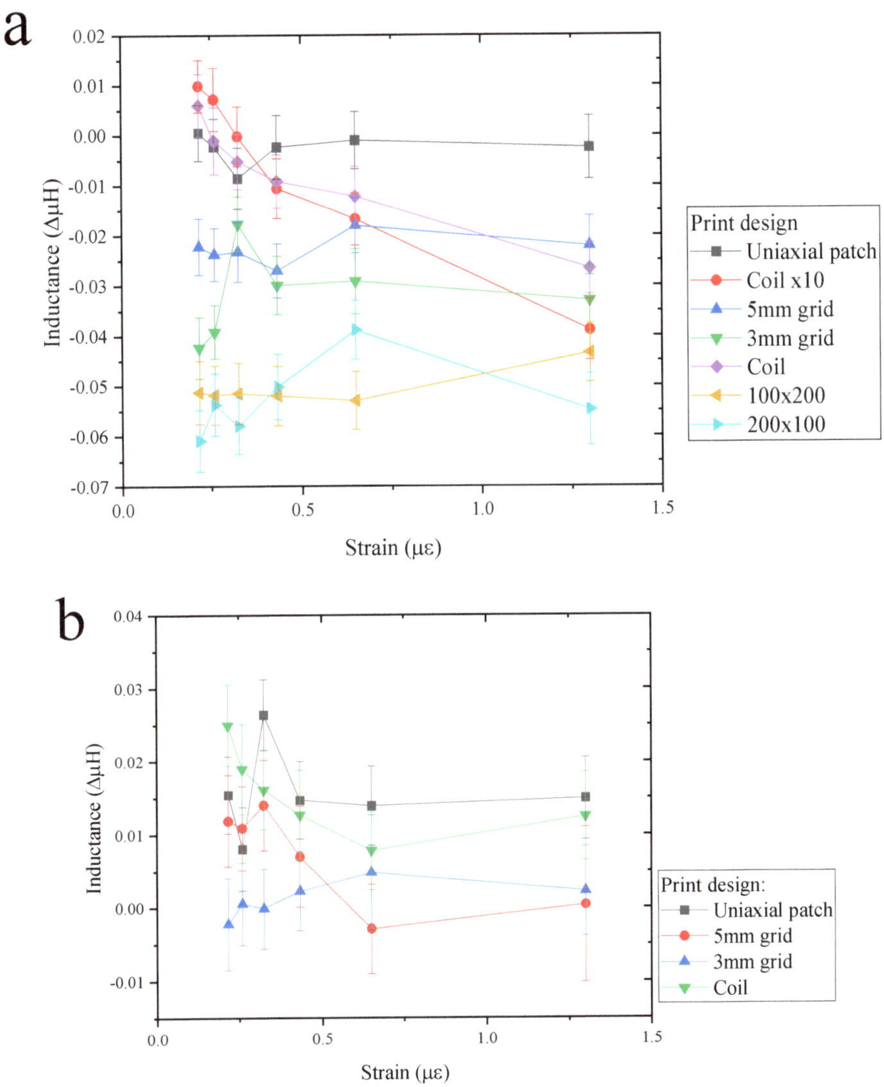

Figure 9. Change in inductance as a function of strain from 0.22 με to 1.3 με for (**a**) magnetite print designs and (**b**) nickel print designs.

A linear fit was performed on the data where Figure 10 shows the calculated linear fitting of the magnetite print for the graph (a) intercept, (b) slope and (c) R-squared values. The gradient is the change in the magnetic field as a function of strain applied; therefore, the greater the change, the better the sensor design performs as a function of strain. The intercept shows the initial inductance value where no strain is applied, which may be significant to compare with other designs such as increased layers. The R-squared value shows how well the data fits the linear fitting and if increasing strain changes the magnetic field; for example, if the data is erratic, then the R-squared value is low. For a reliable sensor, an R-squared value should be close to 1.

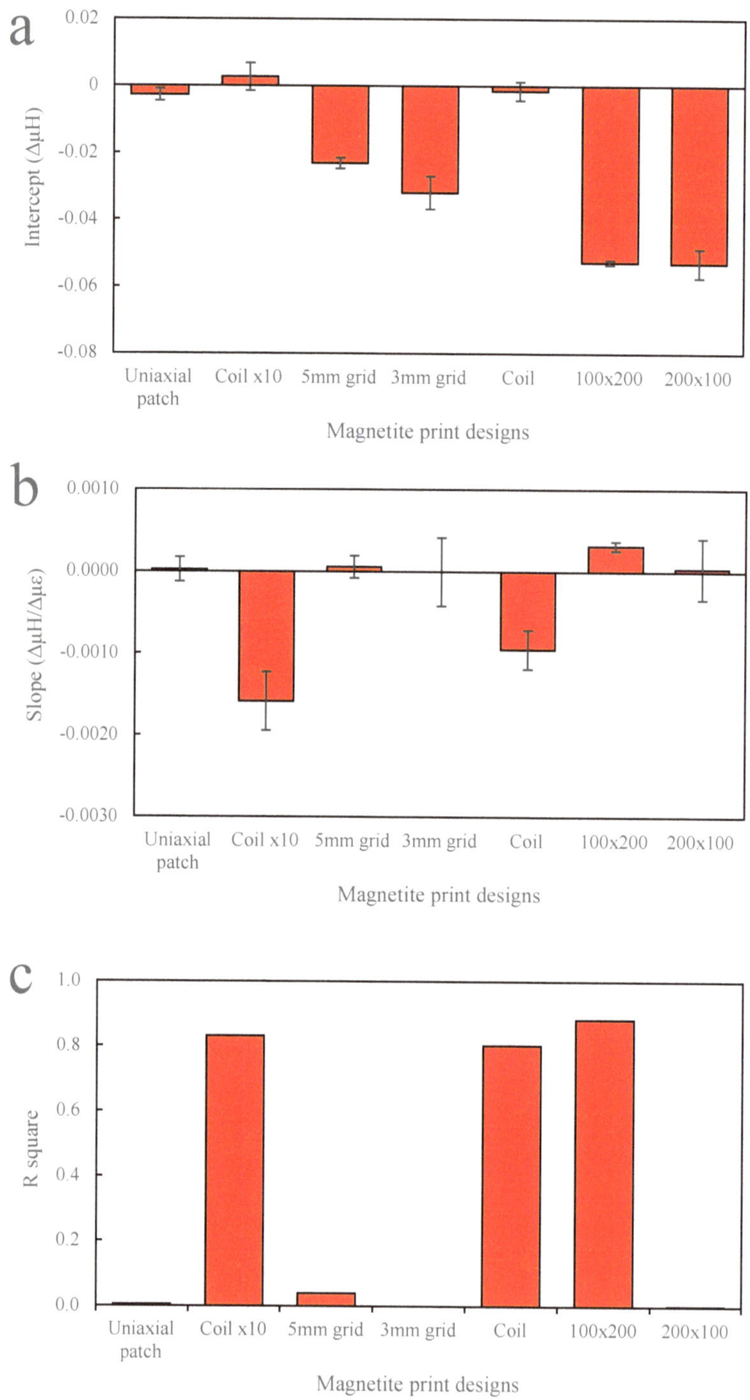

Figure 10. Magnetite print designs linear fitting showing (**a**) intercept, (**b**) gradient and (**c**) R-squared value.

The magnetite design that is most sensitive to strain is the coil design, as shown in Figure 10b. In comparison with other designs tested, the gradient of the coil design is lower, which may be due to the design itself. The R square value (COD), as seen in (c), shows that 80% of the data is due to the applied strain. In comparison, the R-value for a 5 mm grid, a 3 mm grid and a uniaxial patch are close to 0, which suggests that the change in strain does not affect the change in inductance. Data could be improved by gradually reducing the radius of curvature while measuring the inductance. For example, the intercept shown in (a) should show a positive value due to the demagnetising field of the printed NPs. The negative intercept could be due to the initial applied strain; in other words, the starting strain could be too high, resulting in a negative reduction in inductance.

To improve the change in the field with strain, different designs, the number of layers and print directions were explored, as seen in Figure 10b. The effect of changing the design from 5 mm to 3 mm and the uniaxial patch does not show a significant change. There is a slight increase in gradient and R squared value for the 5 mm grid. However, there is a larger deviation in the 3 mm grid than the 5 mm grid design, which may benefit from additional data points; the same can be said with the uniaxial patch. In comparison with the desktop printing grid design, the change in field to strain (cantilever) did show a difference between the 5 mm and 3 mm track gap design, as seen in [6]. Printing multiple layers does make a difference to the magnetoelastic performance. A coil with 10 layers shows an increased slope in the field as a function of strain compared to the single-layered coil, which is expected due to a greater amount of material deposited on the substrate. Even though the gradient for both single and multiple coil layers shows a negative tread, the addition of multiple layering causes an increased gradient to the magnetoelastic performance.

The print that is perpendicular to the strain direction shows the greatest sensitivity to inductance as a function of strain, as seen in Figure 10. The 100 × 200 px rectangular print (perpendicular to strain) has a higher inductance to strain sensitivity than the 200 × 100 px rectangle (parallel to strain) print. The intercept in (a) is shown to be almost the same for both print in parallel and perpendicular lines. As strain is applied, the perpendicular print produces a positive gradient, whereas the parallel print remains at zero, as seen in (b). It is true that the perpendicular gives a higher gradient but at lower to higher strain values, e.g., from 0.65 $\mu\varepsilon$ to 1.3 $\mu\varepsilon$. The parallel print is more sensitive to the field as a function of strain at a lower strain value than the perpendicular direction. This may be due to the shape anisotropy effect where the alignment of print direction in the parallel, the easy direction is along the axis, which is why the inductance peaked much earlier than the perpendicular direction. The perpendicular direction peaked at 1.3 $\mu\varepsilon$, which is due to the hard direction (as the alignment is already aligned towards the perpendicular direction). In Figure 6a, the microscope image of the printed magnetite on photo paper shows that individual drops of the magnetite ink are near-spherical shape, which suggests that the shape anisotropy is down to the direction of print and not due to the shape of individual droplets.

The graphs in Figure 11 show the fitting for nickel with parameters of (a) intercept, (b) gradient and (c) R squared value. Figure 11a shows that even though the intercept is higher, all the designs have smaller gradients in comparison to magnetite print, as shown in Figure 10b. For example, the nickel 5 mm grid and coil designs gave a good response to inductance as a function of strain. However, they have a lower gradient than magnetite coil and a 5 mm grid design. Although the gradient is very small for all designs, it was expected that the coil design would be more sensitive in nickel ink based on the design change rather than the material change. The change in magnetic field can change depending on the design and applied strain, which could be useful in various applications. For example, from Figure 9b, the coil design shows a much steadier and more gradual slope than the 5 mm grid, similar to a magnetite coil as strain is increased. For example, in the 5 mm grid design, from 0.22 $\mu\varepsilon$ to 0.33 $\mu\varepsilon$, there is a small positive change in inductance, but when the strain is increased to 0.65 $\mu\varepsilon$, there is a steep reduction in inductance. In comparison, the coil design shows a reduction of inductance from 0.22 $\mu\varepsilon$ to 0.65 $\mu\varepsilon$. Therefore, the coil design has a better range of sensing for small and larger strains. The R-square values in

(c) in both Figures 10 and 11 show that for both magnetite and nickel, the uniaxial patch is the least effective design, as the data is not influenced by the strain size. This could be due to the print direction (perpendicular to strain) and gaps within the design, which could reduce the change in the field.

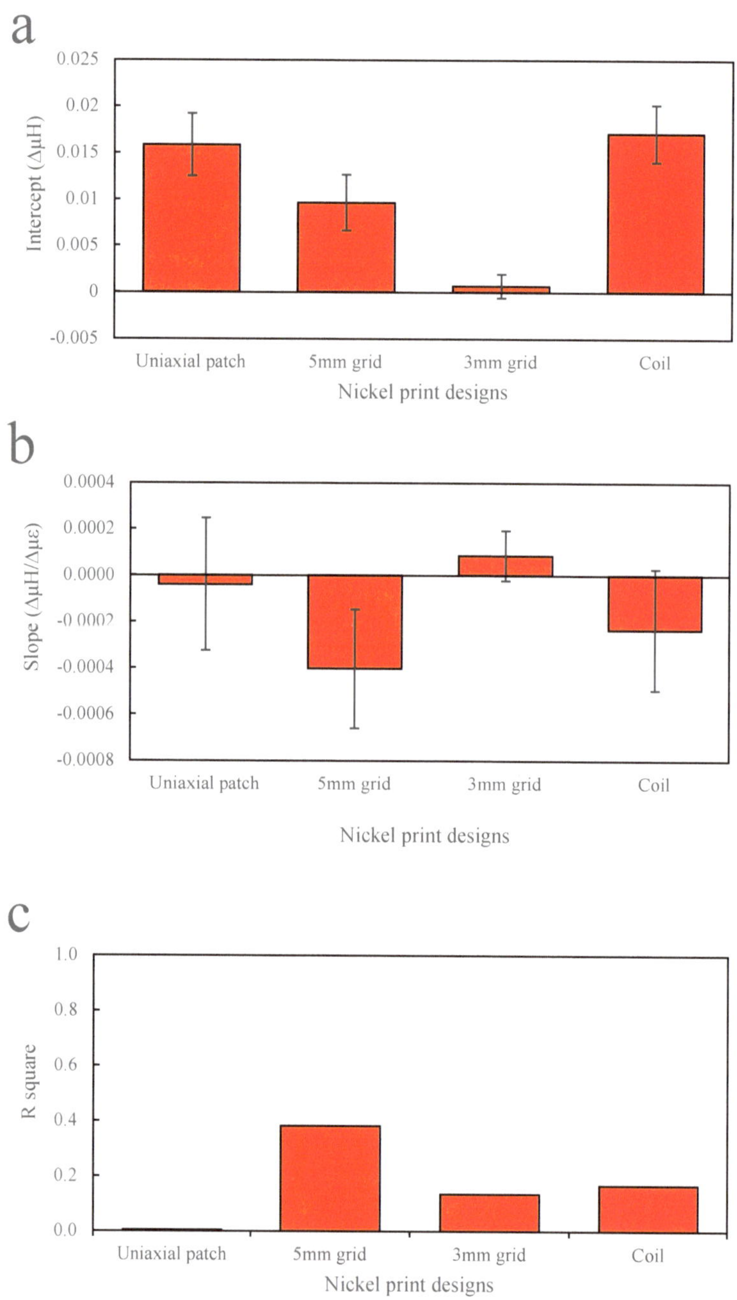

Figure 11. Nickel print designs linear fitting graph showing (**a**) intercept, (**b**) gradient and (**c**) R-squared value.

Overall, the design that gives the best magnetoelastic response is the coil design. While it does show a negative inductance gradient in both nickel and magnetite material, both do show a good trend of change in inductance to strain. The trend shows a gradual change, whereas other designs give a sharp and unpredictable change, which can distort the results. Therefore, this makes it ideal to select it as a sensor in future work. The rate of change in field as a function of strain is higher in magnetite than nickel, even though nickel has a higher intercept (demagnetising field) than magnetite print. Therefore, the material of choice in this work is magnetite.

3.4.1. Additive Layering

Printing magnetite in layers has been shown to improve the NPs change in the field as a function of strain. However, this was performed with one design, comparing the single and 10-layer printed magnetite coil on photo paper. Printing a simple design, such as a square shape (5×5 mm^2), makes it easier to print in layers of 10, 20 and 30.

Figure 12 shows the magnetite square design (5×5 mm^2) in (a) 30, (b) 20 and (c) 10 layers. Significant differences can be seen in each layering print. For example, the 10 layers show clear print direction across the square design, whereas the 30 layers show erratic print direction, thus reducing the printing accuracy. This may be due to the excess satellites and clogging forming during printing, where the excess droplets can be seen on the edges of the square design. Therefore, when printing beyond 30 layers, the jetting parameters may need to be recalibrated in order to prevent satellites or excess droplets. The 20-layered print has been shown to have good definition and density to form an accurate square patch on the photo paper substrate.

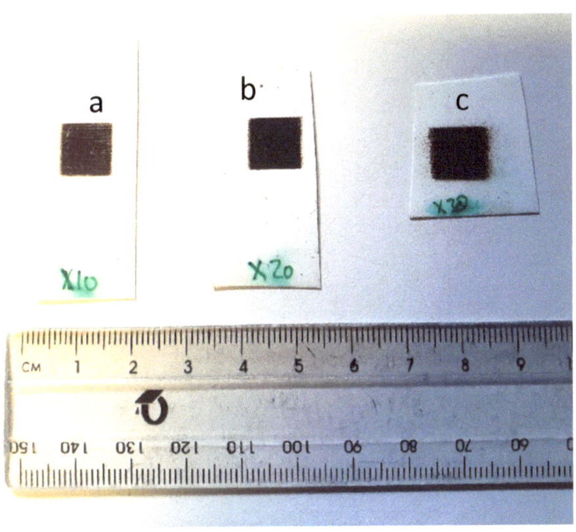

Figure 12. Magnetite print in (**a**) 10, (**b**) 20, (**c**) 30 layered square design.

Figure 13 shows the bending test performed for each layered magnetite sample in Figure 12. The radius of curvature was reduced from 600 mm to 1000 mm; this was used to see the effect of applying a smaller strain to the magnetite designs and whether there was any change in inductance. Not all square prints show a change in inductance or reliability as a function of strain. From all three sample square prints, on average, the change in inductance increased from an applied strain of 0.13 $\mu\varepsilon$ to 0.2 $\mu\varepsilon$, and then the change in inductance reduced as the applied strain increased from 0.25 $\mu\varepsilon$. This pattern was present in all print layers. However, the error seen in 10 and 30 layers exceeds the change in inductance, whereas in 20-layered square patches, the error is lower than the change in inductance to strain. All three prints have shown different levels of noise and sensitivity to strain. In the 30-layered magnetite print, the sensitivity is the largest; however, by using the same inductor coil, the level of signal-to-noise ratio for 30 layers is larger compared to other layers. For example, the error/noise in a 30-layered print is around ±4.2 μH, whereas the error/noise in a 20-layered print is around ±2.2 μH. This could be due to the excess porosity and voids present in the print, which restricts magnetic moment orientation. In the 20-layered square design, the sensitivity is greatest beyond 0.2 με and saturates at 0.4 με. The level of error in the 20-layered print is the best compared to other layered designs. The 10-layered square design shows the least sensitivity to strain and the lowest measured magnetic field. This is expected, as there is less magnetite material on the surface; hence, it will have a lower field and reduced sensitivity. The level of error is quite large, which is surprising, as it would contain less porosity than the other designs, such as the 20-layered magnetite. There is a difference in maximum inductance saturation as more material is layered on top. For the 10 and 20 layers, the inductance is saturated at around 0.2 με whereas 30-layered magnetite has an inductance maximum at around 0.225 με. This is due to the amount of material on top, as more strain is needed to orientate the magnetic moments to reach maximum strain when the coil is placed on top. Coating a thin layer of PDMS or acrylic polymer on the 30-layered magnetite could help with reaching the maximum inductance.

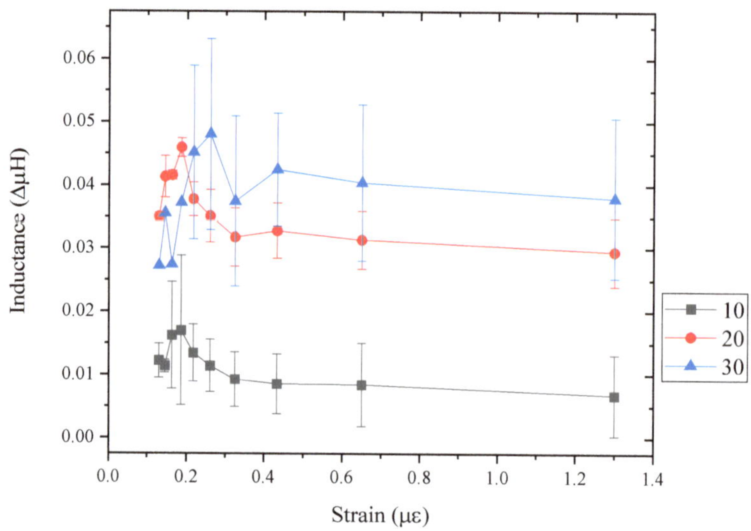

Figure 13. Bending test of magnetite square designs of 10, 20 and 30 layers measuring the inductance as a function of strain.

3.4.2. Bending Test

For structural health monitoring of CFRP, the inkjet-printed sensor will be directly printed onto the CFRP rather than paper. Therefore, a magnetite coil design was directly printed on top (Figure 14). The change in inductance as a function of applied strain was measured, as previously performed for the sensors on paper, but this time, it was found that smaller bend radii were required to achieve a higher strain.

The inductance measurements were carried out using a magnetite coil sensor, printed on CFRP using a 200 coil turns inductor and a 10 μF capacitor in series. While the results in Figure 15 show that there is a change in inductance as a function of radii, the initial measurement (baseline) showed that there was a drift over time; therefore, an average was taken. The change in inductance is higher compared to the magnetite coil sensor on a photo paper substrate. In comparison, ref. [9] have reported a resolution of 0.211 μStrain with CoFeB ribbon, while this work shows that the coil has a resolution of 0.12 μStrain. While different types of inductors were used, such as the number of turns, it does show that the sensor used here is much lower compared to literature due to a higher magnetostriction of CoFeB than magnetite.

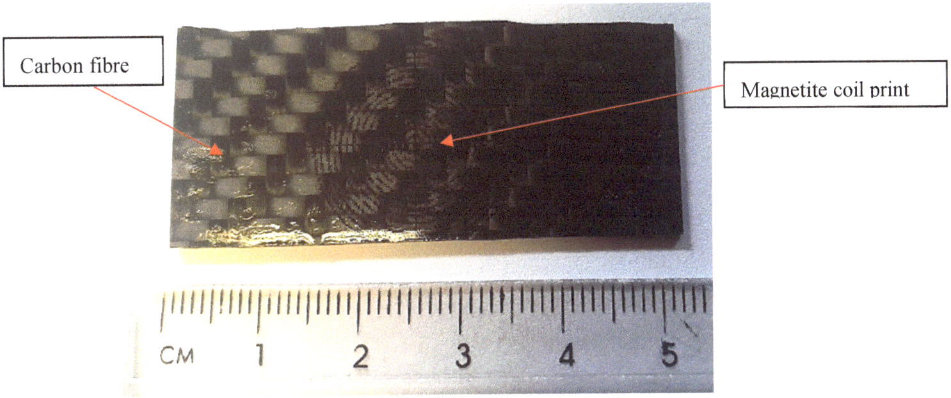

Figure 14. Magnetite coil print on CFRP.

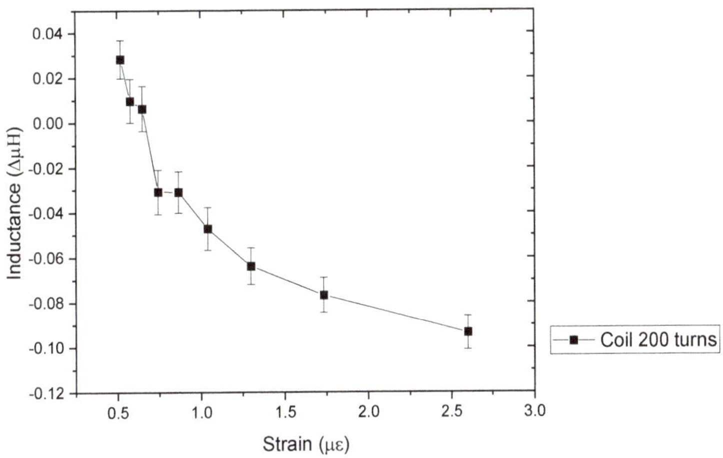

Figure 15. Inductance of coil print on CFRP with 200 coil inductor.

4. Conclusions

Magnetic sensors were made from 20% magnetite NP and DMP solvent, and the designs were made and selected to print using the JetLab 4 inkjet printing system. These included a coil, uniaxial patch, grid and rectangular block to test the inductance sensitivity as a function of strain.

The print was analysed by optical microscope, SEM and EDS. Print analysis shows that the print had a resolution of 132 DPI. This was lower than in literature studies, where they achieved a resolution of 180 DPI. This is because the jetting parameters were set to print larger droplets to ensure magnetic NPs were jetted and not to let the NPs sink to the bottom of the reservoir. EDS shows magnetite and nickel composition, where iron and oxygen were present in magnetite droplets, and nickel was present in nickel droplets.

The coil design was selected, and the best ink for strain bending was magnetite. The coil design far exceeded other designs that were thought to have anisotropy due to the print design such as in uniaxial patch design. This may be due to the print pattern where the coil design contained larger gaps between the coil turns; therefore, it may improve the field through increased demagnetisation. It was evident that the greater number of layers o magnetite increased inductance as a function of strain. For example, the 20-layered square patch had a greater response to strain than the 10-layered square magnetite patch. However beyond 20 layers, the print quality was reduced even though the magneto-mechanica response was slightly higher. This increased the level of noise that would not be suitable for SHM applications.

By continuing from inkjet printing on photo paper, strain by bending over a known curvature was performed on a CFRP sample with a magnetite coil printed directly on top. The results showed that printing directly on the surface of CFRP, performed better than on paper substrate. This could be due to the paper substrate absorption of the magnetite and higher wetting angle on kapton than CFRP. Acrylic spray or PDMS coating can be used to protect the magnetite print on CFRP and prevent damage to the environment.

As magnetostrictive materials are known to provide SHM of CFRP, this paper has contributed to developing a sensor via inkjet printing magnetostrictive material and shown how changing the design and number of layers affects the change in inductance under applied strain. This work provides scope for future research and development into SHM and printing magnetic materials.

Author Contributions: N.A.: Writing—original draft, Writing—review & editing, Investigation. P.J.S. Writing—review & editing. N.A.M.: Conceptualization, methodology, writing—review & editing, Supervision, project administration, funding acquisition. All authors have read and agreed to the published version of the manuscript.

Funding: This research was funded under the DSTL Anglo-French studentship scheme and the Henry Royce Institute for Advanced Materials, funded through EPSRC grants EP/R00661X/1 EP/S019367/1, EP/P02470X/1 and EP/P025285/1, for the SQUID access at The University of Sheffield. The APC was funded by the University of Nottingham.

Institutional Review Board Statement: Not applicable.

Informed Consent Statement: Not applicable.

Data Availability Statement: Unavailable due to privacy.

Conflicts of Interest: The authors declare no conflict of interest.

Appendix A

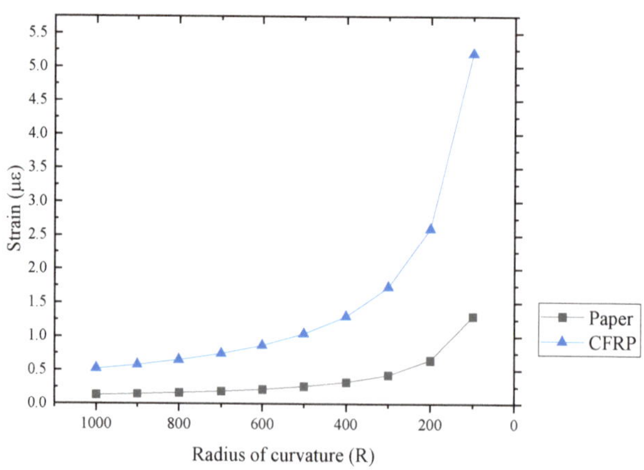

Figure A1. Strain calculated values for radius of curvature from 1000 to 100 for paper and CFRP.

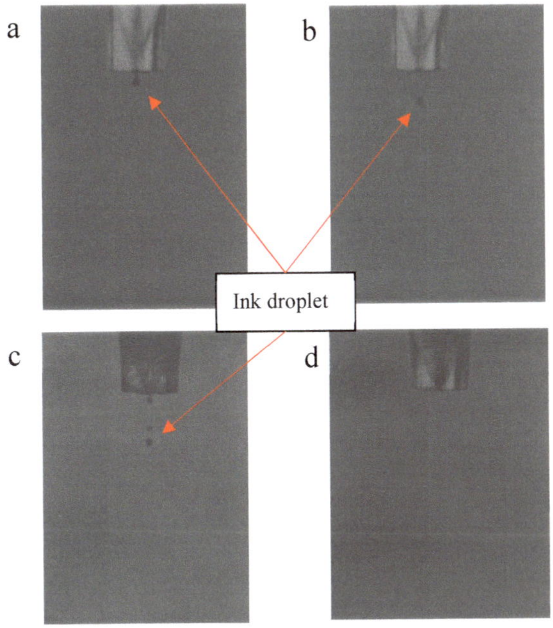

Figure A2. Print head nozzle during jetting metal NP when jetting from nozzle (**a**), single droplet (**b**), satellite (**c**) and clogged (**d**).

Table A1. Viscosities of water at various temperatures (James et al., 1984 [27]).

		Flow Times (S)		
Temperature (K)	**Density (g cm^{-3})**	**Upper Bulb (cP)**	**Lower Bulb (cP)**	**Average (cP)**
273.15	0.99984	1.789	1.7889	1.7891
278.15	0.99996	1.518	1.5187	1.5187
293.15	0.9982	(1.002)	(1.002)	(1.002)
298.15	0.99704	0.89	0.8903	0.8904
303.15	0.99565	0.7978	0.7977	0.7978
308.15	0.99403	0.7193	0.7193	0.719,
313.15	0.9922	0.6531	0.6531	0.6531
323.15	0.988	0.5475	0.547	0.5475
333.15	0.9832	0.467	-	0.467

References

1. Bastola, A.K.; Hossain, M. The shape—Morphing performance of magnetoactive soft materials. *Mater. Des.* **2021**, *211*, 110172. [CrossRef]
2. Güemes, A.; Fernandez-Lopez, A.; Pozo, A.R.; Sierra-Pérez, J. Structural Health Monitoring for Advanced Composite Structures: A Review. *J. Compos. Sci.* **2020**, *4*, 13. [CrossRef]
3. Han, S.; Li, Q.; Cui, Z.; Xiao, P.; Miao, Y.; Chen, L.; Li, Y. Non-destructive testing and structural health monitoring technologies for carbon fiber reinforced polymers: A review. *Nondestruct. Test. Evaluation* **2024**, *39*, 725–761. [CrossRef]
4. Hassani, S.; Dackermann, U. A Systematic Review of Advanced Sensor Technologies for Non-Destructive Testing and Structural Health Monitoring. *Sensors* **2023**, *23*, 2204. [CrossRef] [PubMed]
5. Sharma, R.S.; Vijayakumar, M.N. Recent Developments in Damage Detection of CFRP Composites Using Non-Destructive Techniques—A Review. *J. Eng. Res. Rep.* **2023**, *25*, 51–65. [CrossRef]
6. Ahmed, N.; Deffley, R.; Kundys, B.; Morley, N. 3D printing of magnetostrictive property in 17/4 ph stainless steel. *J. Magn. Magn. Mater.* **2023**, *585*, 171115. [CrossRef]

7. Jones, N.J.; Yoo, J.-H.; Ott, R.T.; Lambert, P.K.; Petculescu, G.; Simsek, E.; Schlagel, D.; Lograsso, T.A. Magnetostrictive performanc of additively manufactured CoFe rods using the LENSTM system. *AIP Adv.* **2018**, *8*, 56403. [CrossRef]
8. Leong, Z.; Holmes, W.; Clarke, J.; Padki, A.; Hayes, S.; Morley, N.A. Magnetostrictive Sensors for Composite Damage Detectio and Wireless Structural Health Monitoring. *IEEE Trans. Magn.* **2019**, *55*, 4003006. [CrossRef]
9. Vincent, J.D.S.; Rodrigues, M.; Leong, Z.; Morley, N.A. Design and Development of Magnetostrictive Actuators and Sensors fo Structural Health Monitoring. *Sensors* **2020**, *20*, 711. [CrossRef]
10. Gullapalli, A.; Beedasy, V.; Vincent, J.D.S.; Leong, Z.; Smith, P.; Morley, N. Flat Inkjet-Printed Copper Induction Coils fo Magnetostrictive Structural Health Monitoring: A Comparison with Bulk Air Coils and an anisotropic magnetoresistive senso (AMR) Sensor. *Adv. Eng. Mater.* **2021**, *23*, 2100313. [CrossRef]
11. Voit, W.; Belova, L.; Zapka, W.; Rao, K.V. Application of inkjet technology for the deposition of magnetic nanoparticles to forn micron-scale structures. *IEE Proc.-Sci. Meas. Technol.* **2003**, *150*, 252–256. [CrossRef]
12. Raut, N.C.; Al-Shamery, K. Inkjet printing metals on flexible materials for plastic and paper electronics. *J. Mater. Chem. C* **2018**, 1618–1641. [CrossRef]
13. Song, H.; Spencer, J.; Jander, A.; Nielsen, J.; Stasiak, J.; Kasperchik, V.; Dhagat, P. Inkjet printing of magnetic materials with alignec anisotropy. *J. Appl. Phys.* **2014**, *115*, 17E308. [CrossRef]
14. Grabham, N.J.; Beeby, S.P.; White, N.M. The formulation and processing of a thick-film magnetostrictive material Related conten the formulation and processing of a thick-film magnetostrictive material. *Meas. Sci. Technol.* **2001**, *13*, 59. [CrossRef]
15. Cabrera, L.; Gutierrez, S.; Menendez, N.; Morales, M.; Herrasti, P. Magnetite nanoparticles: Electrochemical synthesis anc characterization. *Electrochim. Acta* **2008**, *53*, 436–3441. [CrossRef]
16. Kolchanov, D.S.; Slabov, V.; Keller, K.; Sergeeva, E.; Zhukov, M.V.; Drozdov, A.S.; Vinogradov, A.V. Sol–gel magnetite inks fo inkjet printing. *J. Mater. Chem. C* **2019**, *7*, 6426–6432. [CrossRef]
17. Li, D.; Sutton, D.; Burgess, A.; Graham, D.; Calvert, P.D. Conductive copper and nickel lines via reactive inkjet printing. *J. Mate Chem.* **2009**, *19*, 3719–3724. [CrossRef]
18. Marjanović, N.; Chiolerio, A.; Kus, M.; Ozel, F.; Tilki, S.; Ivanović, N.; Rakočević, Z.; Andrić, V.; Barudžija, T.; Baumann, R.R Magnetite nanoparticles: Synthesis, thin film properties and inkjet printing of magnetic cores for inductor applications. *Thin Solic Films* **2014**, *570 Pt A*, 38–44. [CrossRef]
19. Park, S.-H.; Kim, H.-S. Flash light sintering of nickel nanoparticles for printed electronics. *Thin Solid Films* **2014**, *550*, 575–581 [CrossRef]
20. Pasquarelli, R.; Curtis, C.; Van Hest, M. Inkjet printing of nickel and silver metal solar cell contacts. *J. Undergrad. Res.* **2008**, *8* 91–96. Available online: https://www.osti.gov/servlets/purl/1052098 (accessed on 22 February 2021).
21. Tan, H.W.; An, J.; Chua, C.K.; Tran, T. Metallic Nanoparticle Inks for 3D Printing of Electronics. *Adv. Electron. Mater.* **2019**, *5* 1800831. [CrossRef]
22. Tiberto, P.; Barrera, G.; Celegato, F.; Coïsson, M.; Chiolerio, A.; Martino, P.; Pandolfi, P.; Allia, P. Magnetic properties of jet-printe inks containing dispersed magnetite nanoparticles. *Eur. Phys. J. B* **2013**, *86*, 173. [CrossRef]
23. Vaseem, M.; Ghaffar, F.A.; Farooqui, M.F.; Shamim, A. Iron Oxide Nanoparticle-Based Magnetic Ink Development for Fully Printed Tunable Radio-Frequency Devices. *Adv. Mater. Technol.* **2018**, *3*, 1700242. [CrossRef]
24. Altay, B.N. Development and Characterization of Nano Nickel-Based Conductive Inks for Flexographic Printing of Electronics and New Interpretations of Surface Energies of Solids. Ph.D. Thesis, Western Michigan University, Kalamazoo, MI, USA, 2018.
25. Liu, Y.; Derby, B. Experimental study of the parameters for stable drop-on-demand inkjet performance. *Phys. Fluids* **2019**, *31* 032004. [CrossRef]
26. Zhang, F.; Tuck, C.; Hague, R.; He, Y.; Saleh, E.; Li, Y.; Sturgess, C.; Wildman, R. Inkjet printing of polyimide insulators for the 3D printing of dielectric materials for microelectronic applications. *J. Appl. Polym. Sci.* **2016**, *133*, 43361. [CrossRef]
27. James, C.J.; Mulcahy, D.E.; Steel, B.J. Viscometer calibration standards: Viscosities of water between 0 and 60 degrees C and of selected aqueous sucrose solutions at 25 degrees C from measurements with a flared capillary viscometer. *J. Phys. D Appl. Phys* **1984**, *17*, 225–230. [CrossRef]
28. SHD. Advanced Composite Prepreg Materials by SHD. 2022. Available online: https://shdcomposites.com/ (accessed on 5 May 2024).
29. TDK Corp. Inductors—Part 1 The Basics of Inductors. 2020. Available online: https://www.tdk.com/en/tech-mag/electronics_primer/1 (accessed on 22 August 2023).
30. Mahajan, C.G.; Alfadhel, A.; Irving, M.; Kahn, B.E.; Borkholder, D.A.; Williams, S.A.; Cormier, D. Magnetic Field Patterning of Nickel Nanowire Film Realized by Printed Precursor Inks. *Materials* **2019**, *12*, 928. [CrossRef] [PubMed]
31. Zhang, H.T.; Wu, G.; Chen, X.H.; Qiu, X.G. Synthesis and magnetic properties of nickel nanocrystals. *Mater. Res. Bull.* **2006**, *41*, 495–501. [CrossRef]
32. Deepak Dixit, D.; Pattamatta, A. Effect of uniform external magnetic-field on natural convection heat transfer in a cubical cavity filled with magnetic nano-dispersion. *Int. J. Heat Mass Transf.* **2020**, *146*, 118828. [CrossRef]

sensors

MDPI

Article

Theoretical Study of Microwires with an Inhomogeneous Magnetic Structure Using Magnetoimpedance Tomography

Nikita A. Buznikov [1,*] and Galina V. Kurlyandskaya [2]

[1] Institute for Theoretical and Applied Electromagnetics, Russian Academy of Sciences, Moscow 125412, Russia
[2] Institute of Natural Sciences and Mathematics, Ural Federal University, Ekaterinburg 620002, Russia; galinakurlyandskaya@urfu.ru
* Correspondence: n_buznikov@mail.ru

Abstract: The recently proposed magnetoimpedance tomography method is based on the analysis of the frequency dependences of the impedance measured at different external magnetic fields. The method allows one to analyze the distribution of magnetic properties over the cross-section of the ferromagnetic conductor. Here, we describe the example of theoretical study of the magnetoimpedance effect in an amorphous microwire with inhomogeneous magnetic structure. In the framework of the proposed model, it is assumed that the microwire cross-section consists of several regions with different features of the effective anisotropy. The distribution of the electromagnetic fields and the microwire impedance are found by an analytical solution of Maxwell equations in the particular regions. The field and frequency dependences of the microwire impedance are analyzed taking into account the frequency dependence of the permeability values in the considered regions. Although the calculations are given for the case of amorphous microwires, the obtained results can be useful for the development of the magnetoimpedance tomography method adaptation for different types of ferromagnetic conductors.

Keywords: magnetoimpedance; amorphous microwires; modeling; magnetic structure; magnetic anisotropy; permeability

Citation: Buznikov, N.A.; Kurlyandskaya, G.V. Theoretical Study of Microwires with an Inhomogeneous Magnetic Structure Using Magnetoimpedance Tomography. *Sensors* **2024**, *24*, 3669. https://doi.org/10.3390/s24113669

Academic Editor: Christer Johansson

Received: 27 April 2024
Revised: 26 May 2024
Accepted: 3 June 2024
Published: 5 June 2024

1. Introduction

The magnetoimpedance (MI) effect consists in significant change in the complex impedance of a ferromagnetic conductor under application of an external magnetic field [1–3]. MI is promising for the development of magnetic sensors and magnetometers with very high sensitivity with respect to applied magnetic fields [4–8], including devices for the detection of biomagnetic signals and magnetic biomarkers [9–12]. The nature of MI is attributed to the dependence of the skin penetration depth of a magnetic conductor on the external magnetic field. A significant MI effect can be observed in different soft magnetic amorphous and nanocrystalline as well as in complex composites [13–16].

In addition, the microwave properties of glass-coated microwires have attracted particular interest. In some cases, the developed metamaterials have shown extraordinary characteristics that are interesting for high-frequency applications in various devices [17]. Figure 1 shows selected examples of different types of magnetic wires demonstrating promise for MI applications. The varieties of their magnetic, magnetoelastic and microwave properties open the possibility to create modern magnetic sensors of multifunctional type adapted to different temperature ranges and environmental conditions. For example, water-quenched amorphous wires of selected composition and glass-coated microwires (Figure 1a–d) can be very good candidates for applications in biological and harsh chemical conditions. Figure 1 shows images obtained by scanning electron microscopy (SEM) representing different features of the samples. The details of the SEM technique [18] or preparation of the rapidly quenched wires [2,19], glass-coated microwires [20] and electroplated wires [21] can be found elsewhere. However, it is important to emphasize that the wire samples used for imaging are not exactly the same as those used for the magnetic and MI measurements or in a particular sensor device. Magnetic wires of all types have a certain disadvantage: it is necessary to use a special procedure in order to cut them without deformation of the edge part and install them into electronic circuit. Figure 1a,b show purposely deformed rapidly quenched wire in a way that one can appreciate the evidence of the amorphous structure. Figure 1c,d show a glass-coated microwire with broken glass shell, making it possible both to estimate the geometrical parameter of each part of the composite sample and evaluate the quality of the metal surface. Figure 1e is the same, but in the case of an electroplated wire.

There are two different types of magnetic biosensors: one for magnetic label detection and an electrochemical type of detector that works on the label-free principle [22,23]. For magnetic label detection, the MI-sensitive element should be stable in biological and harsh chemical conditions. This can be achieved not only by the selection of appropriate composition, but also by coating/protecting the surface of the sensitive element with an additional metallic, polymer or carbon layer to make it biocompatible [24–28]. Even more, magnetic biosensors are usually designed with an additional layer either for functionalization of the sensitive element or for surface protection [24,25]. According to traditional classification of magnetic biosensors, the first type is for the the measurement of parameters of electrical and magnetic properties of living systems related to their functionality; the biocompatibility does not require the direct contact of the biosystem with the surface of the sensitive element [29–31]. The second type of magnetic biosensors (for testing the bioanalytes) is described above. In addition, electroplated wires having a nanostructured magnetic layer might be suitable for aerospace applications or other rugged radiation conditions [16].

One of the most-studied MI materials is glass-coated amorphous microwires produced by the Taylor–Ulitovsky technique [32–38]. In these microwires, magnetocrystalline anisotropy is negligible and the magnetoelastic anisotropy is governed by the magnetostriction and spatial distribution of internal stresses arising in amorphous metal during the glass-coated microwire production [32,34–36]. Uneven temperature distribution during the microwire fabrication and the difference in the thermal expansion coefficients of the glass covering and metal core parts leads to a non-uniform distribution of internal stresses over the microwire cross-section [32,35]. The magnetoelastic anisotropy in amorphous microwires can be modified by different post-annealing treatments [36,38,39], which results in a relaxation of the internal stresses.

Several models based on the theory of viscoelasticity have been proposed in order to describe the internal stress distribution in glass-coated amorphous microwires (see, for example, [35,40–43]). Usually, the magnetic structure of the amorphous part of glass-coated microwire is described in terms of the core–shell model [32]. It is assumed that the internal core has an axial magnetic anisotropy, whereas the external shell has a circular or radial easy axis of magnetization depending on the sign of the magnetostriction coefficient. The core–shell representation of the magnetic domain structure of glass-coated microwires allows one to explain peculiarities of their magnetic behavior. However, amorphous microwires may have a more complex domain structure. In particular, the magneto-optical Kerr effect studies for glass-coated microwires with a nearly-zero magnetostriction coefficient have shown that a helical anisotropy does exist in a surface region [44,45]. In addition, it has been demonstrated that the domain wall between the core and shell regions may significantly affect the magnetic behavior of microwires with a nearly-zero magnetostriction coefficient [46].

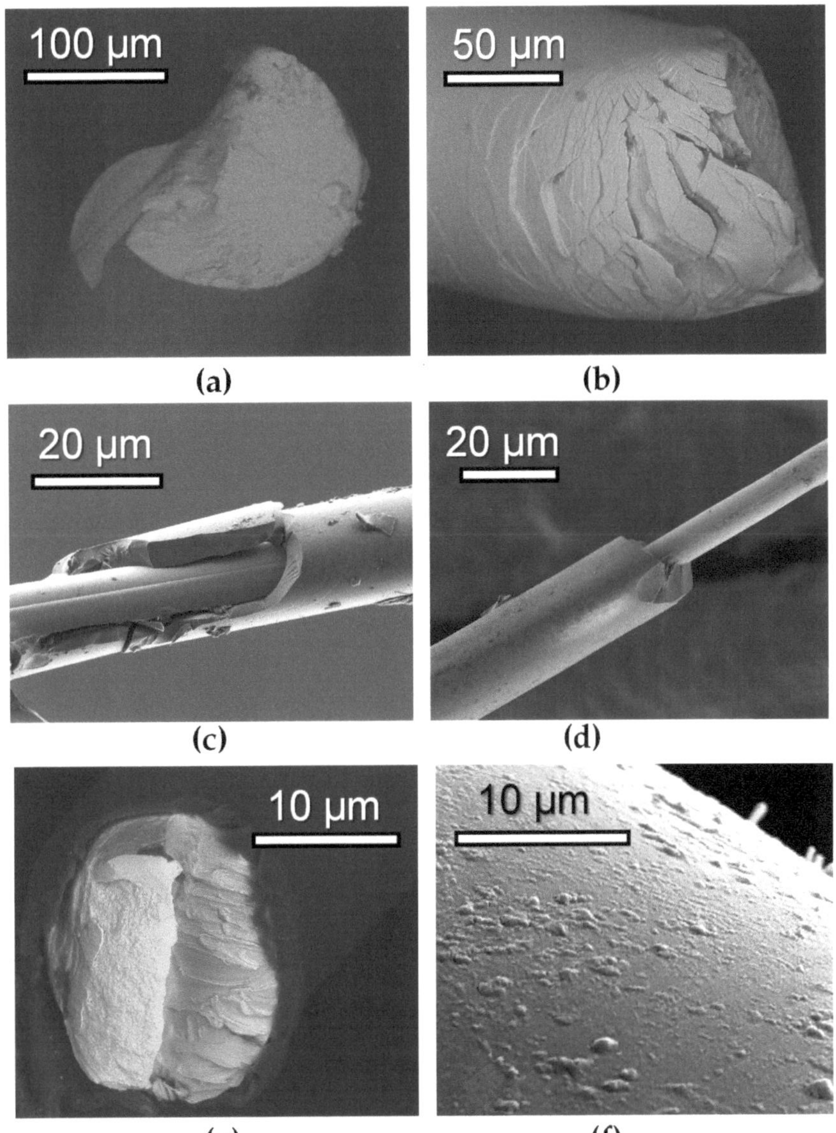

Figure 1. SEM images showing the general views of different types of magnetic wires. Cold-drawn in the rotating water amorphous wires with following compositions: $Fe_{75}Si_{10}B_{15}$ (**a**) and $(Co_{94}Fe_6)_{72\cdot5}Si_{12\cdot5}B_{15}$ (**b**). Glass-coated amorphous microwires obtained by Taylor–Ulitovsky technique: $(Co_{94}Fe_6)_{72\cdot5}Si_{12\cdot5}B_{15}$ (**c**) and $(Co_{50}Fe_{50})_{72\cdot5}Si_{12\cdot5}B_{15}$ (**d**). Composite CuBe/CoFeNi electroplated wires: cross-section showing both the central conductive base wire and magnetic coating (**e**), general features of the surface properties of CoFeNi layer showing different types of the surface defects appearing daring the electroplating process (**f**).

The use of a glass-coated amorphous microwire as an MI-sensitive element requires sufficient understanding of its magnetic anisotropy and magnetic structure, since the MI response depends on the microwire geometry, composition and post-preparation annealing/cooling or other type of additional treatment. In the first approximation, the impedance of a magnetic microwire can be described as follows [47]:

$$Z/R_{\mathrm{dc}} = (ka/2)J_0(ka)/J_1(ka). \quad (1)$$

Here, $R_{\mathrm{dc}} = l/\pi\sigma a^2$ is the microwire resistance in the direct current mode; l and a are the length and radius of the amorphous part of the microwire; σ is the conductivity; J_0 and J_1 are the Bessel functions of the first kind of zero and first orders; $k = (1 + i)/\delta$; i is the imaginary unit; $\delta = c/(2\pi\sigma\omega\mu)^{1/2}$; c is the speed of light in vacuum; ω is the angular frequency and μ is the circumferential permeability.

Equation (1) was obtained assuming that the circumferential permeability does not change over the microwire cross-section. Moreover, the tensor form of the permeability is neglected in Equation (1). The models for MI responses taking into account the tensor form of the permeability were developed for the geometry of a cylindrical conductor with circular or helical anisotropy [48–50]. It was shown that these models are applicable in cases of asymmetric MI and off-diagonal MI effect.

The influence of the axially magnetized core on the MI in the rapidly quenched amorphous wires and glass-coated microwires was investigated in the framework of different models [51–55]. It was demonstrated that in some cases, the core–shell models allow one to explain specific features of the frequency dependences of the MI response.

It should be noted that there are no methods for the direct study of the magnetic structure in the cross-section of glass-coated amorphous microwires, so it is usually interpreted on the basis of magnetostatic measurements. Recently, a new method to evaluate magnetic structure in amorphous materials of different types exhibiting a MI effect was developed [56–62]. The method is based on measured data on the frequency dependence of the MI, which allows one to analyze the distribution of the permeability over the conductor cross-section. This method is referred to as MI tomography.

In Ref. [56], the radial distribution of the permeability in glass-coated amorphous microwires was restored by using Equation (1). The permeability was reconstructed by averaging over a surface layer of the microwire to which part the major intensity of the excitation current corresponded. However, this approach has some limitations when the permeability has strongly non-uniform distribution over the microwire cross-section [56].

Another approach is based on a comparison of the experimental frequency dependences of the MI with numerical simulation [57]. The conductor cross-section is divided into several regions with constant electromagnetic properties. The Maxwell equations were solved numerically by the finite element method by using Comsol Multiphysics commercial software (https://www.comsol.com/products). The method allowed description of the experimental data on the MI frequency dependences in the composite CuBe/FeCoNi wires [57–59], amorphous rapidly quenched wires [57,60,62] and amorphous ribbons [61].

Note that there are some assumptions in the method developed in Ref. [57]. First, the field dependence of the permeability in each region was not specified explicitly. Second, it was assumed that the permeability is independent of the frequency. In addition, the imaginary part of the permeability was neglected. Although, in many cases, these approximations are well applicable, it is of interest to develop a model that does not have these restrictions.

In this work, we propose a theoretical model for the description of the MI effect in amorphous microwires with inhomogeneous magnetic structure. The microwire cross-section is divided into several regions, where the magnetic properties are assumed to be constant. The approach is based on the calculation of electromagnetic fields in the regions of the microwire by using an analytical solution of Maxwell equations. In the framework of the model, the frequency dependence and complex nature of the permeability are taken into account, and the field dependence of the permeability is expressed in an explicit form.

2. Model

Let us consider amorphous microwire exciting by the alternating current $I(t) = I_0\exp(-i\omega t)$. The external magnetic field H_e is directed along the microwire axis. Following the approach proposed previously [57], we divide the amorphous nucleus of the microwire into n coaxial regions. It is assumed that each region has uniaxial magnetic anisotropy, and the anisotropy axis makes an angle ψ_j with the azimuthal direction (hereinafter, the index $j = 1, \ldots n$ denotes the region number in the model).

To find the magnetization distribution, we neglect the exchange and magnetostatic coupling between the selected regions. This approximation significantly simplifies the calculations. However, it should be noted that the coupling between the regions can be taken into account by introducing effective interaction fields similar to the core–shell models proposed previously [53,54].

The static equilibrium magnetization distribution within the regions can be found by the minimization of the free energy. The free energy density U for each region can be presented as a sum of the magnetic anisotropy term and Zeeman energy [13]:

$$U = (MH_j/2)\sin^2(\theta_j - \psi_j) - MH_e \sin\theta_j. \tag{2}$$

Here, M is the saturation magnetization, H_j is the anisotropy field in the region j and θ_j is the equilibrium magnetization angle with respect to the azimuthal direction.

The minimization of the free energy density results in the following equations for the equilibrium magnetization angles θ_j:

$$H_j \sin(\theta_j - \psi_j)\cos(\theta_j - \psi_j) = H_e \cos\theta_j. \tag{3}$$

In the following analysis, we neglect the contribution of the domain-wall motion to the permeability and assume that the values of the permeability in the regions are governed by the magnetization rotation. This approximation is valid at not-too-low frequencies, when the domain-wall motion is damped by eddy currents. In soft magnetic amorphous materials exhibiting the MI effect, the domain-wall motion is negligible in the frequency range from several hundred kHz to a few MHz [63,64]. It is assumed also that the permeability tensor has a quasi-diagonal form, and the MI response depends only on the values of the circumferential permeability in the regions. For simplicity, we neglect the contribution of the exchange-conductivity effect to the permeability in the regions. This contribution is relatively low at high frequencies, and it could be taken into account by means of the methods developed previously [49,65]. Under the above assumptions, the circumferential permeability μ_j in the region j is expressed as [13,50,63]:

$$\mu_j = 1 + \frac{\gamma 4\pi M(\gamma 4\pi M + \omega_j^* - i\kappa\omega)\sin^2\theta_j}{(\gamma 4\pi M + \omega_j^* - i\kappa\omega)(\omega_j^{**} - i\kappa\omega) - \omega^2}. \tag{4}$$

Here, γ is the gyromagnetic constant, κ is the Gilbert damping parameter and the characteristic frequencies are given by [50]

$$\begin{aligned} \omega_j^* &= \gamma[H_j\cos^2(\theta_j - \psi_j) + H_e\sin\theta_j], \\ \omega_j^{**} &= \gamma[H_j\cos\{2(\theta_j - \psi_j)\} + H_e\sin\theta_j]. \end{aligned} \tag{5}$$

Taking into account the cylindrical symmetry, the distribution of electromagnetic fields within the microwire is described by Maxwell equations [47]:

$$\begin{aligned} -\frac{\partial e_j}{\partial \rho} &= \frac{i\omega\mu_j}{c}h_j, \\ \frac{1}{\rho}\frac{\partial}{\partial \rho}(\rho h_j) &= \frac{4\pi\sigma}{c}e_j. \end{aligned} \tag{6}$$

Here, ρ is the radial coordinate; e_j and h_j are the amplitudes of the longitudinal electric field and the circular magnetic field, respectively.

The solution of Equation (6) for each region is expressed as follows:

$$\begin{aligned} e_j &= (ck_j/4\pi\sigma)[A_j J_0(k_j\rho) + B_j Y_0(k_j\rho)], \\ h_j &= A_j J_1(k_j\rho) + B_j Y_1(k_j\rho). \end{aligned} \tag{7}$$

Here, $k_j = (1 + i)/\delta_j$; $\delta_j = c/(2\pi\sigma\omega\mu_j)^{1/2}$; Y_0 and Y_1 are the Bessel functions of the second kind of zero and first orders; A_j and B_j are the constants. Note that due to the symmetry the constant B_1 is equal to zero in the central region, $j = 1$.

The constants A_j and B_j can be found from the continuity conditions for the amplitudes of the electric and magnetic fields at the interfaces between different regions, $\rho = r_j$ ($j < n - 1$),

$$\begin{aligned} e_j(r_j) &= e_{j+1}(r_j), \\ h_j(r_j) &= h_{j+1}(r_j). \end{aligned} \tag{8}$$

Using Equation (7), these conditions can be rewritten as

$$\begin{aligned} A_j J_0(k_j r_j) + B_j Y_0(k_j r_j) &= (\mu_{j+1}/\mu_j)^{1/2}[A_{j+1} J_0(k_{j+1} r_j) + B_{j+1} Y_0(k_{j+1} r_j)], \\ A_j J_1(k_j r_j) + B_j Y_1(k_j r_j) &= A_{j+1} J_1(k_{j+1} r_j) + B_{j+1} Y_1(k_{j+1} r_j). \end{aligned} \tag{9}$$

The boundary condition at the surface of amorphous part of the microwire is obtained from the current excitation condition. For the surface region, $j = n$, we have

$$A_n J_1(k_n a) + B_n Y_1(k_n a) = 2\, I_0/ca. \tag{10}$$

The field distribution within the microwire is completely determined by Equations (7), (9) and (10). The microwire impedance Z is expressed as [48–50]

$$Z = \frac{l e_n(a)}{I_0} = \frac{2l}{ca} \times \frac{e_n(a)}{h_n(a)}, \tag{11}$$

where $e_n(a)$ and $h_n(a)$ are the amplitudes of the electrical and magnetic field at the surface of the amorphous part of the microwire.

Equation (11) can be rewritten as follows:

$$Z/R_{\mathrm{dc}} = (k_n a/2) \times \frac{A_n J_0(k_n a) + B_n Y_0(k_n a)}{A_n J_1(k_n a) + B_n Y_1(k_n a)}. \tag{12}$$

Note that for a uniform distribution of the anisotropy over the microwire cross-section $n = 1$, we have $B_n = 0$ from the symmetry conditions. Then, Equation (9) transfers to a standard expression for the microwire impedance (see Equation (1)).

Thus, the procedure for the calculation of the MI response in the microwire with inhomogeneous magnetic structure can be summarized as follows. The cross-section of the metallic part of the microwire is divided into n regions. In principle, there are no restrictions on the selection of the number of regions. For each region, the anisotropy field H_j and the anisotropy axis ψ_j can be specified. The equilibrium magnetization angles θ_j and the circumferential permeability μ_j in the region j are found from Equations (3) and (4), respectively. After that, the field distribution within the microwire is calculated using Equations (7), (9) and (10), and the impedance Z is obtained by means of Equation (12).

3. Results and Discussion

The proposed model allows one to analyze the MI effect in amorphous microwires with inhomogeneous magnetic structure. A comparison of modeling results with measured data on the MI in microwires is out of the scope of this paper, and below we present some results of the MI calculations to illustrate the applicability of the model. For calculations, we use the following parameters of amorphous microwire: the diameter of metallic nucleus $2a = 20$ μm, the saturation magnetization $M = 600$ G, the conductivity $\sigma = 10^{16}$ s^{-1} and the Gilbert damping parameter $\kappa = 0.1$.

We divide the cross-section of the amorphous part of the microwire into five regions, $n = 5$. It is assumed that in the central region, the easy magnetization axis is along the microwire axis. In the surface region, the anisotropy is close to the circular one. In the intermediate regions, the anisotropy axes change gradually from the axial to transverse directions. The radii of the regions r_j, the anisotropy fields H_j and the anisotropy axis angles ψ_j are presented in Table 1.

Table 1. Parameters of the five-region model, $n = 5$.

j	Region Radius r_j, μm	Anisotropy Field H_j, Oe	Anisotropy Axis Angle ψ_j, rad
1	2	5	0.5π
2	4	8	0.4π
3	5	10	0.2π
4	7	15	0.1π
5	10	20	0.05π

The dependences of the values of the static permeability in the regions on the external field are shown in Figure 2. The values of μ_j are calculated by means of Equations (3)–(5) at $\omega = 0$ using the parameters presented in Table 1.

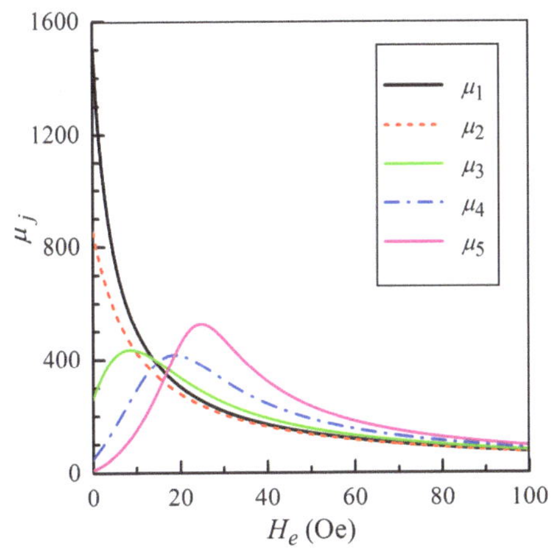

Figure 2. The values of the static permeability in the five-region model, $n = 5$, versus the external field H_e.

The permeability in two central regions, $j = 1$ and $j = 2$, has a maximum at zero external field and decreases with a growth of H_e. In other regions, the permeability has non-monotonic behavior and achieves a peak at some non-zero external field. The difference in the field dependences of the permeability values significantly affect the MI response of the microwire. Note that at high external fields, the permeability values for different regions approach each other (see Figure 2). This is due to the fact that the magnetization directions in each region are close to the direction of the external field. The magnetization distribution over the microwire cross-section becomes almost uniform, and the permeability values depend weakly on the anisotropy field.

Let us return to Equation (4) and results shown in Figure 2. The developed model for calculation of the frequency dependence and complex nature of the magnetic permeability were taken into account. However, Figure 2 presents values of the static permeability to illustrate the changes in the permeability in the regions with external field just for the simplicity of visual representation.

Further, we compare the results of the MI modeling by using the five-region approach, $n = 5$, with two other models for distribution of the magnetic anisotropy within the microwire. In the first model, it is assumed that the anisotropy distribution is uniform over the microwire cross-section, $n = 1$. Then, the MI response can be found by using Equation (1). It is assumed that in this case, the anisotropy field and the anisotropy axis angle coincide with those for the surface region in the five-region model, that is, $H_1 = 20$ Oe and $\psi_1 = 0.05\pi$.

In the second model, we assume that the magnetic structure of the microwire is described in the framework of the core–shell model, $n = 2$. Note that for $n = 2$, the microwire impedance Z is expressed in the following form [51]:

$$\begin{aligned} Z/R_{\rm dc} &= (k_2 a/2) \times \frac{J_0(k_2 a)+PY_0(k_2 a)}{J_1(k_2 a)+PY_1(k_2 a)}, \\ P &= \frac{(\mu_2/\mu_1)^{1/2} J_1(k_1 r_1) J_0(k_2 r_1) - J_0(k_1 r_1) J_1(k_2 r_1)}{J_0(k_1 r_1) Y_1(k_2 r_1) - (\mu_2/\mu_1)^{1/2} J_1(k_1 r_1) Y_0(k_2 r_1)}. \end{aligned} \quad (13)$$

Here, r_1 is the radius of the inner core, and the subscripts 1 and 2 correspond to the core and shell regions. Further, we take the following values of parameters for calculations in the core–shell model: $2r_1 = 10$ μm, $H_1 = 5$ Oe, $H_2 = 20$ Oe, $\psi_1 = 0.5\pi$ and $\psi_2 = 0.05\pi$.

The studied magnetic structure with axial anisotropy in the inner region of the microwire and close to the circular anisotropy in the surface region is typical for Co-rich glass-coated amorphous microwires with nearly-zero or negative magnetostriction [13,32]. These microwires exhibit the highest values of the MI effect. A similar structure exists in stress-annealed Fe-rich glass-coated amorphous microwires with positive magnetostriction, since the circular anisotropy appears in the surface region after the annealing [36,38,55,66,67]. Note also that the proposed model is applicable for microwires with a regular domain structure in the surface layer consisting of ring domains (so-called bamboo domain structure), since Equation (4) is valid for description of the effective circumferential permeability after averaging over the surface domain structure.

Figure 3 shows the field dependences of the MI ratio $\Delta Z/Z$ calculated at several frequencies $f = \omega/2\pi$ for different models. Note that the dependences of $\Delta Z/Z$ on H_e is presented only for positive values of the external magnetic field, since in the framework of the model, the field dependences of the MI ratio are symmetric with respect to the sign of the external field. The MI ratio is defined as follows:

$$\Delta Z/Z = [Z(H_e) - Z(H_0)]/Z(H_0), \quad (14)$$

where $H_0 = 150$ Oe is the external field sufficient for magnetic saturation of the microwire.

It follows from Figure 3 that at relatively low frequencies of the excitation current the discrepancy between the calculated results for the field dependence of $\Delta Z/Z$ ratio in the models with different number of regions n is quite high. The model with uniform magnetic anisotropy distribution, $n = 1$, predicts that the field dependence of the MI ratio exhibits two-peak behavior. For the core–shell model, $n = 2$, the MI ratio has an additional maximum at zero field (see Figure 3a,b). This fact is related to the contribution of the core region with high permeability at low fields. The five-region model, $n = 5$, provides the results intermediate between $n = 1$ and $n = 2$ at low fields.

With an increase in the frequency, all the models give almost identical results, with the exception of the field region close to zero magnetic field (Figure 3c,d). This is due to the fact that at high frequencies, the thickness of the skin layer decreases, and the main contribution to the MI comes from the surface region, which has the same magnetic properties for all the models under consideration.

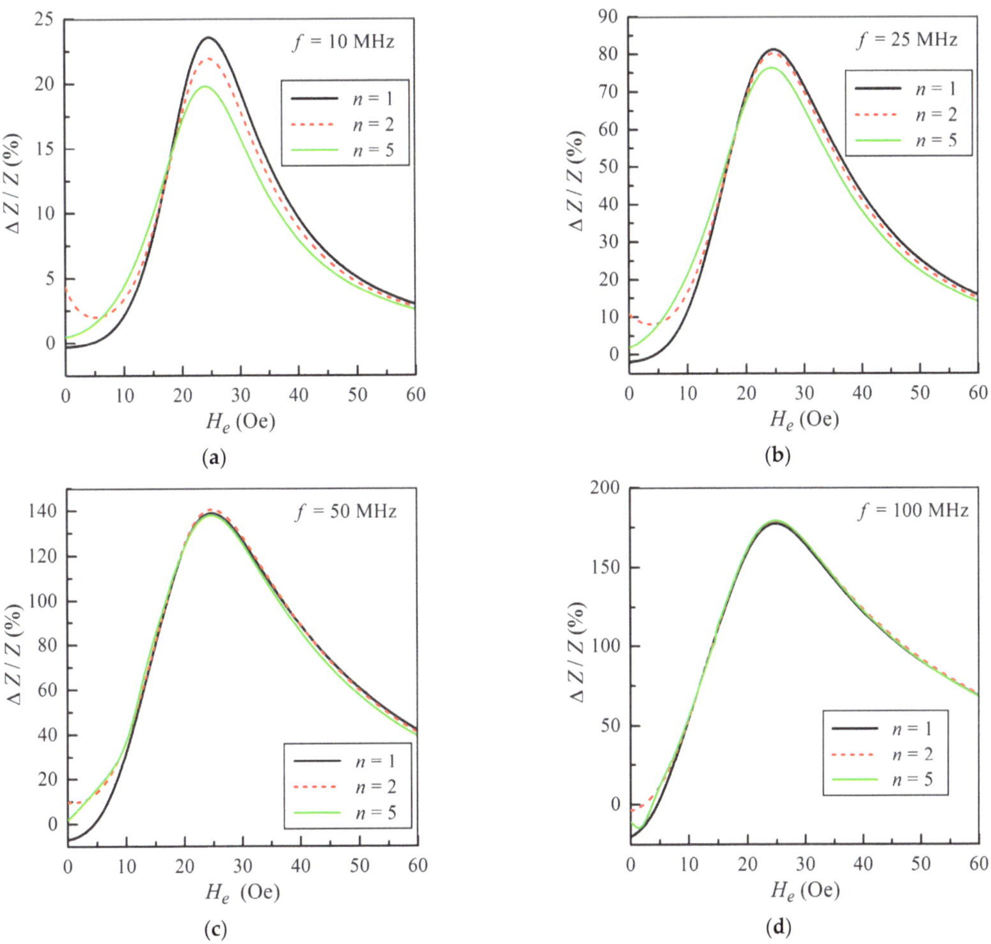

Figure 3. The field dependences of the MI ratio $\Delta Z/Z$ calculated for different numbers of regions n at several frequencies: f = 10 MHz (**a**); f = 25 MHz (**b**); f = 50 MHz (**c**); f = 100 MHz (**d**).

The MI tomography method is based on a comparison of calculations of the frequency dependence of the MI with measurement results. The calculation results are very sensitive to the distribution of the permeability over the microwire cross-section. Figure 4 presents the frequency dependences of the impedance Z calculated at several external fields for different models. At zero external field, the difference between the three models is of significance at all frequencies (Figure 4a). The discrepancy between the calculated impedance values for different numbers of regions decreases with an increase in the external field. At $H_e > 10$ Oe, all the studied models provide very similar impedance values within the whole frequency range, since the permeability values are approximately the same for all regions at high external fields. The obtained results are in qualitative agreement with numerical calculations using the finite element method [57]. Thus, to study the anisotropy distribution within the microwire by the MI tomography method, the frequency dependence of the impedance at low external fields should be analyzed.

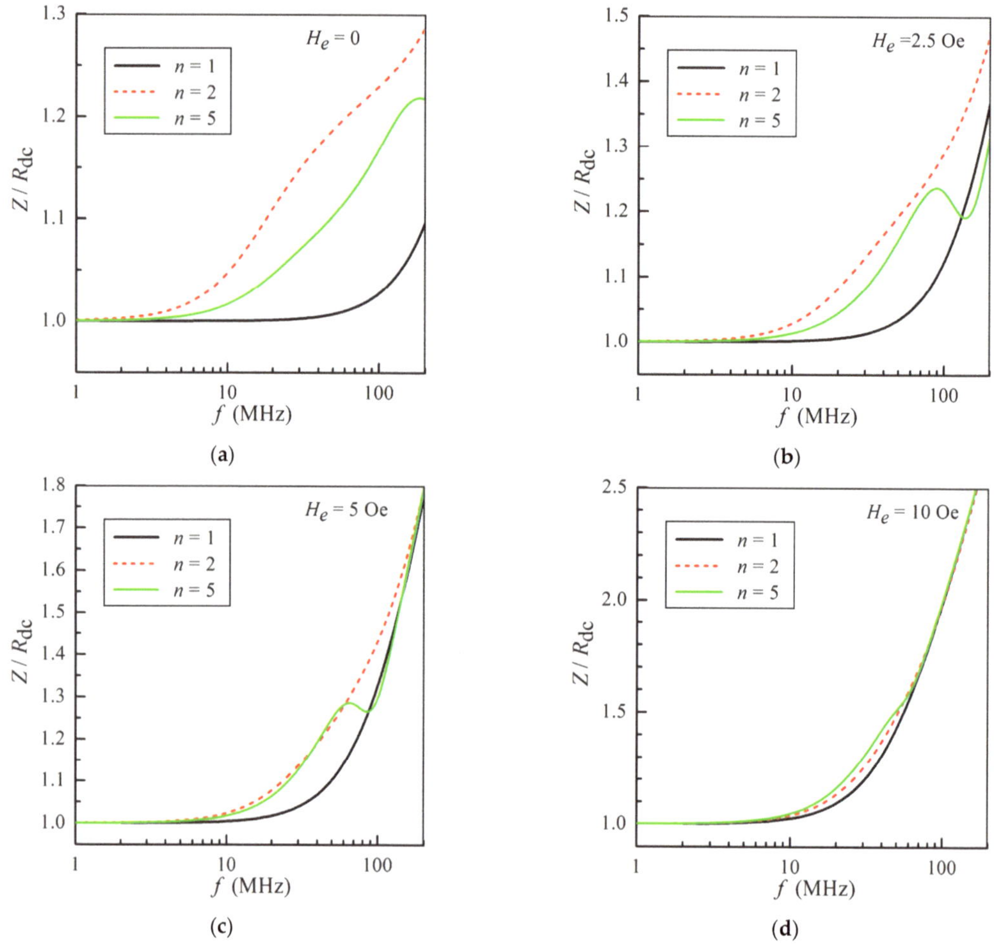

Figure 4. The frequency dependence of the impedance Z calculated for different numbers of regions n at several external magnetic fields: $H_e = 0$ (**a**); $H_e = 2.5$ Oe (**b**); $H_e = 5$ Oe (**c**); $H_e = 10$ Oe (**d**).

In conclusion of this section, let us discuss the frequency dependence the MI and the effect of the imaginary part of magnetic permeability on the MI response. In the framework of the model proposed, the absence of the frequency dependence of the circumferential permeability can be obtained by setting $\omega = 0$ in Equation (4). The imaginary part of the permeability is neglected if we assume in Equation (4) that $\omega \neq 0$ and $\kappa = 0$.

An analysis shows that at all frequencies, the highest difference between the proposed model and simplified approaches is observed near the maximum of the field dependence of the impedance. In this regard, we analyze further the frequency dependence of the maximum MI ratio $(\Delta Z/Z)_{\max}$ calculated for the cases when the frequency dependence of the permeability is neglected, $\omega = 0$, and when the imaginary part of the permeability equals zero, $\kappa = 0$. The calculated frequency dependences of $(\Delta Z/Z)_{\max}$ are shown in Figure 5 for the five-region model, $n = 5$.

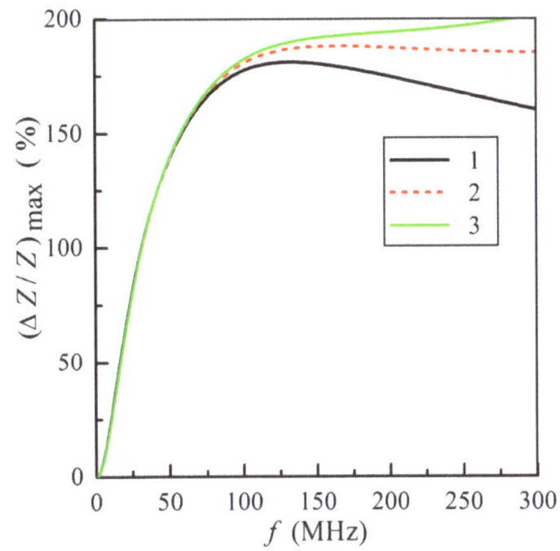

Figure 5. Frequency dependence of maximum MI ratio $(\Delta Z/Z)_{max}$ for the number of regions $n = 5$: curve 1, the proposed model; curve 2, model with the permeability independent of the frequency; curve 3, model with the imaginary part of the permeability equal to zero.

Figure 5 shows that for the studied microwire at frequencies below 70 MHz, the difference in the calculated values of $(\Delta Z/Z)_{max}$ is negligible. The significant difference between different models appears at the frequencies higher than 100 MHz, and it increases with the frequency. Note that the frequency dependence of maximum MI ratio exhibits different behavior in the proposed model and simplified approaches.

In the model proposed, the maximum MI ratio has a peak and decreases at high frequencies (see curve 1 in Figure 5). If the frequency dependence of the permeability is not taken into account, then the maximum MI ratio is almost constant within a wide frequency range (see curve 2 in Figure 5). When we take into account the frequency dependence of the permeability and neglect the imaginary part of the permeability, then $(\Delta Z/Z)_{max}$ increases with the frequency (see curve 3 in Figure 5). Thus, the analysis demonstrates that the simplified approach developed previously [57–62] is valid within a wide frequency range. However, at high frequencies, the frequency dependence and the complex nature of the circumferential permeability should be taken into account in order to describe the MI response of amorphous microwires correctly.

Despite the main focus of the present study being on glass-coated magnetic microwires, the obtained results are not limited by this particular kind of MI materials, and the developed model can be useful for MI tomography as an additional instrument complementing existing methodology. For example, a recently proposed type of biphase glass-coated microwires, i.e., composites based on magnetic microwire having an additional magnetic coating deposited by the sputtering technique would be of special interest to researchers and engineers. Such a phenomenon as low field microwave absorption or low field MI [16,37,68–70] to some extent could be rethought and may be even better understood.

In the present work we made a focus on the study of glass-coated microwires. Future research directions can be connected to the development of this approach for the other cylindrical ferromagnetic conductors and even more complex composites.

4. Conclusions

The MI tomography method consists in a comparison of the calculated and measured frequency dependences of the MI at different external fields. The method can be used to restore the distribution of the magnetic permeability and magnetic anisotropy over the conductor cross-section.

In this work, we designed, proposed and tested a theoretical model for the description of the MI effect in amorphous microwires with inhomogeneous magnetic structure over their cross-section. This approach allows one to analyze the distribution of magnetic properties over the cross-section of the conductor, including the cases of composite conducting materials. The microwire cross-section is divided in several regions, in which the magnetic properties are assumed to be constant. The approach is based on the calculation of electromagnetic fields in the regions of the microwire by using an analytical solution of the Maxwell equations. The frequency dependence and complex nature of the permeability are taken into account, and the field dependence of the permeability is expressed in an explicit form in the framework of the model. The results obtained can be useful for the application and further development of the MI tomography method.

Author Contributions: Conceptualization, N.A.B. and G.V.K.; methodology, N.A.B. and G.V.K.; validation, N.A.B. and G.V.K.; formal analysis, N.A.B. and G.V.K.; investigation, N.A.B. and G.V.K.; writing—original draft preparation, N.A.B. and G.V.K.; writing—review and editing, N.A.B. and G.V.K.; visualization, N.A.B. and G.V.K. All authors have read and agreed to the published version of the manuscript.

Funding: This research received no external funding.

Institutional Review Board Statement: Not applicable.

Informed Consent Statement: Not applicable.

Data Availability Statement: Data available from the corresponding author upon reasonable request.

Acknowledgments: We thank Hector Garcia-Miquel for special collaboration and providing selected data obtained by scanning electron microscopy.

Conflicts of Interest: The authors declare no conflicts of interest.

References

1. Makhotkin, V.E.; Shurukhin, B.P.; Lopatin, V.A.; Marchukov, P.Y.; Levin, Y.K. Magnetic field sensors based on amorphous ribbons. *Sens. Actuators A* **1991**, *21*, 759–762. [CrossRef]
2. Beach, R.S.; Berkowitz, A.E. Giant magnetic field dependent impedance of amorphous FeCoSiB wire. *Appl. Phys. Lett.* **1994**, *64*, 3652–3654. [CrossRef]
3. Panina, L.V.; Mohri, K. Magneto-impedance effect in amorphous wires. *Appl. Phys. Lett.* **1994**, *65*, 1189–1191. [CrossRef]
4. Mohri, K.; Uchiyama, T.; Panina, L.V.; Yamamoto, M.; Bushida, K. Recent advances of amorphous wire CMOS IC magnetoimpedance sensors: Innovative high-performance micromagnetic sensor chip. *J. Sens.* **2015**, *2015*, 718069. [CrossRef]
5. Nakai, T. Sensitivity of thin film magnetoimpedance sensor in 0.3 T surface normal magnetic field. *IEEJ Trans. Electr. Electron. Eng.* **2020**, *15*, 1230–1235. [CrossRef]
6. Riveros, P.A.D.; Silva, E.C.; Pacheco, S.; Cabrera, L.S.B.; Barbosa, C.R.H. Design, implementation and experimental characterisation of a high sensitivity GMI gradiometer with an interference compensation system. *IET Sci. Meas. Technol.* **2020**, *14*, 688–694. [CrossRef]
7. Traoré, P.S.; Asfour, A.; Yonnet, J.-P. Noise analysis of a high sensitivity GMI sensor based on a Field-Programmable-Gate-Array. *Sens. Actuators A* **2021**, *331*, 112972. [CrossRef]
8. Yao, R.; Takemura, Y.; Uchiyama, T. High precision MI sensor with low energy consumption driven by low-frequency Wiegand pulse. *AIP Adv.* **2023**, *13*, 025201. [CrossRef]
9. Blanc-Béguin, F.; Nabily, S.; Gieraltowski, J.; Turzo, A.; Querellou, S.; Salaun, P.Y. Cytotoxicity and GMI bio-sensor detection of maghemite nanoparticles internalized into cells. *J. Magn. Magn. Mater.* **2009**, *321*, 192–197. [CrossRef]
10. Buznikov, N.A.; Safronov, A.P.; Orue, I.; Golubeva, E.V.; Lepalovskij, V.N.; Svalov, A.V.; Chlenova, A.A.; Kurlyandskaya, G.V. Modelling of magnetoimpedance response of thin film sensitive element in the presence of ferrogel: Next step toward development of biosensor for in tissue embedded magnetic nanoparticles detection. *Biosens. Bioelectr.* **2018**, *117*, 366–372. [CrossRef]
11. Uchiyama, T.; Ma, J. Development of pico tesla resolution amorphous wire magneto-impedance sensor for bio-magnetic field measurements. *J. Magn. Magn. Mater.* **2020**, *514*, 167074. [CrossRef]
12. Pei, C.; Zhang, B.; Xie, J.; Kou, Z.; Li, X.; Feng, T.; Sun, B.; Wang, W. Superlattice-shelled nanocrystalline core structural design for highly sensitive GMI sensors. *Acta Mater.* **2023**, *255*, 119088. [CrossRef]
13. Knobel, M.; Vázquez, M.; Kraus, L. Giant magnetoimpedance. In *Handbook of Magnetic Materials*; Buschow, K.H.J., Ed.; Elsevier: Amsterdam, The Netherlands, 2003; Volume 15, pp. 497–563.
14. Zhukov, A.; Ipatov, M.; Zhukova, V. Advances in giant magnetoimpedance of materials. In *Handbook of Magnetic Materials*; Buschow, K.H.J., Ed.; Elsevier: Amsterdam, The Netherlands, 2015; Volume 24, pp. 139–236.

15. Antonov, A.S.; Rakhmanov, A.L.; Buznikov, N.A.; Prokoshin, A.F.; Granovsky, A.B.; Perov, N.S.; Usov, N.A. Magnetic properties and magneto-impedance of cold-drawn permalloy-copper composite wires. *IEEE Trans. Magn.* **1999**, *35*, 3640–3642. [CrossRef]
16. Jantaratana, P.; Bebenin, N.G.; Kurlyandskaya, G.V. Magnetoimpedance and magnetization processes of FeCoNi electroplated tubes. *J. Appl. Phys.* **2009**, *105*, 013908. [CrossRef]
17. García-Miquel, H.; Carbonell, J.; Boria, V.E.; Sánchez-Dehesa, J. Experimental evidence of left handed transmission through arrays of ferromagnetic microwires. *Appl. Phys. Lett.* **2009**, *94*, 054103. [CrossRef]
18. Smith, K.C.A.; Oatley, C.W. The scanning electron microscope and its fields of application. *Br. J. Appl. Phys.* **1955**, *6*, 391–399. [CrossRef]
19. Ogasawara, I.; Ueno, S. Preparation and properties of amorphous wires. *IEEE Trans. Magn.* **1995**, *31*, 1219–1223. [CrossRef]
20. Marin, P.; Marcos, M.; Hernando, A. High magnetomechanical coupling on magnetic microwire for sensors with biological applications. *Appl. Phys. Lett.* **2010**, *96*, 262512. [CrossRef]
21. Shcherbinin, S.V.; Pérez, R.; Vazquez, M.; Kurlyandskaya, G.V. Ferromagnetic resonance in electroplated CuBe/FeCoNi and amorphous CoFeSiB wires Ferromagnetic resonance in electroplated CuBe/FeCoNi and amorphous CoFeSiB wires. *IEEE Trans. Magn.* **2020**, *56*, 2800110. [CrossRef]
22. Baselt, D.R.; Lee, G.U.; Natesan, M.; Metzger, S.W.; Sheehan, P.E.; Colton, R. A biosensor based on magnetoresistance technology. *Biosens. Bioelectron.* **1998**, *13*, 731–739. [CrossRef]
23. Kurlyandskaya, G.V.; Fal Miyar, V. Surface modified amorphous ribbon based magnetoimpedance biosensor. *Biosens. Bioelectron.* **2007**, *22*, 2341–2345. [CrossRef]
24. Cerdeira, M.A.; Kurlyandskaya, G.V.; Fernandez, A.; Tejedor, M.; Garcia-Miquel, H. Giant magnetoimpedance effect in surface modified CoFeMoSiB amorphous ribbons. *Chin. Phys. Lett.* **2003**, *20*, 2246–2249. [CrossRef]
25. Fal-Miyar, V.; Kumar, A.; Mohapatra, S.; Shirley, S.; Frey, N.A.; Barandiaránd, J.M.; Kurlyandskaya, G.V. Giant magnetoimpedance for biosensing in drug delivery. *Appl. Phys. Lett.* **2007**, *91*, 143902.
26. Volchkov, S.O.; Pasynkova, A.A.; Derevyanko, M.S.; Kozlov, N.V.; Svalov, A.V.; Semirov, A.V. Magnetoimpedance of CoFeCrSiB ribbon-based sensitive element with FeNi covering: Experiment and modeling. *Sensors* **2021**, *21*, 6728. [CrossRef] [PubMed]
27. Yang, Z.; Lei, C.; Sun, X.C.; Zhou, Y.; Liu, Y. Enhanced GMI effect in tortuous-shaped Co-based amorphous ribbons coated with graphene. *J. Mater. Sci. Mater. Electron.* **2016**, *27*, 3493–3498. [CrossRef]
28. Semirov, A.V.; Derevyanko, M.S.; Bukreev, D.A.; Moiseev, A.A.; Kudryavtsev, V.O.; Safronov, A.P. Magnetoimpedance of cobalt-based amorphous ribbons/polymer composites. *J. Magn. Magn. Mater.* **2016**, *415*, 97–101. [CrossRef]
29. Kurlyandskaya, G.V.; Blyakhman, F.A.; Makarova, E.B.; Buznikov, N.A.; Safronov, A.P.; Fadeyev, F.A.; Shcherbinin, S.V.; Chlenova, A.A. Functional magnetic ferrogels: From biosensors to regenerative medicine. *AIP Adv.* **2020**, *10*, 125128. [CrossRef]
30. Dolabdjian, C.; Ménard, D. Giant magneto-impedance (GMI) magnetometers. In *High Sensitivity Magnetometers*; Grosz, A., Haji-Sheikh, M.J., Mukhopadhyay, S.C., Eds.; Springer: Berlin/Heidelberg, Germany, 2017; pp. 103–126.
31. Uchiyama, T.; Mohri, K.; Honkura, Y.; Panina, L.V. Recent advances of pico-Tesla resolution magneto-impedance sensor based on amorphous wire CMOS IC MI Sensor. *IEEE Trans. Magn.* **2012**, *48*, 3833–3839. [CrossRef]
32. Chiriac, H.; Óvári, T.A. Amorphous glass-covered magnetic wires: Preparation, properties, applications. *Prog. Mater. Sci.* **1996**, *40*, 333–407. [CrossRef]
33. Vázquez, M. Advanced magnetic microwires. In *Handbook of Magnetism and Advanced Magnetic Materials*; Kronmüller, H., Parkin, S.S.P., Eds.; Wiley: Chichester, UK, 2007; pp. 2193–2226.
34. Zhukov, A.; González, J.; Vázquez, M.; Larin, V.; Torcunov, A. Nanocrystalline and amorphous magnetic microwires. In *Encyclopedia of Nanoscience and Nanotechnology*; Nalwa, H.S., Ed.; American Scientific: Stevenson Ranch, CA, USA, 2014; Volume 6, pp. 365–367.
35. Baranov, S.A.; Larin, V.S.; Torcunov, A.V. Technology, preparation and properties of the cast glass-coated magnetic microwires. *Crystals* **2017**, *7*, 136. [CrossRef]
36. Zhukova, V.; Ipatov, M.; Talaat, A.; Blanco, J.M.; Churyukanova, M.; Zhukov, A. Effect of stress annealing on magnetic properties and GMI effect of Co- and Fe-rich microwires. *J. Alloys Compd.* **2017**, *707*, 189–194. [CrossRef]
37. El Kammouni, R.; Vázquez, M.; Lezama, L.; Kurlyandskaya, G.; Kraus, L. Temperature dependence of microwave absorption phenomena in single and biphase soft magnetic microwires. *J. Magn. Magn. Mater.* **2014**, *368*, 126–132. [CrossRef]
38. Gonzalez, A.; Zhukova, V.; Corte-Leon, P.; Chizhik, A.; Ipatov, M.; Blanco, J.M.; Zhukov, A. Tuning of magnetoimpedance effect and magnetic properties of Fe-rich glass-coated microwires by Joule heating. *Sensors* **2022**, *22*, 1053. [CrossRef] [PubMed]
39. Zhukova, V.; Corte-Leon, P.; Talaat, A.; Ipatov, M.; García-Gomez, A.; González, A.; Blanco, J.M.; Zhukov, A. Optimization of giant magnetoimpedance effect of amorphous microwires by postprocessing. *Processes* **2024**, *12*, 556. [CrossRef]
40. Chiriac, H.; Óvári, T.A.; Pop, G. Internal stress distribution in glass-covered amorphous magnetic wires. *Phys. Rev. B* **1995**, *52*, 10104–10113. [CrossRef] [PubMed]
41. Antonov, A.S.; Borisov, V.T.; Borisov, O.V.; Prokoshin, A.F.; Usov, N.A. Residual quenching stresses in glass-coated amorphous ferromagnetic microwires. *J. Phys. D Appl. Phys.* **2000**, *33*, 1161–1168. [CrossRef]
42. Larin, V.S.; Torcunov, A.V.; Zhukov, A.; González, J.; Vazquez, M.; Panina, L. Preparation and properties of glass-coated microwires. *J. Magn. Magn. Mater.* **2002**, *249*, 39–45. [CrossRef]
43. Baranov, S.A. Magnetic models of cast microwires. *Surf. Eng. Appl. Electrochem.* **2011**, *47*, 316–330. [CrossRef]

44. Chizhik, A.; Garcia, C.; Zhukov, A.; Gawronski, P.; Kulakowski, K.; Gonzalez, J.; Blanco, J.M. Investigation of helical magneti structure in Co-rich amorphous microwires. *J. Magn. Magn. Mater.* **2007**, *316*, 332–336. [CrossRef]
45. Chizhik, A.; Blanco, J.M.; Zhukov, A.; Gonzalez, J.; Garcia, C.; Gawronski, P.; Kulakowski, K. Magneto-optical determination o helical magnetic structure in amorphous microwires. *Physica B* **2008**, *403*, 289–292. [CrossRef]
46. Chiriac, H.; Óvári, T.-A.; Corodeanu, S.; Ababei, G. Interdomain wall in amorphous glass-coated microwires. *Phys. Rev. B* **2007**, *76*, 214433. [CrossRef]
47. Landau, L.D.; Lifshitz, E.M. *Electrodynamics of Continuous Media*; Pergamon: London, UK, 1975.
48. Usov, N.A.; Antonov, A.S.; Lagar'kov, A.N. Theory of giant magneto-impedance effect in amorphous wires with different types o magnetic anisotropy. *J. Magn. Magn. Mater.* **1998**, *185*, 159–173. [CrossRef]
49. Ménard, D.; Yelon, A. Theory of longitudinal magnetoimpedance in wires. *J. Appl. Phys.* **2000**, *88*, 379–393. [CrossRef]
50. Makhnovskiy, D.P.; Panina, L.V.; Mapps, D.J. Field-dependent surface impedance tensor in amorphous wires with two types o magnetic anisotropy: Helical and circumferential. *Phys. Rev. B* **2001**, *63*, 144424. [CrossRef]
51. Chen, D.-X.; Pascual, L.; Fraga, E.; Vazquez, M.; Hernando, A. Magnetic and transport eddy-current anomalies in cylinders wit core-and-shell regions. *J. Magn. Magn. Mater.* **1999**, *202*, 385–396. [CrossRef]
52. Usov, N.A.; Antonov, A.S.; Lagar'kov, A.N.; Granovsky, A.B. GMI Spectra of amorphous wires with different types of magneti anisotropy in the core and the shell regions. *J. Magn. Magn. Mater.* **1999**, *203*, 108–110. [CrossRef]
53. Melo, L.G.C.; Ménard, D.; Ciureanu, P.; Yelon, A.; Cochrane, R.W. Coupled core–shell model of magnetoimpedance in wires. *J. Appl. Phys.* **2004**, *95*, 1331–1335. [CrossRef]
54. Popov, V.V.; Berzhansky, V.N.; Gomonay, H.V.; Qin, F.X. Stress-induced magnetic hysteresis in amorphous microwires probed by microwave giant magnetoimpedance measurements. *J. Appl. Phys.* **2013**, *113*, 17A326. [CrossRef]
55. Buznikov, N.A.; Popov, V.V. A core–shell model for magnetoimpedance in stress-annealed Fe-rich amorphous microwires. *J. Supercond. Nov. Magn.* **2021**, *34*, 169–177. [CrossRef]
56. Alekhina, I.; Kolesnikova, V.; Rodionov, V.; Andreev, N.; Panina, L.; Rodionova, V.; Perov, N. An indirect method of micromagneti structure estimation in microwires. *Nanomaterials* **2021**, *11*, 274. [CrossRef]
57. Bukreev, D.A.; Derevyanko, M.S.; Moiseev, A.A.; Svalov, A.V.; Semirov, A.V. The study of the distribution of electrical and magnetic properties over the conductor cross-section using magnetoimpedance tomography: Modeling and experiment. *Sensors* **2022**, *22*, 9512. [CrossRef]
58. Bukreev, D.A.; Derevyanko, M.S.; Moiseev, A.A.; Kudryavtsev, V.O.; Kurlyandskaya, G.V.; Semirov, A.V. Modeling and an experimental study of the frequency dependences of the impedance of composite wires. *Phys. Met. Metallogr.* **2022**, *123*, 887–892 [CrossRef]
59. Bukreev, D.A.; Derevyanko, M.S.; Moiseev, A.A.; Matsyuk, I.M.; Ballesteros, A.; Svalov, A.V.; Semirov, A.V. Magneto- impedance tomography of composite CuBe/FeCoNi wires. *SPIN* **2023**, *13*, 2340004. [CrossRef]
60. Bukreev, D.A.; Derevyanko, M.S.; Moiseev, A.A.; Semirov, A.V. Magnetoimpedance tomography of amorphous CoFeTaSiB wires *Phys. Met. Metallogr.* **2023**, *124*, 781–786. [CrossRef]
61. Bukreev, D.A.; Derevyanko, M.S.; Semirov, A.V. Magnetoimpedance effect in cobalt-based amorphous ribbons with an inhomogeneous magnetic structure. *Sensors* **2023**, *23*, 8283. [CrossRef]
62. Bukreev, D.A.; Derevyanko, M.S.; Moiseev, A.A.; Kurlyandskaya, G.V.; Semirov, A.V. The influence of relaxation annealing on the magnetic properties and magnetic impedance of amorphous Co-based wires. *Phys. Met. Metallogr.* **2023**, *124*, 1159–1164 [CrossRef]
63. Panina, L.V.; Mohri, K.; Ushiyama, T.; Noda, M.; Bushida, K. Giant magneto-impedance in Co-rich amorphous wires and films *IEEE Trans. Magn.* **1995**, *31*, 1249–1260. [CrossRef]
64. Kraus, L. GMI modeling and material optimization. *Sens. Actuators A* **2003**, *106*, 187–194. [CrossRef]
65. Kraus, L. The theoretical limits of giant magneto-impedance. *J. Magn. Magn. Mater.* **1999**, *196–197*, 354–356. [CrossRef]
66. Zhukova, V.; Blanco, J.M.; Ipatov, M.; Gonzalez, J.; Churyukanova, M.; Zhukov, A. Engineering of magnetic softness and giant magnetoimpedance effect in Fe-rich microwires by stress-annealing. *Scr. Mater.* **2018**, *142*, 10–14. [CrossRef]
67. Zhukova, V.; Blanco, J.M.; Ipatov, M.; Churyukanova, M.; Taskaev, S.; Zhukov, A. Tailoring of magnetoimpedance effect and magnetic softness of Fe-rich glass-coated microwires by stress-annealing. *Sci. Rep.* **2018**, *8*, 3202. [CrossRef] [PubMed]
68. Varga, R.; Vojtanik, P.; Davies, H.A. Low-field magnetoimpedance of amorphous CoFeSiB alloy wire. *J. Magn. Magn. Mater.* **2003**, *261*, 360–368. [CrossRef]
69. Garcia-Beneytez, J.M.; Vinai, F.; Brunetti, L.; Garcia-Miquel, H.; Vázquez, M. Study of magneto impedance effect in the microwave frequency range for soft magnetic wires and microwires. *Sens. Actuators A* **2000**, *81*, 78–81. [CrossRef]
70. Lofland, S.E.; Garcia-Miquel, H.; Vázquez, M.; Bragat, S.M. Microwave magnetoabsortion in glass-coated amorphous microwires with radii close to skin depth. *J. Appl. Phys.* **2002**, *92*, 2058–2063. [CrossRef]

sensors MDPI

Communication

Monitoring the Velocity of Domain Wall Motion in Magnetic Microwires

Alexander Chizhik [1,*], Paula Corte-Leon [1,2], Valentina Zhukova [1,2], Juan Mari Blanco [2] and Arcady Zhukov [1,2,3]

1 Department Advanced Polymers and Materials: Physics, Chemistry and Technology, University of Basque Country UPV/EHU, 20018 San Sebastian, Spain; paula.corte@ehu.eus (P.C.-L.); valentina.zhukova@ehu.eus (V.Z.); arkadi.joukov@ehu.eus (A.Z.)
2 Department of Applied Physics, University of Basque Country EIG, UPV/EHU, 20018 San Sebastian, Spain; juanmaria.blanco@ehu.eus
3 IKERBASQUE, Basque Foundation for Science, 48011 Bilbao, Spain
* Correspondence: oleksandr.chyzhyk@ehu.es

Abstract: An approach was proposed to control the displacement of domain walls in magnetic microwires, which are employed in magnetic sensors. The velocity of the domain wall can be altered by the interaction of two magnetic microwires of distinct types. Thorough investigations were conducted utilizing fluxmetric, Sixtus–Tonks, and magneto-optical techniques. The magneto-optical examinations revealed transformation in the surface structure of the domain wall and facilitated the determination of the mechanism of external influence on the movement of domain walls in magnetic microwires.

Keywords: soft magnetic materials; amorphous magnetic microwires; magnetic domains; magneto-optic Kerr effect; magnetic anisotropy

Citation: Chizhik, A.; Corte-Leon, P.; Zhukova, V.; Blanco, J.M.; Zhukov, A. Monitoring the Velocity of Domain Wall Motion in Magnetic Microwires. *Sensors* **2024**, *24*, 1326. https://doi.org/10.3390/s24041326

Academic Editor: Galina V. Kurlyandskaya

Received: 30 January 2024
Revised: 14 February 2024
Accepted: 17 February 2024
Published: 19 February 2024

1. Introduction

Amorphous soft magnetic materials, such as glass-coated microwires and ribbons, play a fundamental role in numerous technological applications [1–7]. The most advanced application of these soft magnetic materials is their utilization in magnetometers and magnetic sensors [8–13]. This utilization is made possible by the exceptional magnetic properties and good mechanical properties of the materials, as well as the existence of a well-established and validated production and quality control system.

The perfectly cylindrical shape of magnetic wires presents the opportunity to observe magnetic properties that are quite unusual, such as spontaneous magnetic bistability and/or the giant magnetoimpedance (GMI) effect [1,14–19]. These properties are intrinsically linked to the distinctive domain structure of magnetic wires, which consists of an inner axially magnetized core surrounded by an outer domain shell [14,20]. Consequently, the high GMI effect of Co-rich magnetic wires is a result of the high circumferential magnetic permeability of Co-rich amorphous wires [14,20–24]. On the other hand, spontaneous magnetic bistability is attributed to the remagnetization process within the axially magnetized core brought about by the rapid propagation of domain walls (DWs) [1,20].

The observation of rapid single-domain wall (DW) propagation in magnetic wires has garnered significant attention from the perspective of fundamental physics. This includes investigating the origins of DW nucleation and propagation fields, as well as the remarkably high DW velocities (v) and DW mobility (S) [1,20,25–30]. Conversely, various potential applications, such as racetrack memories, magnetic logic, and electronic surveillance, have been developed by leveraging the magnetic bistability of fast and controllable DW propagation [31–33].

In the majority of applications, the efficient regulation of individual domain walls (DWs) through injection, the management of controllable DW propagation, and the act of pinning are of utmost significance [31,32,34].

Generally, this study aligns with the overarching concept of interlayer interaction present in magnetic multilayer structures featuring non-magnetic separators surpassing exchange lengths [35,36]. Notably, oscillation periods of interlayer exchange were noted contingent upon the non-magnetic material type and thickness. In our investigation the glass coating of two microwires functioned as the non-magnetic layer. Additionally within this framework lies the notion of employing two magnetic microwires with distinct chemical compositions, hence differing magnetic characteristics.

One of the primary focal points in our previous research, within the context of the specified orientations of employing magnetic microwires, entailed a comprehensive examination of the dynamics of domain walls (DWs), both internally within the microwires and on their surfaces [34,37]. Particularly noteworthy to us was the matter of regulating DW propagation [31,34,37–39]. In the course of developing diverse methodologies for such regulation, we pursued the concept of reciprocal influence between closely positioned microwires. This concept was explored not only through our own research endeavors [37] but also by other scientific communities [40–42], thereby enabling us to assess the potential level of this reciprocal influence in relation to the distance separating the microwires.

In the majority of prior publications, researchers have investigated the magnetostatic interaction among microwires that possess similar compositions and geometry [37,40–44]. In terms of the dynamics of domain walls, our previous work from several years ago demonstrated that the relationship between the velocity of the DW and the applied magnetic field could be influenced by the magnetostatic interaction with another microwire with identical properties [37]. Furthermore, only a small number of publications have reported on the distinctive magnetic properties observed in a linear array consisting of microwires with varying chemical compositions and magnetic properties [45].

As an outcome, the notion of situating two microwires with distinct characteristics in close proximity to each other appeared innovative and promising to us. In order to achieve a located influence, one of the microwires was deliberately chosen to be considerably shorter than the other. Consequently, one of the designated wires was a lengthy Fe-rich microwire. In this microwire, which possesses the magnetic bistability phenomenon, we examined the displacement of an individual DW, the presence of which we were already aware of. The second microwire, with which we aimed to exert a localized influence, was a short Co-rich microwire.

We opted for a specific set of research techniques that appeared most appropriate for this study, including the fluxmetric method [46], the Sixtus–Tonks method [37], and the magneto-optical Kerr effect technique [47].

2. Experimental Details

In our studies, we used the following two microwires: an as-prepared 12 cm long $Fe_{75}B_9Si_{12}C_4$ microwire (diameter of the metallic nucleus d = 19.8 μm, total diameter D = 28.6 μm) and a 1 cm long $Co_{64.6}Fe_{5.8}B_{16.8}Si_{11}Cr_{3.4}$ microwire (d = 80 μm, D = 92.3 μm). The Co-rich microwire was located in close proximity to the Fe-rich microwire (Figure 1).

The magnetic hysteresis loops were obtained utilizing the fluxmetric technique, which has been previously employed for the characterization of magnetically soft microwires. The investigation of the magnetization reversal process in the surface region of the microwire was conducted using a MOKE loop tracer. The polarized light emitted by a He–Ne laser was directed towards the detector after being reflected from the microwire (Figure 1).

The DW velocity was determined using a modified Sixtus–Tonks experiment, which has previously proven effective in the examination of DW dynamics in magnetic microwires [20,26]. The microwire was positioned within an extended solenoid, establishing a magnetic field.

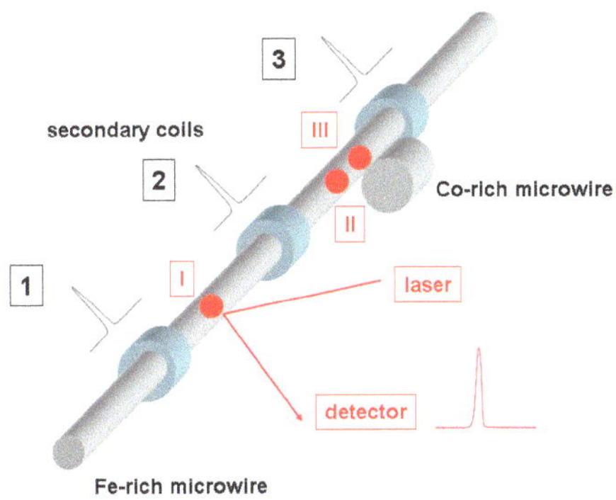

Figure 1. Schematic picture of the experimental set-up: 1, 2, and 3—pick-up coils; I, II, and III, points of laser reflection during the MOKE experiment. The position of the Co-rich microwire relative to the pick-up coils and the Fe-rich microwire is demonstrated.

Three pickup coils were positioned coaxially within the solenoid (Figure 1), encircling the sample and kept at an equal distance apart, for the purpose of evaluating the velocity of the DW. The velocity of the DW was determined by observing the difference in time between the electromotive force (*EMF*) peaks caused in the pick-up coils by the displacement of the DW. The pick-up coils were separated by the distance *L*.

In addition to the phenomenon of surface hysteresis, the MOKE method facilitates the detection of MOKE peaks that correspond to the displacement of a domain wall across the surface of the specimen. By acquiring knowledge about the characteristics of the MOKE peaks at various positions on the surface of the elongated microwire sample (designated as locations I, II, and III in Figure 1), we employed the MOKE technique to infer the structure of the domain wall at the different aforementioned locations within the investigated microwire.

3. Results and Discussion

As an initial investigation, magnetic and magneto-optical hysteresis were acquired from the two samples utilized in the experiments. Figure 2 shows the hysteresis loops for the magnetic (represented by black lines) and MOKE (represented by red line) methods. The Fe-rich microwire, which displayed positive magnetostriction, exhibited a volume hysteresis loop with a perfectly rectangular shape (Figure 2a, black line). Conversely, the Co-rich microwire, with a vanishing value of magnetostriction, exhibited an inclined hysteresis loop with significantly lower coercivity (Figure 2b). The shape of the volume hysteresis loop observed for the Fe-rich microwire is associated with the so-called "magnetic bistability effect".

Figure 2 depicts diagrams illustrating the domain structures. The black arrows represent magnetization in a schematic manner. When the external magnetic field reached a sufficient magnitude, both types of microwires reached saturation (top and bottom inserts). This is denoted by the black arrows that align with the microwire axis. In Fe-rich microwires, magnetization reversal occurs through the swift movement of a flat domain wall separating domains with opposite magnetization (as shown in the left inset). Generally, in Co-rich microwires, various domain structures, such as axial, helical, or circular structures, may exist. However, in this instance, magnetization reversal occurs through the rotation of magnetization without forming a stable domain structure (as shown in the right inset).

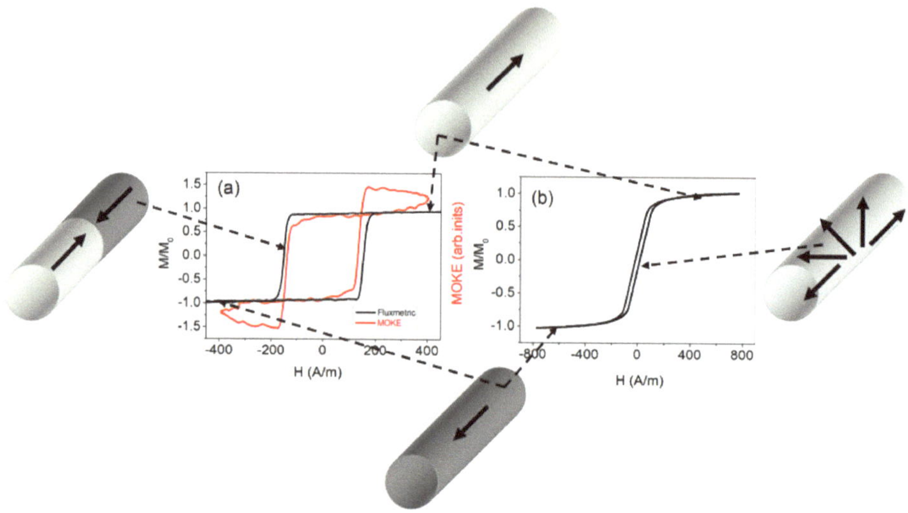

Figure 2. Fluxmetric (black lines) and MOKE (red line) hysteresis loops obtained from Fe-rich (**a**) and Co-rich (**b**) samples.

The process of magnetization reversal in Fe-rich microwires takes place when a single DW is detached from one of the closure end domains and is subsequently displaced along the microwire [20,26,48,49]. On the other hand, the magnetization reversal process in Co-rich microwires is characterized by the rotation of magnetization. As a result, these microwires typically exhibit low coercivity and high magnetic permeability [50–53].

It was equally important to obtain pertinent details regarding the process of magnetization reversal within the surface layer of a Fe-rich microwire. The confirmation of our assumptions regarding the magnetic bistability, referred to as the "surface bistability effect", was made evident by the rectangular shape observed in the MOKE hysteresis loop. This observation was made both in the sample's volume and on its surface. Therefore we deduced that during the magnetization reversal, there was swift displacement of a solitary domain wall on the surface of the Fe-rich sample, as well as within the bulk. It is important to highlight the slight deviation of the signal from a linear path, as observed in the MOKE hysteresis loop. This occurred because the laser beam reflected not from a flat surface, but from a cylindrical surface of the sample. Nonetheless, this did not hinder our ability to detect magnetization jumps on the sample's surface that were linked to the rapid movement of the domain wall. The slight disparity in the coercivity magnitude observed in the magnetic and MOKE hysteresis loops arose from the fact that the process of magnetization reversal initiates from the surface of the Fe-rich sample.

After obtaining initial insights into the bulk and surface magnetization reversal process, we initiated an investigation into the propagation of the domain wall (DW) using the Sixtus-Tonks method. In Figure 3, the magnetic field dependencies of velocity in the Fe-rich sample are illustrated. The placement of the Co-rich short wire aligns with the diagram depicted in Figure 1.

The Co-rich microwire was situated in close proximity to the surface of the long Fe-rich microwire, as well as in the space between the secondary coils 2 and 3. The black line (Figure 3) represents the velocity that was measured between coils 1 and 2, whereas the red line (Figure 3) was measured in the region of the sample between coils 2 and 3, where the shorter Co-rich microwire was situated. It can be observed that a substantial disparity in the value of the velocity of the domain wall, as measured in various regions of the sample, exists. In simpler terms, a controlled deceleration in the motion of the domain wall is apparent.

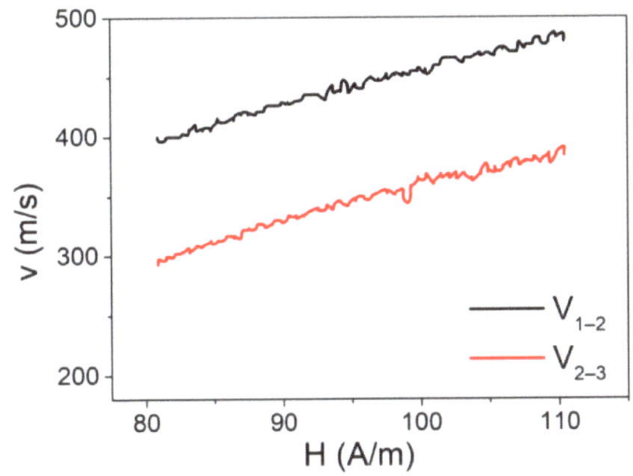

Figure 3. V(H) dependencies obtained from the Fe-rich microwire. V_{1-2} represents the velocity dependence obtained between secondary coils 1 and 2. V_{2-3} represents the velocity dependence obtained between secondary coils 2 and 3.

To examine the operation of regulated alterations in the speed of the domain wall, we investigated the conversion of the MOKE peaks acquired at different positions on the surface of the microwire.

The initial MOKE measurement (designated as point I in Figure 1) was carried out between coils 1 and 2. A comparison of the EMF peak registered by coil 1 and the MOKE peak registered at point I is shown in Figure 4. It is evident that the shapes of these two signals bear significant resemblance. This observation suggests the rapid, homogeneous, and compact motion of the domain wall within the volume of the microwire and on its surface. In this particular instance, the MOKE peak exhibited a somewhat narrower width compared to the peak acquired using the Sixtus–Tonks technique. Presumably, this discrepancy arose from the fact that the width of the domain wall on the sample's surface was slightly smaller than within its interior.

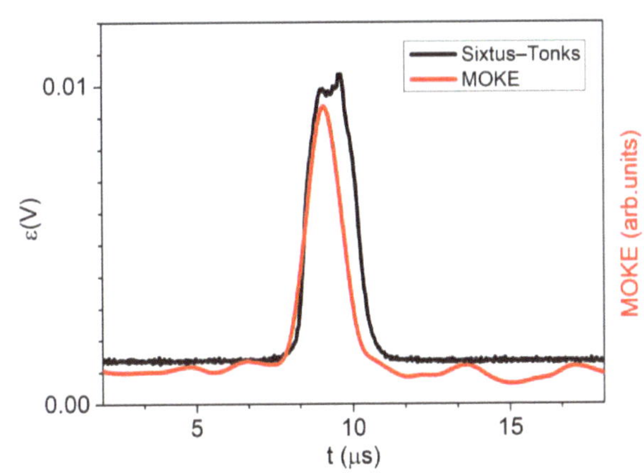

Figure 4. V(H) dependencies obtained in a linear array consisting of long Fe-rich and short Co-rich microwires. V_{1-2} represents the velocity dependence obtained between secondary coils 1 and 2. V_{2-3} represents the velocity dependence obtained between secondary coils 2 and 3.

In the second phase of the MOKE investigation, our focus shifted towards the region between coils 2 and 3, which exhibited a lower speed of the domain wall. The MOKE peaks were obtained at the designated locations II and III, as illustrated in Figure 1. Point II can be identified on the surface of the Fe-rich microwire, positioned in close proximity to th Co-rich microwire. On the other hand, point III was situated on the surface of the Fe-ric microwire, slightly above the Co-rich microwire. Our belief is that this particular area of th Fe-rich microwire, encompassing points II and III, found itself within the zone of influenc exerted by the Co-rich microwire, albeit to a varying degree at each point.

The findings of these MOKE experiments are depicted in Figure 5. It is evident that th morphology of the MOKE peaks exhibits variability contingent upon the precise locatio of the observation point. As previously mentioned, a consistent peak achieved at point corresponds to the displacement of the compact and regular domain wall. Conversely, th signal received at point II (the blue line in Figure 5) exhibits a modified shape in relatio to the peak obtained at point I. An additional series of smaller peaks and a reduction i the amplitude of the primary peak signified a metamorphosis of the domain wall on th surface of the Fe-rich microwire. The signal received at point III (the green line in Figure 5 underwent even more pronounced alterations. At this juncture, a series of peaks wit varying heights became apparent. Simultaneously, the primary peak continued to decreas

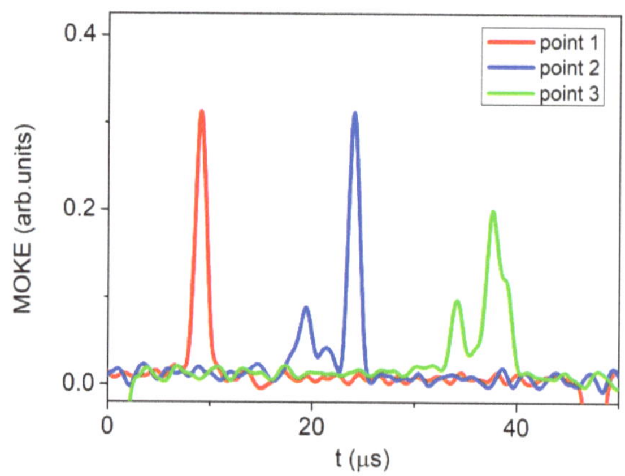

Figure 5. V(H) dependencies obtained in a linear array consisting of long Fe-rich and short Co-rich microwires. V_{1-2} represents the velocity dependence obtained between secondary coils 1 and 2. V_{2-3} represents the velocity dependence obtained between secondary coils 2 and 3.

Henceforth, the alterations observed in the MOKE signal imply a metamorphosis of the surface segment of the domain wall within the sphere of impact of the Co-rich microwire. Evidently, the impact of the Co-rich microwire is associated with its stray fields (Figure 6); however, certain peculiarities exist.

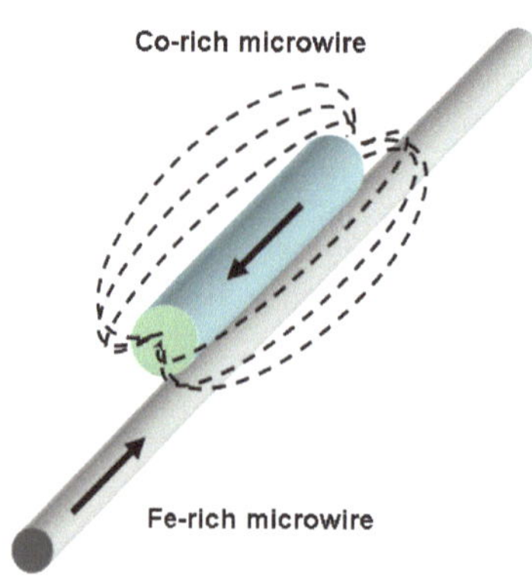

Figure 6. Schematic sketch of stray field distribution.

We maintain the belief that the impact of stray fields is non-uniform, both in terms of the microwire's length and diameter. In the vicinity of the Co-rich microwire (designated as point II in our experimental setup), the configuration of domains was primarily influenced by the stray magnetic field that was perpendicular to the microwire's axis. As we traversed the surface of the microwire (designated as point III in our experiment), the contribution of the longitudinal stray magnetic field became increasingly prominent. In this particular scenario, the surface region of the microwire, which we examined using the MOKE method, was situated within a region exhibiting significant variations in the amplitude of the magnetic field.

The observed transformation of the domain wall manifested through the emergence of a specific form of non-uniformity. In the region where the perpendicular stray fields exerted their effect (identified as point II), local areas with a perpendicular component of magnetization appeared on the domain wall. Based on the minor peak count, there were two such regions. At point III, two minor peaks transitioned into a singular peak; however, simultaneously, the primary peak widened and diminished in magnitude, thereby indicating an augmentation in the influence of the stray field that aligned parallel to the sample's axis.

Generally speaking, we posit that the primary impact of a closely positioned additional microwire lies in the induced complexity of the moving domain wall's shape. The key elements of this mechanism involve the creation and enlargement of regions with uneven magnetization rotation within the domain wall, together with its general broadening.

4. Conclusions

We proposed a novel method for controlling the movement of domain boundaries in magnetic microwires. Our investigation focused on understanding the impact of the Co-rich microwire's proximity on the velocity of the domain wall in the Fe-rich microwire. Various experimental techniques, including the fluxmetric method, Sixtus–Tonks method, and magneto-optical Kerr effect method, were employed. These complementary methods allowed us to ascertain the key properties of the influence of the Co-rich microwire and, more importantly, comprehend the underlying mechanism. Our findings revealed a noticeable alteration in the domain wall velocity within the region directly adjacent to the additional Co-rich microwire in the studied sample.

During these experiments, we observed a notable alteration in the magneto-optical peaks linked to the motion of the domain wall in the surface region of the microwire. We interpreted this change in MOKE peaks as a complexity emerging in the structure of the surface segment of the domain wall as it neared the influence zone of the additional microwire.

Ultimately, alterations in the moving domain wall's structure resulted from the combination of longitudinal and transverse projections of the stray magnetic field, enabling local control over its velocity. We posit that when applied to a system comprising numerous microwires with diverse properties, this method holds promising potential.

Author Contributions: Conceptualization, A.C. and A.Z.; methodology, P.C.-L., J.M.B., A.C. and V.Z.; validation, A.C. and P.C.-L.; formal analysis, P.C.-L., J.M.B., A.C. and V.Z.; investigation, P.C.-L., J.M.B., A.C. and V.Z.; resources, P.C.-L. and V.Z.; data curation, P.C.-L., J.M.B., A.C. and V.Z.; writing—original draft preparation, A.C. and A.Z.; writing—review and editing, A.C. and A.Z.; funding acquisition, A.Z. and V.Z. All authors have read and agreed to the published version of the manuscript.

Funding: Funding was provided by the EU under the "INFINITE" (HORIZON-CL5-2021-D5-01-06) project, by the Spanish MICIN under the PID2022-141373NB-I00 project, and by the Government of the Basque Country under PUE_2021_1_0009, the Elkartek (MINERVA, MAGAF, and ZE-KONP) projects, and under the scheme of "Ayuda a Grupos Consolidados" (ref. IT1670-22). The authors are grateful for the technical and human support provided by SGIker of UPV/EHU (Medidas Magneticas Gipuzkoa) and European funding (ERDF and ESF). The authors wish to acknowledge the useful discussions held with Prof. A. Fert. We also wish to thank the administration team from the University of the Basque Country, which not only provided very limited funding, but also expropriated the resources received by the research group from private companies for the research activities of the group. Such assistance helped to keep us on our toes.

Institutional Review Board Statement: Not applicable.

Informed Consent Statement: Not applicable.

Data Availability Statement: Data is available upon request due to restrictions related to development

Conflicts of Interest: The authors declare no conflicts of interest.

References

1. Zhukova, V.; Corte-Leon, P.; González-Legarreta, L.; Talaat, A.; Blanco, J.M.; Ipatov, M.; Olivera, J.; Zhukov, A. Review of Domain Wall Dynamics Engineering in Magnetic Microwires. *Nanomaterials* **2020**, *10*, 2407. [CrossRef]
2. Makhotkin, V.; Shurukhin, B.; Lopatin, V.; Marchukov, P.; Levin, Y. Magnetic field sensors based on amorphous ribbons. *Sens. Actuators A Phys.* **1991**, *27*, 759–762. [CrossRef]
3. Morón, C.; Cabrera, C.; Morón, A.; García, A.; González, M. Magnetic Sensors Based on Amorphous Ferromagnetic Materials: A Review. *Sensors* **2015**, *15*, 28340–28366. [CrossRef]
4. Phan, M.; Peng, H. Giant magnetoimpedance materials: Fundamentals and applications. *Prog. Mater. Sci.* **2008**, *53*, 323–420. [CrossRef]
5. Nakayama, S.; Uchiyama, T. Real-time Measurement of Biomagnetic Vector Fields in Functional Syncytium Using Amorphous Metal. *Sci. Rep.* **2015**, *5*, 8837. [CrossRef]
6. Sabol, R.; Klein, P.; Ryba, T.; Hvizdos, L.; Varga, R.; Rovnak, M.; Sulla, I.; Mudronova, D.; Galik, J.; Polacek, I.; et al. Novel Applications of Bistable Magnetic Microwires. *Acta Phys. Pol. A* **2017**, *131*, 1150. [CrossRef]
7. Qin, F.; Peng, H.X. Ferromagnetic microwires enabled multifunctional composite materials. *Prog. Mater. Sci.* **2013**, *58*, 183–259. [CrossRef]
8. Mohri, K.; Uchiyama, T.; Panina, L.V.; Yamamoto, M.; Bushida, K. Recent Advances of Amorphous Wire CMOS IC Magneto-Impedance Sensors: Innovative High-Performance Micromagnetic Sensor Chip. *J. Sens.* **2015**, *2015*, 718069. [CrossRef]
9. Correa, M.A.; Ferreira, A.; Souza, A.L.R.; Dantas Neto, J.M.; Bohn, F.; Vaz, F.; Kurlyandskaya, G.V. Anomalous Nernst Effect in Flexible Co-Based Amorphous Ribbons. *Sensors* **2023**, *23*, 1420. [CrossRef]
10. Andò, B.; Baglio, S.; Bulsara, A.R.; Trigona, C. Design and characterization of a microwire fluxgate magnetometer. *Sens. Actuators A Phys.* **2009**, *151*, 145–153. [CrossRef]
11. Herrero-Gómez, C.; Marín, P.; Hernando, A. Bias free magnetomechanical coupling on magnetic microwires for sensing applications. *Appl. Phys. Lett.* **2013**, *103*, 142414. [CrossRef]
12. Goleman, K.; Sasada, I. High Sensitive Orthogonal Fluxgate Magnetometer Using a Metglas Ribbon. *IEEE Trans. Magn.* **2006**, *42*, 3276–3278. [CrossRef]
13. Edelstein, A. Advances in magnetometry. *J. Phys. Condens. Matter* **2007**, *19*, 165217. [CrossRef]
14. Vazquez, M.; Chen, D.-X. The magnetization reversal process in amorphous wires. *IEEE Trans. Magn.* **1995**, *31*, 1229–1238. [CrossRef]
15. Varga, R.; Torrejon, J.; Kostyk, Y.; Garcia, K.L.; Infantes, G.; Badini, G.; Vazquez, M. Single-wall dynamics and power law in bistable magnetic microwires. *J. Phys. Condens. Matter* **2008**, *20*, 445215. [CrossRef]
16. Taysioglu, A.A.; Peksoz, A.; Kaya, Y.; Derebasi, N.; Kaynak, G. GMI effect in CuO coated Co-based amorphous ribbons. *J. Alloys Compd.* **2009**, *487*, 38–41. [CrossRef]
17. Malátek, M.; Kraus, L. Off-diagonal GMI sensor with stress-annealed amorphous ribbon. *Sens. Actuators A Phys.* **2010**, *164*, 41–45. [CrossRef]
18. Fodil, K.; Denoual, M.; Dolabdjian, C.; Harnois, M.; Senez, V. Dynamic sensing of magnetic nanoparticles in microchannel using GMI technology. *IEEE Trans. Magn.* **2013**, *49*, 93–96. [CrossRef]
19. Corodeanu, S.; Ovari, T.A.; Lupu, N.; Chiriac, H. Magnetization process and GMI effect in as-cast nanocrystalline microwires. *IEEE Trans. Magn.* **2010**, *46*, 380–382. [CrossRef]
20. Zhukova, V.; Blanco, J.M.; Chizhik, A.; Ipatov, M.; Zhukov, A. AC-current-induced magnetization switching in amorphous microwires. *Front. Phys.* **2018**, *13*, 137501. [CrossRef]
21. Ciureanu, P.; Melo, L.G.C.; Yelon, A. Circumferential and longitudinal 1 GHz permeabilities in Co-rich melt-extracted amorphous wires. *J. Magn. Magn. Mater.* **2002**, *242–245*, 224–228. [CrossRef]
22. Rouabhi, M.; Cochrane, R.W.; Ménard, D.; Britel, M.; Ciureanu, P.; Yelon, A. Anisotropy and magnetization processes in Co-rich amorphous wires. *J. Appl. Phys.* **1999**, *85*, 5441–5443. [CrossRef]
23. Zhao, Y.; Hao, H.; Zhang, Y. Preparation and giant magneto-impedance behavior of Co-based amorphous wires. *Intermetallics* **2013**, *42*, 62–67. [CrossRef]
24. Mu, C.; Hu, L.; Zhu, F.; Cui, Z.; He, A.; Dong, Y.; Man, Q.; Ma, H.; Li, J. Effects of drawing and tension stress annealing on the structure and magnetic properties of Co-based amorphous wire. *J. Magn. Magn. Mater.* **2024**, *589*, 171569. [CrossRef]

25. Klein, P.; Varga, R.; Badini-Confalonieri, G.A.; Vazquez, M. Study of domain structure and magnetization reversal after thermal treatments in $Fe_{40}Co_{38}Mo_4B_{18}$ microwires. *J. Magn. Magn. Mater.* **2011**, *323*, 3265–3270. [CrossRef]
26. Ziman, J.; Onufer, J.; Kladivová, M. Dynamics of domain wall depinning from closure domain structure at the end of bistable glass coated microwire. *J. Magn. Magn. Mater.* **2020**, *514*, 167233. [CrossRef]
27. Vereshchagin, M.; Baraban, I.; Leble, S.; Rodionova, V. Structure of head-to-head domain wall in cylindrical amorphous ferromagnetic microwire and a method of anisotropy coefficient estimation. *J. Magn. Magn. Mater.* **2020**, *504*, 166646. [CrossRef]
28. Beck, F.; Rigue, J.N.; Carara, M. The profile of the domain walls in amorphous glass-covered microwires. *J. Magn. Magn. Mater.* **2017**, *435*, 21–25. [CrossRef]
29. Richter, K.; Vahovsky, O.; Varga, R. Direct visualization of a domain wall pinning by time-resolved microscopy in amorphous glass-coated microwires. *IEEE Trans. Magn.* **2022**, *58*, 6500505. [CrossRef]
30. Klein, P.; Varga, R.; Vázquez, M. Stable and fast domain wall dynamics in nanocrystalline magnetic microwire. *J. Alloys Compd.* **2013**, *550*, 31–34. [CrossRef]
31. Allwood, D.A.; Xiong, G.; Faulkner, C.C.; Atkinson, D.; Petit, D.; Cowburn, R.P. Magnetic domain-wall logic. *Science* **2005**, *309*, 1688–1692. [CrossRef]
32. Parkin, S.S.P.; Hayashi, M.; Thomas, L. Magnetic Domain-Wall Racetrack Memory. *Science* **2008**, *320*, 190. [CrossRef]
33. Zhukova, V.; Corte-Leon, P.; Blanco, J.M.; Ipatov, M.; Gonzalez, J.; Zhukov, A. Electronic Surveillance and Security Applications of Magnetic Microwires. *Chemosensors* **2021**, *9*, 100. [CrossRef]
34. Corte-León, P.; Zhukova, V.; Blanco, J.M.; Chizhik, A.; Ipatov, M.; Gonzalez, J.; Fert, A.; Alonso, A.; Zhukov, A. Engineering of domain wall propagation in magnetic microwires with graded magnetic anisotropy. *Appl. Mater. Today* **2022**, *26*, 101263. [CrossRef]
35. Dieny, B.; Speriosu, V.S.; Metin, S.; Parkin, S.S.; Gurney, B.A.; Baumgart, P.; Wilhoit, D.R. Magnetotransport properties of magnetically soft spin-valve structures. *J. Appl. Phys.* **1991**, *69*, 4774. [CrossRef]
36. Rifai, D.; Abdalla, A.N.; Ali, K.; Razali, R. Giant Magnetoresistance Sensors: A Review on Structures and Non-Destructive Eddy Current Testing Applications. *Sensors* **2016**, *16*, 298. [CrossRef]
37. Gawronski, P.; Zhukova, V.; Zhukov, A.; Gonzalez, J. Manipulation of domain propagation dynamics with the magnetostatic interaction in a pair of Fe- rich amorphous microwires. *J. Appl. Phys.* **2023**, *114*, 043903. [CrossRef]
38. Atkinson, D.; Allwood, D.A.; Xiong, G.; Cooke, M.D. Magnetic domain-wall dynamics in a submicrometre ferromagnetic structure. *Nat. Mater.* **2003**, *2*, 85–87. [CrossRef]
39. Chiriac, H.; Óvári, T.A.; Tibu, M. Domain wall propagation in nearly zero magnetostrictive amorphous microwires. *IEEE Trans. Magn.* **2008**, *44*, 3931–3933. [CrossRef]
40. Kolesnikova, V.; Martínez-García, J.C.; Rodionova, V.; Rivas, M. Study of bistable behaviour in interacting Fe-based microwires by first order reversal curves. *J. Magn. Magn. Mater.* **2020**, *508*, 166857. [CrossRef]
41. Pereira, A.; Denardin, J.C.; Escrig, J. How do magnetic microwires interact magnetostatically? *J. Appl. Phys.* **2009**, *105*, 083903. [CrossRef]
42. Knobel, M.; Sampaio, L.C.; Sinnecker, E.H.C.P.; Vargas, P.; Altbir, D. Dipolar magnetic interactions among magnetic microwires. *J. Magn. Magn. Mater.* **2002**, *249*, 60–72. [CrossRef]
43. Raposo, V.; Garcia, J.M.; Gonzalez, J.M.; Vazquez, M. Long-range magnetostatic interactions in arrays of nanowires. *J. Magn. Magn. Mater.* **2000**, *222*, 227–232. [CrossRef]
44. Escrig, J.; Allende, S.; Altbir, D.; Bahiana, M.; Torrejón, J.; Badini, G.; Vázquez, M. Magnetostatic bias in multilayer microwires: Theory and experiments. *J. Appl. Phys.* **2009**, *105*, 023907. [CrossRef]
45. Rodionova, V.; Ipatov, M.; Ilyn, M.; Zhukova, V.; Perov, N.; Gonzalez, J.; Zhukov, A. Tailoring of Magnetic Properties of Magnetostatically-Coupled Glass-Covered Magnetic Microwires. *J. Supercond. Nov. Magn.* **2011**, *24*, 541–547. [CrossRef]
46. Gonzalez-Legarreta, L.; Corte-Leon, P.; Zhukova, V.; Ipatov, M.; Blanco, J.M.; Gonzalez, J.; Zhukov, A. Optimization of magnetic properties and GMI effect of Thin Co-rich Microwires for GMI Microsensors. *Sensors* **2020**, *20*, 1588. [CrossRef]
47. Chizhik, A.; Gonzalez, J. *Magnetic Microwires. A Magneto-Optical Study*; Pan Stanford Publishing: Singapore, 2014; ISBN 978-981-4411-25-7.
48. Chen, D.X.; Dempsey, N.; Vázquez, M.; Hernando, A. Propagating domain wall shape and dynamics in iron-rich amorphous wires. *IEEE Trans. Magn.* **1995**, *31*, 781–790. [CrossRef]
49. Kladivová, M.; Ziman, J. Properties of a domain wall in a bi-stable magnetic microwire. *J. Magn. Magn. Mater.* **2019**, *480*, 193–198. [CrossRef]
50. Liu, B.B.; Liu, C.; Ma, L.; Han, C.; Zhen, S.; You, L.; Ye, F. Enhancing magnetic properties of the Co66Fe6Si13B15 metallic glass through DC annealing. *J. Magn. Magn. Mater.* **2022**, *544*, 168676. [CrossRef]
51. Marín, P.; Cortina, D.; Hernando, A. High-frequency behavior of amorphous microwires and its applications. *J. Magn. Magn. Mater.* **2005**, *290–291*, 1597–1600. [CrossRef]

52. Tuan, L.A.; Huy, N.T.; Huy, P.T. Giant magnetoimpedance in layered composite micro-wires for high-sensitivity magnetic sens applications. *J. Phys. Conf. Ser.* **2009**, *187*, 012044. [CrossRef]
53. Han, L.; Maccari, F.; Souza Filho, I.R.; Peter, N.J.; Wei, Y.; Gault, B.; Gutfleisch, O.; Li, Z.; Raabe, D. A mechanically strong an ductile soft magnet with extremely low coercivity. *Nature* **2022**, *608*, 310–316. [CrossRef]

sensors

MDPI

Article

Relationship of Magnetic Domain and Permeability for Clustered Soft Magnetic Narrow Strips with In-Plane Inclined Magnetization Easy Axis on Distributed Magnetic Field

Tomoo Nakai

Industrial Technology Institute, Miyagi Prefectural Government, Sendai 981-3206, Japan; nakai-to693@pref.miyagi.lg.jp

Abstract: A unique functionality was reported for a thin-film soft magnetic strip with a certain angle of inclined magnetic anisotropy. It can switch magnetic domain by applying a surface normal field with a certain distribution on the element. The domain switches between a single domain and a multi-domain. Our previous study shows that this phenomenon appears even in the case of the adjacent configuration of multiple narrow strips. It was also reported that the magnetic permeability for the alternating current (AC) magnetic field changes drastically in the frequency range from 10 kHz to 10 MHz as a function of the strength of the distributed magnetic field. In this paper, the correspondence of AC permeability and the magnetic domain as a function of the intensity of the distributed field is investigated. It was confirmed that the extension of the area of the Landau–Lifshitz-like multi-domain on the clustered narrow strips was observed as a function of the intensity of the distributed magnetic field, and this domain extension was matched with the permeability variation. The result leads to the application of this phenomenon to a tunable inductor, electromagnetic shielding, or a sensor for detecting and memorizing the existence of a distributed magnetic field generated by a magnetic nanoparticle in the vicinity of the sensor.

Keywords: distributed magnetic field; magnetic domain transition; permeability; soft magnetic thin film; amorphous; narrow strip; magnetization easy axis; surface normal field

Citation: Nakai, T. Relationship of Magnetic Domain and Permeability for Clustered Soft Magnetic Narrow Strips with In-Plane Inclined Magnetization Easy Axis on Distributed Magnetic Field. *Sensors* **2024**, *24*, 706. https://doi.org/10.3390/s24020706

Academic Editor: Galina V. Kurlyandskaya

Received: 11 December 2023
Revised: 16 January 2024
Accepted: 18 January 2024
Published: 22 January 2024

1. Introduction

The magnetic domain is a region within a magnetic material [1,2]. The magnetization within each domain points in a uniform direction, but the magnetization of different domains may point in a different direction. The regions separating magnetic domains are called domain walls. The magnetic domains and the domain walls are generated in order to reduce the overall free energy, mainly the magneto-static energy, in magnetic systems. Thus, without the application of external magnetic fields, the magnetic domain and domain walls are unavoidable in patterned soft-magnetic material systems. As they can have different shapes and widths, the magnetic domains are an exciting playground for fundamental research and have become in recent years the subject of intense studies [3]; these studies have mainly been focused on controlling, manipulating, and moving their internal magnetic configuration; their usage in information storage [4–6]; their performance in logic operations [7–9]; high frequency inductors [10]; and sensor applications.

The domain structure has been studied in the research field of magnetic inductors and sensors using soft magnetic materials. The relationship between the magnetic domain and permeability has been extensively studied for ferromagnetic materials. The basic magnetization process consists of domain wall movement, magnetization rotation, and magnetization reversal. They are intricately intertwined in a manner that depends on the inner structure, such as crystal structure, anisotropy property, magnetostriction, defects, and grain size [11].

For the study of the inductor, the relationship between permeability and the magnetic domain was recognized as indispensable. It was initiated in the study of silicon iron [12–14]. The improvement of magnetization properties was studied based on the structural and dimensional change in the magnetic materials, such as in the study of amorphous powder, the size dependence of spherical ferromagnetic particles, and the soft magnetic multilayered films [15–17]. Some recent studies on modeling and simulation had advantageous results such as the improvement of the performance of magnetic shielding [18]. In order to clarify the magnetization process for the actual application, the domain wall propagation is essential; therefore, some attempts were made to achieve it, such as those in the study on domain boundary propagation velocity, the domain wall transverse caused by fast-rising current pulses, and the effect of plastic deformation on the permeability [19–21].

Many studies have also been conducted on magneto-impedance (MI) sensors. The recent performance improvement of the magnetic domain observation has contributed to the investigation of the sensor from the viewpoint of the dependence of high-frequency impedance on the magnetic domain structure of the thin-film element. This electrical response against the alternating current which depends on the magnetic domain is closely related to this study; an application to the MI sensor is one of the expected candidates. The physical fundamentals of the MI sensor have been studied to clarify the sensing mechanisms [22–24]. Thereafter, the study of the thin-film MI sensors was carried out for the purpose of improving the properties of the MI sensor based on the control of the magnetic domains. The study subjects ranged from sensor structures, materials, and treatment methods to biasing [25–34]. Regarding the magnetic domain simulation, an original method was developed which was adaptive for the actual size of the MI element which are difficult to apply using the conventional micro-magnetic simulations due to the larger dimensions of the element [35,36].

This paper considers a narrow rectangular element made of amorphous soft magnetic thin film with an inclined in-plane magnetization easy axis. It is well known that a sputter deposited amorphous thin film is able to induce a uniaxial magnetization easy axis using magnetic field annealing [11]. Our previous study shows that a narrow rectangular strip of amorphous thin film with an in-plane inclined easy axis has two different magnetic domain transition properties. One is a multi-domain with inclined striped domain configuration, in which the width of the stripe gradually changes as a function of the external magnetic field. The other is a single domain with a magnetic moment along the length direction of the strip in which the magnetization reversal appears as a function of the external field. A typical characteristic of magnetic domain transition was reported when the easy axis lay at around 70° [37], which is a transition condition of the easy axis angle between these different magnetic domain variations. This study aims to apply this phenomenon to a functional device with a switching as well as a memorizing property. In this transitional condition, the magnetic domain transition between the single domain and the multi-domain is artificially controlled by applying a surface normal magnetic field with a certain distribution [38,39]. The details are explained in the next paragraph.

Figure 1 shows the schematics of the domain transition of this case and the effect of the applied distributed normal field. This figure shows the cases both with and without the application of the distributed normal field. When the normal field is not applied, the domain apparently switches between the single domain + direction and the − direction (Figure 1a). There is a non-apparent multi-domain state, which is shown by a short line in the middle of this figure. This state has lower energy compared with the single-domain one, thus, this state is a stable one [40]. Figure 1b shows a case when a distributed normal field is applied. In this case, the multi-domain state appears and the transition from the single domain to the multi-domain is available. This was confirmed experimentally [38], and a sensor with a memory function or a three-state memory as an application was proposed.

We also discovered that this phenomenon appears for a clustered many-body element, with an almost 70° easy axis, even in the conditions of the existing magnetic mutual interaction between the multiple elements [41]. It was also reported that the magnetic permeability for the alternating current (AC) magnetic field changes drastically in the frequency range from 10 kHz to 10 MHz as a function of the strength of the distributed magnetic field [42,43]. It was also noticed in a newly reported typical phenomenon, which has recently been reported [44], that there was an appearance of the multi-domain state when a small magnetic particle was placed just above the strip to control the applied surface normal magnetic field. These reports provide hope for the achievement of an integrated device applying this phenomenon. Based on the background, this article's study aims to clarify the basis of the permeability caused by the magnetic domain transition of this phenomenon.

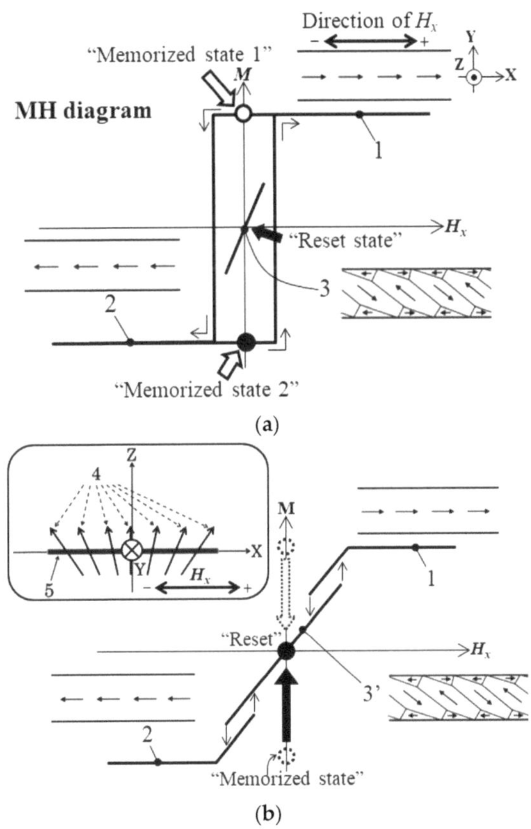

Figure 1. Magnetization diagram with a hidden multi-domain state. (**a**) Without surface normal magnetic field. (**b**) With a distributed normal magnetic field. The numbers in the figure indicate as follows: 1—longitudinal single domain (parallel direction); 2—longitudinal single domain (anti-parallel direction); 3—inclined Landau–Lifshitz domain [45,46] (hidden stable state); 3′—inclined Landau–Lifshitz domain (state of transition available); 4—canted normal field with distribution, where B_z = const. and $\Delta B_x/\Delta x$ = const.; 5—sensor element. The "Memorized state" and the "Reset state" indicates a function of three-state memory.

In this paper, the correspondence between the AC permeability and the structure of the magnetic domain as a function of the intensity of the distributed field is investigated. It was confirmed that the extension of the area of the Landau–Lifshitz-like multi-domain (LLD) [45,46] on the clustered narrow strips was observed as a function of the intensity of the distributed magnetic field. This domain extension was compared with the permeability variation, and it was revealed that they match with each other. The effect of the distributed field was compared with an element which has the easy magnetization axis along the length direction (90°), and it was revealed that the 70° easy axis has the clear transitio to the LLD domain structure and also that it needed a lower distributed field to chang the permeability. The result leads to the application of this phenomenon to a tunabl inductor, electromagnetic shielding, or a sensor for detecting and memorizing the exis tence of distributed magnetic field generated by a magnetic nanoparticle in the vicinity o the sensor.

2. Experimental Procedure

2.1. Magnetic Elements

The magnetic domain of the clustered many-body elements was investigated in re lation to the intensity of the distribution of the applied normal magnetic field. The distri bution means a variation in the inclination angle, which depends on the position of th element. In this study, the normal direction is defined as a surface normal direction agains the flat substrate plane of the thin-film element, and the inclination angle of the field i defined as the angle between the normal direction and the direction of the applied magneti field. The applied magnetic field, in this study, has a certain inclination toward the lengt direction of the element strip, which changes depending on the position. Figure 2 indicate a schematic of the distributed field applied to the element. The intensity of the distributio is defined as a gradient of the partial in-plane field of the inclined normal field, whic varies proportionally as a function of the length displacement of the element strip. In thi study, the gradient is assumed to have a linear relationship.

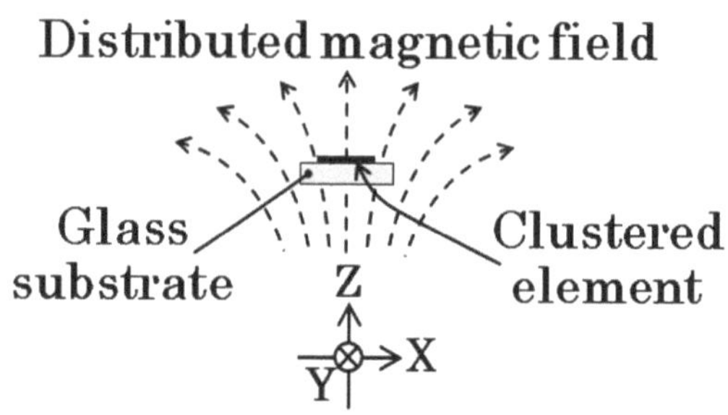

Figure 2. Schematic of distributed surface normal magnetic field.

The clustered element, which is shown in Figure 3, was fabricated by a thin-film process. The process flow was the same as that of our previous study, which is show in [42]. An amorphous $Co_{85}Nb_{12}Zr_3$ film was RF sputter deposited onto a soda glass substrate and then micro-fabricated into a clustered element made of rectangular strips by a lift-off process. A composite metal target with the same composition as $Co_{85}Nb_{12}Zr_3$ and the purity 3N was used. The RF sputter was carried out in a 1.3 Pa Ar atmosphere with the deposition ratio of 1 μm/h and 196 W input power. The substrate plate in the sputter apparatus was water cooled using 21 °C water supply. An XRD analysis of the fabricated film showed a broad hump and no sharp peaks; then, the film was recognized as an amorphous state. The dimensions of the single strip of the element were as follows thousands of μm in length, tens of μm wide, and 2.1 μm thick. The thickness was measured by a stylus-type step profiler. A uniaxial magnetic anisotropy was induced by magnetic field annealing, at 240 kA/m at 673 K for 1 h in a vacuum. The easy magnetization axis of the magnetic anisotropy was controlled along the processing magnetic field. The annealing apparatus used in this study had an angle position accuracy of 0.5°. In this study, it was induced at around θ = 70°, which has a unique magnetic domain transition, as shown in Figure 1. Also, at θ = 90°, the easy magnetization axis element was fabricated and measured for comparison. The definition of the direction θ is shown in Figure 3. Figure 3 is a schematic of the clustered element and an explanation of the direction of the easy magnetization axis. The configuration of the element was the line arrangement adjacent to the many-body elements that had a mutual magnetic interaction with each other. The investigated many-body element in this study had a strip width of 20 μm, and it had the easy magnetization axis directions of θ = 71° and θ = 90°. The individual single strip has a hidden domain state, as shown in Figure 1a, when the element has the easy axis of θ = 70° and a 20 μm strip width.

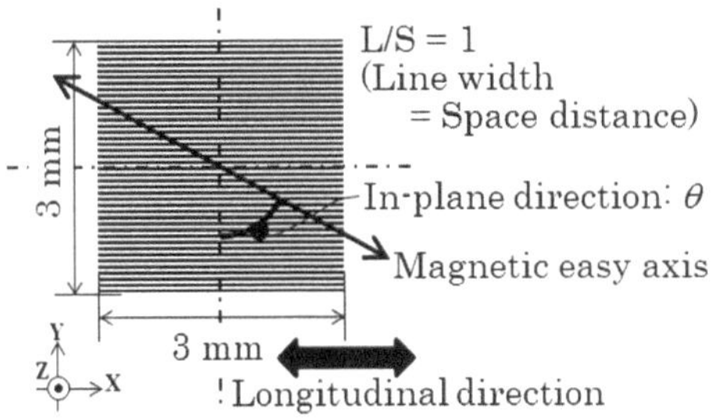

Figure 3. Schematic of the clustered many-body element.

The following is the detailed explanation of the element configuration: The adjacent many-body elements in this study had individual strips which each had lengths of 3000 μm; the width was the previously shown 20 μm, and the thickness was 2.1 μm. The element length was determined according to the knowledge obtained by the previous study, which stated that the element would have a residual domain at both ends of the strip. The length was determined to prevent the effect of the residual domain. In our previous report, the strip length, which was determined for a single element, was set as 2000 μm [38,39], and it indicated a suitable domain transition property. In this study, the individual strip was assembled to form a parallel line arrangement configuration with the line and space (L/S) as 20/20. The "line" corresponds to the strip width, and the "space" corresponds to the gap between the strips. The two-dimensional area of this assembled element was 3000 μm × 3000 μm, which was suitable for a measurement of magnetization loop using a vibrating sample magnetometer (VSM), and it was also suitable for an AC permeability measurement. The element of θ = 90° has a changing property between the single domain − and + [41]. This domain transition appears gradually with the changing of the external magnetic field. Under the condition of an easy axis of θ = 71°, which is the focusing condition of our continuous study, due to the existence of hidden, stable multi-domain state, the apparent domain transition is the same as that of θ = 90°, whereas the inclined LLD domain state appears when a certain distributed field is applied.

The photograph (Figure 4) shows an actual fabricated element. There are different width elements; each element was individually cut and divided using a dicing machine. The image of the element with a 20 μm width, on the left side of the photograph, is hardly recognizable in the individual strip, whereas the diffracted color image can be observed. In this study, the element with a 20 μm width was used, and the direction of the easy magnetization axis was made at θ = 71° and θ = 90°, using the magnetic field annealing.

Figure 4. Photograph of the measured elements on glass substrate. (The distance between the vertical lines on the bottom corresponds to 1 mm).

2.2. Measurement Apparatus

The measurements, which were carried out in this study, consisted of a magnetic domain observation with a control of the intensity of the applied distributed magnetic field. We also carried out an AC permeability measurement. The magnetic domain was observed by a Kerr microscope (BH-762PI-MAE, NEOARK Corporation, Tokyo, Japan). The AC permeability in the frequency from 10 kHz to 13 MHz was measured by a solenoid coil of my own making and an impedance analyzer (HP4192A) for measuring coil impedance. The homemade permeability measurement system was calibrated for the measurement value using the commercially available PMF-3000 (Ryowa Electronics Co., Ltd., Sendai, Japan), which is available for the frequency range from 10 MHz to 1.8 GHz. The details of this permeability measurement are explained in reference [43].

The following explanation is of the magnetic domain observation in this study, which was carried out with the control of the distributed magnetic field.

Figure 5 schematically explains an observation apparatus of the magnetic domain with the application of the distributed surface normal field. This apparatus was set up on a Kerr microscope, and the domain observation was carried out with the control of the intensity of the distributed field. The distributed magnetic field was generated using a combination of a soft magnetic steel rod and a winding coil. The distance between the thin-film element on a glass substrate and the end surface of the steel rod was set as 4 mm. The profile of the distributed magnetic field was designed using a finite element simulator to form a uniform distributed profile on the whole area of the square-shaped element. The center of the distributed field was placed at the center of the 2D square-shaped clustered element. Figure 6 shows a photograph of an actual observation system with the distributed field generator set at the observation focusing point.

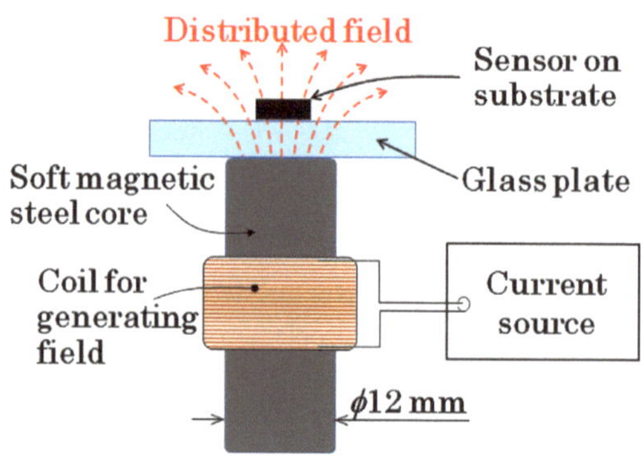

Figure 5. Schematic of the observation apparatus of magnetic domain with application of the distributed surface normal field.

Figure 7 shows a measured variation in the substrate's in-plane magnetic field as a function of the position of the central line of the element area. The horizontal axis indicates a longitudinal position, x, of the central strip element. The zero corresponds to the central position of the square area of the element. The vertical axis indicates the measured magnetic field in the direction of the in-plane X-direction. The measurement result shows that the in-plane field, which corresponds to the X-directional partial vector of the distributed normal field, has a linear variation, and it was controlled by an applied current of the winding coil. The gradient of the linear variation is defined as the intensity of the distributed field, as previously defined in this paper and our related papers. It should be noticed that the gradient in the case of the zero current was not zero; this was due to the residual magnetization of the steel core.

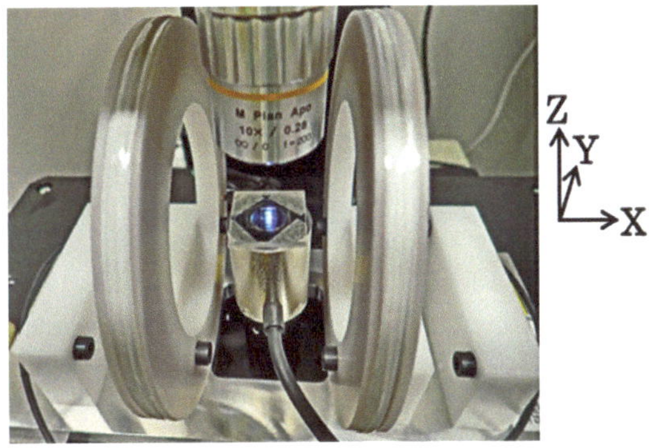

Figure 6. Photograph of the observation apparatus of magnetic domain with application of the distributed surface normal field.

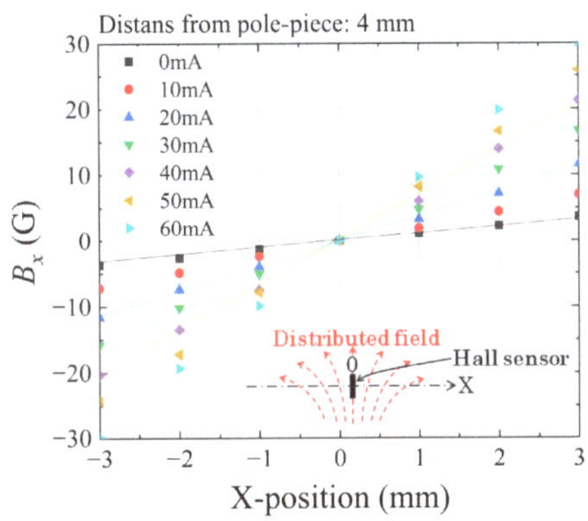

Figure 7. Measured variation in the in-plane magnetic field, B_x, as a function of the position along the length direction of the element, as a parameter of the coil current.

Figure 8 indicates a measured variation in the surface normal field in the Z-direction. The vertical axis shows a magnetic flux density B_z. The horizontal axis indicates a current introduced in the coil. The vertical field was confirmed as being almost uniform in the area of the clustered element, which was placed on the steel core. It is ordinally the case that the vertical magnetic field in the air is proportional to the current. It should be noticed that the Y-intercept value was not zero; this was also due to the residual magnetization of the steel core.

Figure 9 shows the dependence of the intensity of the distribution, dB_x/dx G/mm, on the coil current. Our apparatus, as shown in Figure 6, can generate the intensity of the distribution on the element plane, dB_x/dx, ranging from 0.89 G/mm to 9.89 G/mm, when the coil current is controlled from zero to 60 mA. In this case, the magnetic flux density in the Z-direction changes from 16 G to 142 G. According to the results of our previous study [38], the element is able to tolerate the surface normal field up to 600 G; then, the variation in the magnetic domain is not distorted by the surface normal field under the conditions of this study. Due to the existence of the residual magnetization of the steel core, which means the existence of the hysteresis of the MH loop, our experiment was carried out by keeping the limitation of the current range from zero to 60 mA, in order to keep the linear variation in the magnetic field.

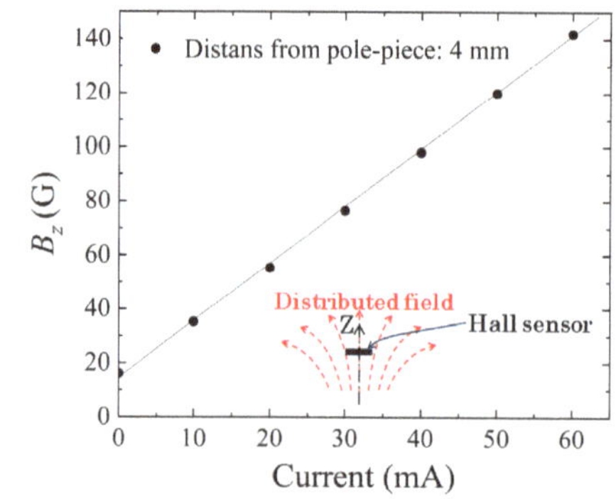

Figure 8. Measured variation in the surface normal field, B_z, as a function of the current of excitation coil.

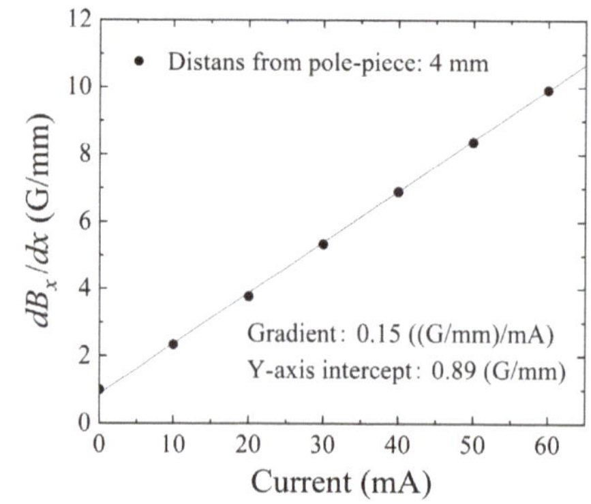

Figure 9. Dependence of the intensity of the distribution, dB_x/dx, on the current of excitation coil.

3. Results

In this section, the results of the magnetic domain variation for two different directions of the magnetization easy axis are shown. One is the case for θ = 71°, and the other is in the direction of almost θ = 90°. The element structures were the same as the structure of the adjacent narrow strips of a 3000 μm length, with L/S = 20 μm/20 μm, which were formed as a 3000 μm × 3000 μm square element. The quasi-steady magnetization curve, which was the MH loop measured by VSM, indicated, along the longitudinal direction, their soft magnetic property, which was reported in our previous papers [41,42]. According to the papers, the saturation magnetization of 0.96T appeared at 3.5 Oe of the applied magnetic field. The coercivity had almost the same value, 0.45 Oe, for both easy axis cases; then they had a soft magnetic property. Under these conditions, the magnetic domain of the individual strips in the clustered element forms a single domain with the magnetic moment along the bi-directional length direction. The number of strips having the same direction as the direction of the external magnetic field increases as a function of the strength of the field. Under the condition of a zero external field, both directional strips are mixed and form the equilibrium state. As shown in the introduction of this paper, the variation in the magnetic domain throughout the whole strip length was investigated, when a distributed surface normal field was applied; the correlation of the permeability of the alternating current magnetic field was discussed in the following study of this paper. Firstly, the condition of θ = 71° is dealt with based on the previously shown typical AC permeability of the element [43]. Secondly, a newly reported magnetic domain of almost θ = 90° and its AC permeability are reported and discussed, depending on the distributed surface normal field.

The magnetic domain was observed from the left end to the right end of the 10 strips placed in the middle of the width direction of the clustered many-body elements. The observation division and their corresponding index symbols are schematically shown in Figure 10. The observed domain photos are continuously connected from one end to the other end. The 3000 μm length elements were divided into six areas of photographs: L1, L2, and L3 from the left end and R1, R2, and R3 from the right end. Due to the observation procedure of the Kerr microscope, in which the magnetically saturated reference image must be obtained in advance of the observation of the domain in a certain external magnetic field, it should be noticed and explained that the structure of the magnetic domain of the same element in the neighboring photos is not always smoothly connected. In the following results, both the actual domain photographs and the schematic illustrations of the tendency of the domain variation are shown, depending on the strength of the distributed field.

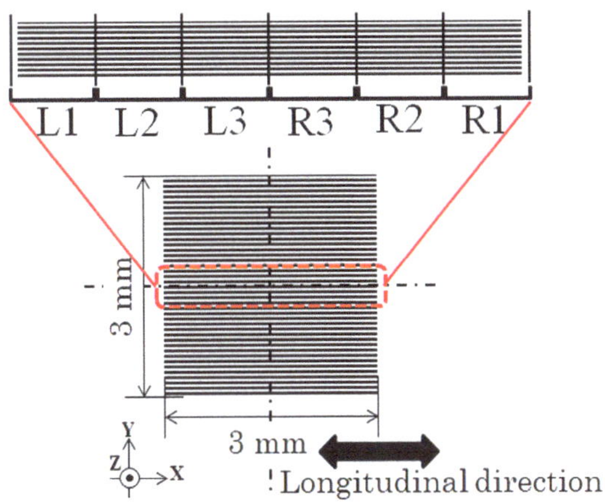

Figure 10. Observation division and corresponding index symbols of the following magnetic domain photographs.

3.1. *Magnetic Domain Variation for θ = 71° Element*

3.1.1. Domain Observation

Figure 11 shows a magnetic domain variation in the case of θ = 71°, as a function of the applied intensity of the distribution of the surface normal magnetic field. The whole lengths of the 10 strips were observed in the neighboring six photographs. The observation layout and the index symbols are shown in Figure 10.

Figure 11a indicates a case when the intensity of the distributed field was 0.89 G/mm. This condition was obtained when the current of the field generating coil was set at zero, due to the existing residual magnetization of the steel core. Figure 11a consists of contiguous bright and dark elements, which indicate an opposite directional single domain. There were some residual LLD stripe domains in the edge area of each strip. There are several photos which have an LLD domain around the middle area of the length of the strip, which is observed in the bottom strip of the photo. The observation magnetic field includes a slightly distributed field, due to the residual magnetization of the core; thus, it is assumed that this is the reason for the appearance of the LLD domain in the middle area of the length of the strip.

Figure 11c indicates a case when the intensity of the distributed field was 5.39 G/mm. In this case, the LLD stripe domain area extends toward the middle area of the strips, and it appears in every other strip in the observation area. L2 and R2 are the typical cases of this. In the neighborhood of the LLD element, a single-domain strip exists; the single domain of the bright ones appear in the L area, which is in the left moment direction, and the dark ones appear in the R area, which is in right moment direction.

Figure 11d indicates a case when the intensity of the distributed field was 9.89 G/mm. It is a relatively large intensity for the element. There is a single-domain area in the outermost edge areas, L1 and R1, and also around the middle position of the next edge areas, L2 and R2. The LLD stripe domain area shrinks and is limited to the vicinity of the edge area, the residual edge domain, and the middle area of strip.

In Figure 12, a schematic of the magnetic domain variation as a tendency of the whole clustered element, which corresponds to Figure 11, is shown.

Figure 12a shows a case which corresponds to Figure 11a. It is a domain structure without a distributed surface normal field for the magnetization easy axis of θ = 71°.

Figure 12b indicates a case which corresponds to Figure 11c, in which the LLD extends toward the middle area of the element strips. In this case, the single strip, which consists of the clustered element, is divided into two areas; one is the LLD stripe domain and the other is the single domain with a magnetic moment in the longitudinal direction. When the left side of the strip has the stripe domain, the right side has a single domain towards right-end direction. In the case that the right side of the strip has the stripe domain, the left side has a single domain towards the left-end direction.

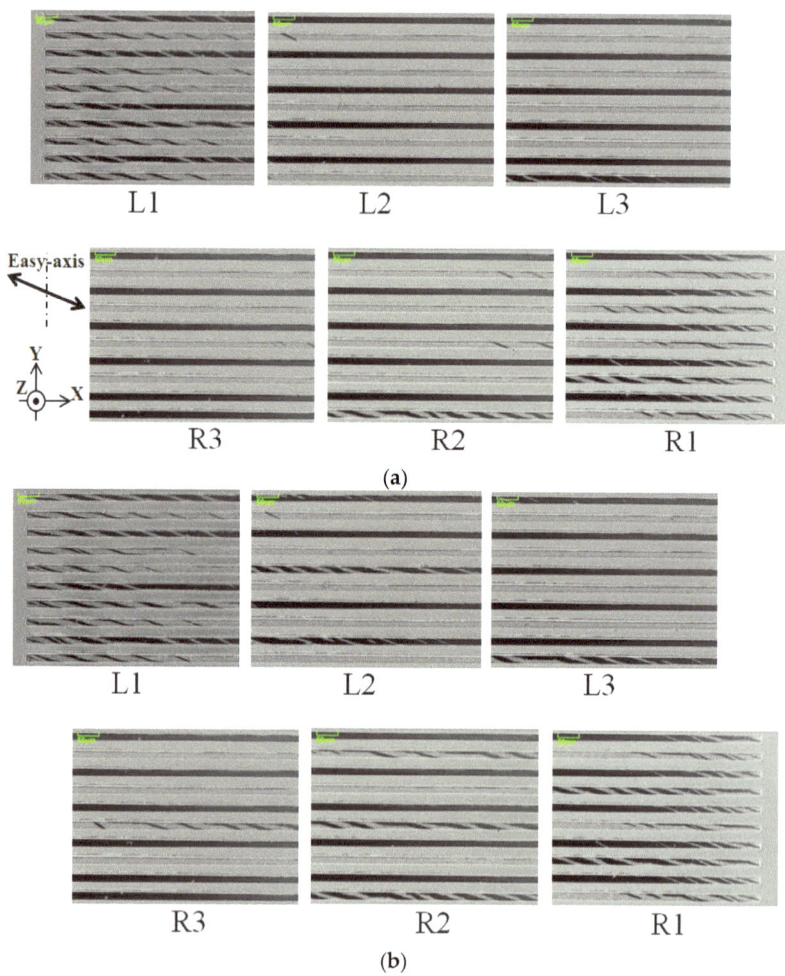

Figure 11. *Cont.*

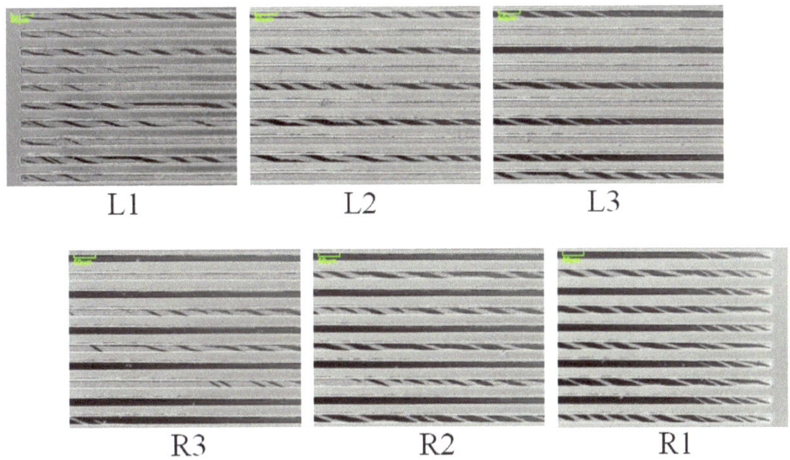

(c)

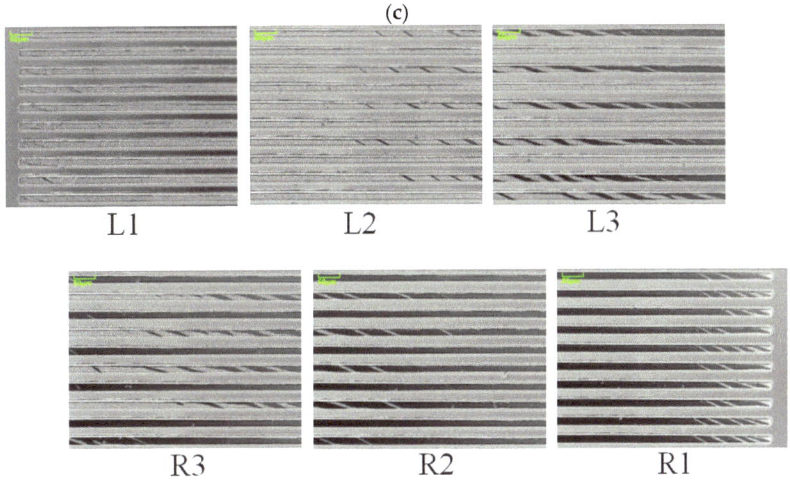

(d)

Figure 11. Magnetic domain variation in the case of θ = 71° as a function of applied intensity of the distribution of the surface normal magnetic field. (**a**) Intensity of the distributed field as 0.89 G/mm, (**b**) the result for 2.39 G/mm, (**c**) the result for 5.39 G/mm, and (**d**) for 9.89 G/mm.

Figure 12c indicates a case which corresponds to Figure 11d. In this case, the intensity of the distribution of the applied magnetic field has a certain large value, then the end-side area of the strip has single domain toward the outer direction, due to the magnetic force toward the outer direction. Even in this case, the edge residual stripe domain was still observed.

To summarize the magnetic domain variation in this sub-section, which is when the element has the magnetization easy axis direction of θ = 71°, the distributed field has the effect of changing the sectional ratio of the LLD stripe domain versus the single domain. Figure 13 shows a schematic explanation of this effect. As the intensity of the distributed field increases, the single-domain area gradually decreases, and the border of the single domain and the multi-domain moves toward the middle of the narrow strip. The red arrow in this figure indicates the variation direction of this phenomenon.

Figure 14 shows a variation in an area ratio of the LLD stripe domain as a function of the intensity of the distributed magnetic field. As the intensity increases, the area of the LLD stripe domain increases until the horizontal value of the intensity is up to 5.4 G/mm. The intensity of the distributed magnetic field, 5.4 G/mm, corresponds to Figures 11c and 12b. When the intensity of the distributed field increases more than 5.4 G/mm, both of the end areas of the narrow strip change to a single domain with a moment direction toward the outer direction in the length of each strip, as shown in Figures 11d and 12c.

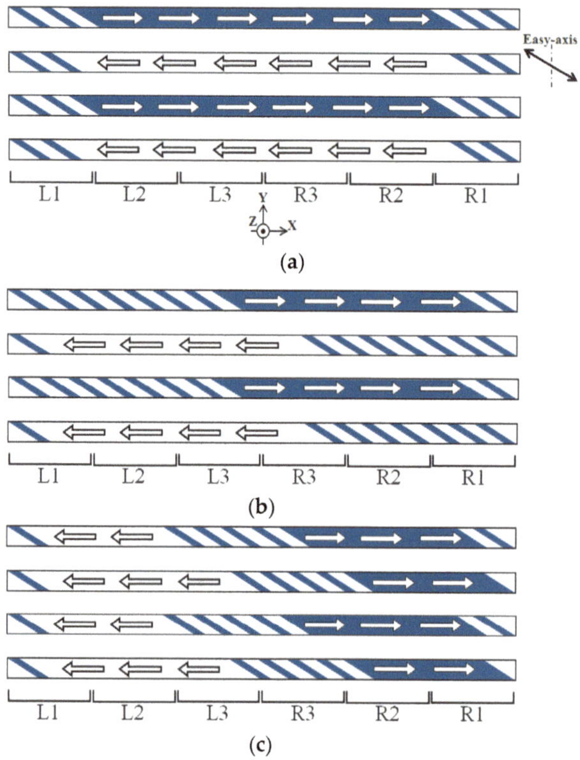

Figure 12. Schematic expression of magnetic domain as a tendency of whole clustered element with correspondence to Figure 11. (**a**) Schematic expression for Figure 11a, (**b**) for Figure 11c, and (**c**) for Figure 11d. The arrow indicates the magnetic moment direction, and the color show the moment direction, which is in much with Figure 11.

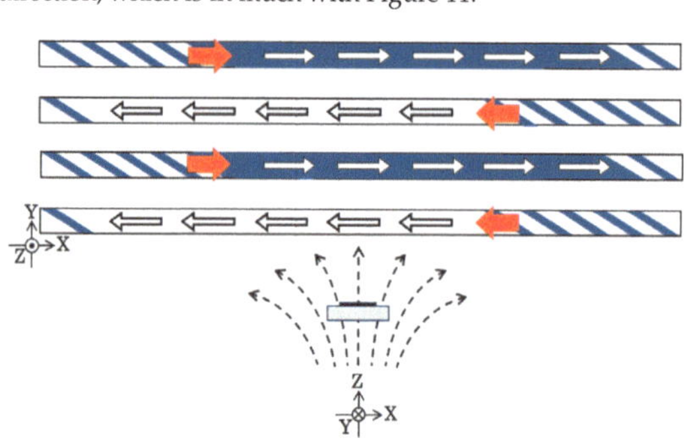

Figure 13. Schematic expression of the effect of the intensity of distributed magnetic field in the case of θ = 71°. The red arrow indicates the variation direction of the transition point.

3.1.2. Comparison of Alternating Current Permeability and Magnetic Domain

Figure 15 indicates our previously reported alternating current (AC) permeability of the clustered element with the magnetization easy axis of θ = 71° [43]. It indicates a measured complex permeability as a parameter of the intensity of the distributed field. Figure 15a shows a real part of the permeability, and Figure 15b shows an imaginary part of the permeability. Both of them are indicated as a function of the frequency of the AC magnetic field.

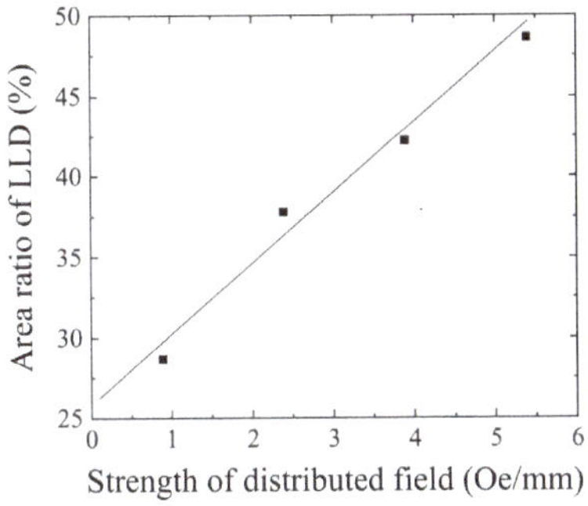

Figure 14. Variation in the area ratio of the LLD stripe domain as a function of the intensity of the distributed magnetic field.

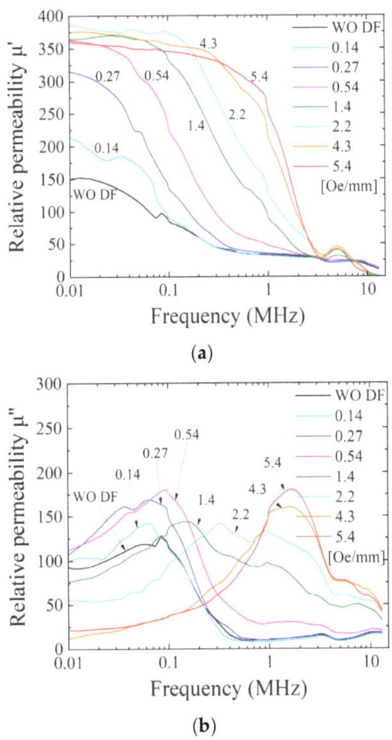

Figure 15. Previously reported measured variation in complex alternating current (AC) permeability as a parameter of the intensity of the distributed field [43]. (**a**) Real part permeability, and (**b**) imaginary part permeability.

Here, a discussion of the effect of the distributed magnetic field is provided, based on the comparison of the permeability and the magnetic domain variation.

As reported in the previous paper [43], which is reposted as Figure 15 in this paper, the real part of the permeability gradually increases its value and then saturates around the relative permeability of 370. This variation in the increment and saturation of the permeability appears initially at a low frequency; then, it extends its frequency range up to 1 MHz. The highest permeability and the wider frequency range were obtained when the applied intensity of the distributed field had the value of 5.4 Oe/mm. The value of the 5.4 Oe/mm distribution intensity was a value in which a wide multi-domain extension was observed experimentally in this paper, as shown in the previous subsection. It is assumed that the higher AC permeability in the frequency range of our experiment comes from the domain wall movement; thus, the extension of the multi-domain area is a reason for this phenomenon. The imaginary part of the AC permeability has a typical tendency, in which the peak frequency shifts from low frequency to high frequency. This phenomenon appeared in accordance with the variation in the distributed fields. The observation shows that there is almost a single domain in the element under the condition of the weak intensity of the distributed field, and the area ratio of the multi-domain increases and extends more widely when the intensity is 5.4 Oe/mm. It is well known that the imaginary permeability means a lost parameter. Then, the phenomenon would be assumed to come from a change in the resonant condition depending on the change in the domain state. A comparison between the AC permeability and the magnetic domain variation shows that it has a certain linear relation, which is assumed to come from a tendency of the proportional increment of the LLD stripe domain sectional ratio, as shown in Figure 14.

3.2. Magnetic Domain Variation for θ = 90° Element

In this subsection, an observation of the magnetic domain and a discussion on the comparison of the permeability, as a function of the intensity of the distributed magnetic field, is given for the element which has the magnetization easy axis of θ = 90°. The experimental methods and the development of the argument are the same as those of the previous subsection for the different easy axis angle, θ = 71°. In our previous study [41], the clustered element with a θ = 90° easy axis has a directional switching property in the magnetic domain. When the distributed magnetic field does not apply to the element, the direction of the magnetic moment in each magnetic narrow strip in the clustered element has the single domain along the length direction and its direction is switched by applying a certain value of the in-plane uniform external magnetic field. In this subsection, the effect of the distributed field for the element of θ = 90° is investigated.

3.2.1. Domain Observation

Figure 16 shows a magnetic domain variation in the case of θ = 90°, as a function of the applied intensity of the distributed magnetic field. Figure 16a indicates a case when the intensity of the distributed field was 0.89 G/mm, which is the value of the coil current at zero. Figure 16a consists of contiguous bright and dark elements, which indicate a single domain with an opposite direction. There were typical residual magnetic domains in the edge area of each strip, which consisted of long triangular domain lengthening along the length direction of the strip, as if the domain wall divided the magnetic strip into two along the length direction.

Figure 16c indicates a case when the intensity of the distributed field was 5.39 G/mm. In this case, the multi-domain area extends towards the inner area of the strip. There are several patterns of multi-domain in this area, which are different from the previous θ = 71° case. Figure 17 shows the schematics of the typical multi-domain structure, which is observed in Figure 16c. These are typical domain patterns for the multi-domain in the case of θ = 90° with the application of the distributed field. The observed index number of the photo and the line number are indicated below each schematic. The multi-domain appeared in the border area between the single domain+ and the single domain−. The multi-domain structure is not clear compared with the previous case of θ = 71°; consequently, a quantitative discussion of the area ratio is difficult to carry out.

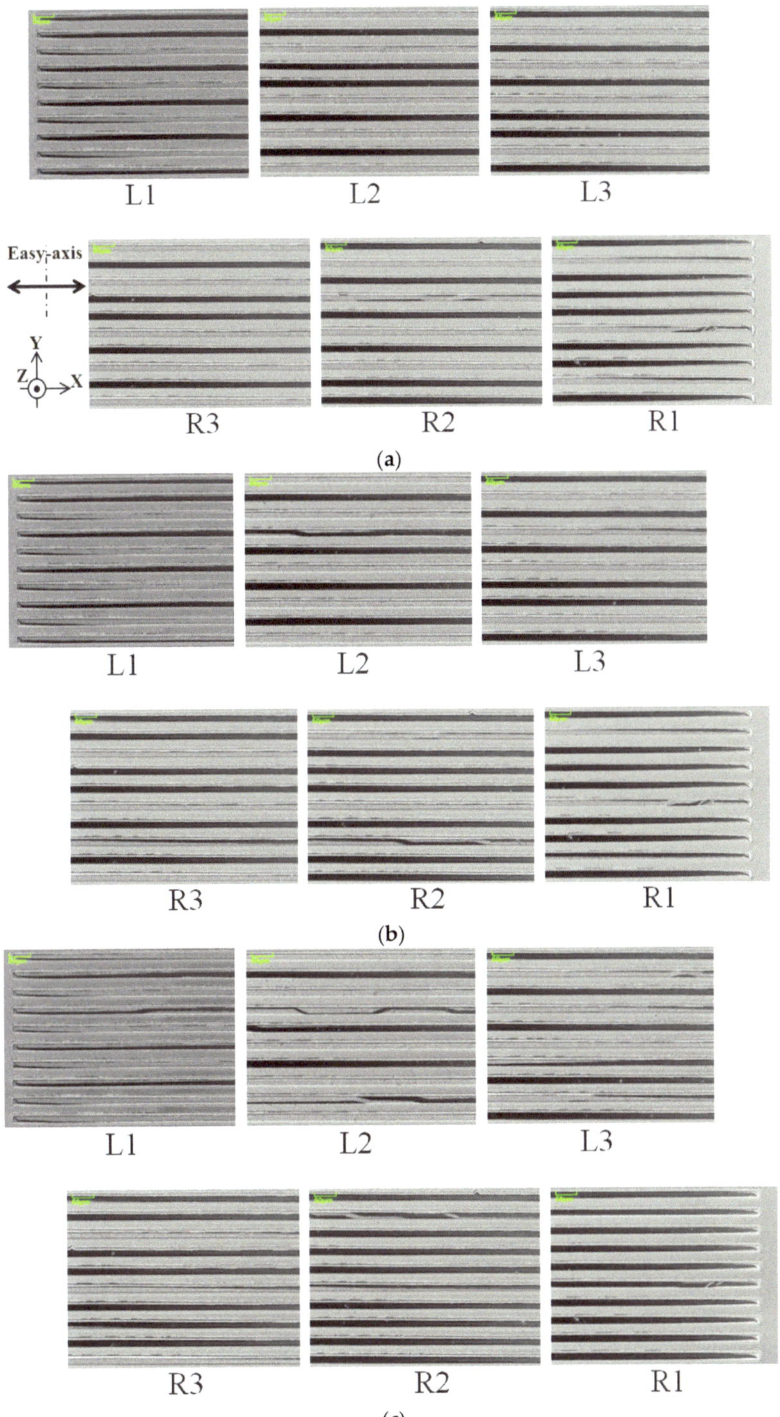

Figure 16. *Cont.*

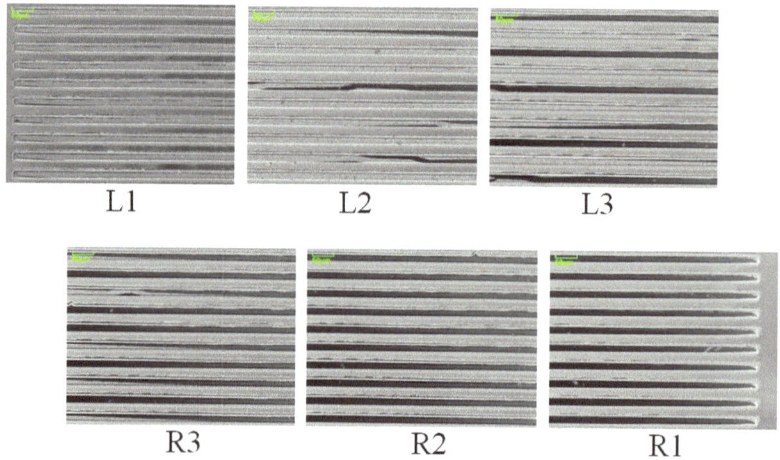

(**d**)

Figure 16. Magnetic domain variation in the case of θ = 90° as a function of applied intensity of the distribution of the surface normal magnetic field. (**a**) Intensity of the distributed field as 0.89 G/mm, (**b**) the result for 2.39 G/mm, (**c**) the result for 5.39 G/mm, and (**d**) for 9.89 G/mm.

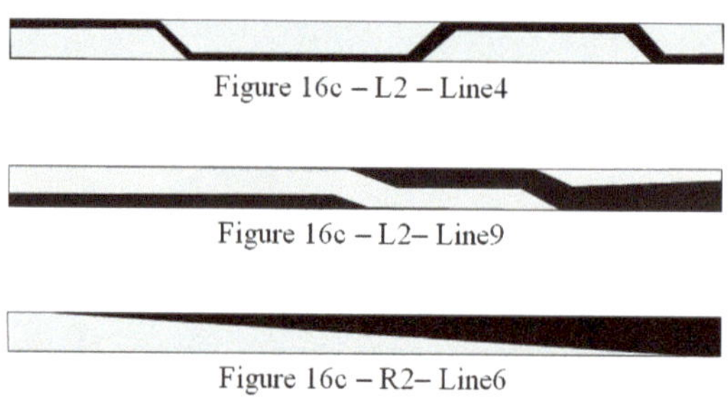

Figure 17. Schematics of the typical multi-domain structure in case of θ = 90°, as indicated in Figure 16.

Figure 16d indicates a case when the intensity of the distributed field was 9.89 G/mm. It is a relatively large intensity for the element. There is a whole single domain in the near area of the edge of L1 and R1. In this case, the multi-domain area seems to shrink and is limited in the vicinity of the middle of the strip.

In Figure 18, a schematic of the magnetic domain variation as a tendency of the whole clustered element, corresponding to Figure 16, is shown. Figure 18a shows a case which corresponds to Figure 16a. It is a domain structure without a distributed surface normal field for the magnetization easy axis of θ = 90°.

Figure 18b indicates a case which corresponds to Figure 16c, in which the multi-domain area extends toward the middle area of the element strips. In this case, the single strip, which consists of the clustered element, is divided into three areas; the first one is the single domain with a magnetic moment in the longitudinal direction +; the second one is the single domain in the longitudinal – direction; and the third one is their border area with the abovementioned typical multi-domain structure in Figure 17. When the left side area of the strip has the single domain −, the right side has single domain + towards the right end direction. The multi-domain area appears between them. In Figure 18, the multi-domain area is indicated as the bottom pattern of Figure 17, the triangular multi-domain, to simplify the expression.

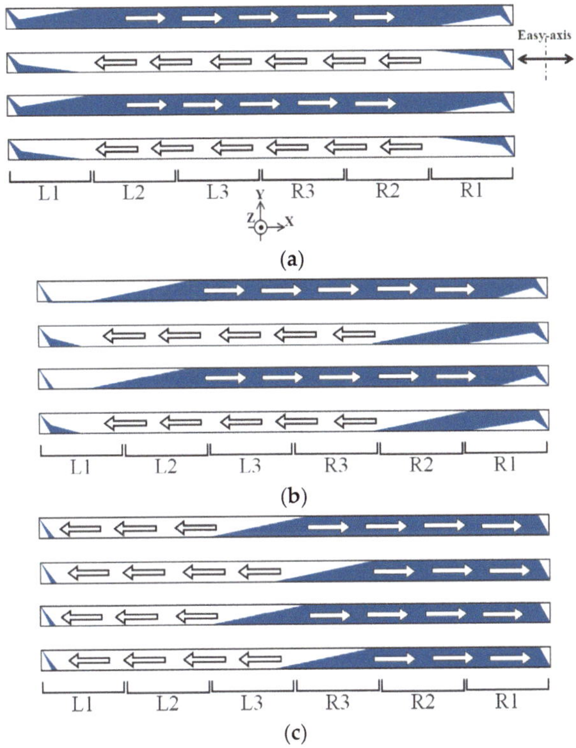

Figure 18. Schematic expression of magnetic domain as a tendency of whole clustered element with correspondence to Figure 16. (**a**) Schematic expression for Figure 16a, (**b**) for Figure 16c, and (**c**) for Figure 16d. The white arrow indicates the magnetic moment direction.

Figure 18c indicates a case which corresponds to Figure 16d. In this case, the intensity of the distribution of the applied magnetic field has a certain large value; then, the end-side area of the strip has a single domain toward the outer direction, due to the magnetic force toward the outer direction. Even in this case, a small edge residual triangular domain was observed, and the multi-domain area also exists between the opposite single-domain areas.

To summarize the magnetic domain variation in this subsection, which concerns the element with an easy axis of θ = 90°, the distributed field has an effect of generating the multi-domain area between the single-domain + and − areas. Figure 19 shows a schematic explanation of this effect. As the intensity of the distributed field increases, the multi-domain area appears around the border area of the single-domain + and − areas. The multi-domain border area moves toward the middle of the narrow strip, as the intensity of the distributed surface normal magnetic field increases. The red arrow in this figure indicates the variation direction of this phenomenon. The multi-domain area seems to widen; then, the area ratio would increase when the intention of the distributed field was around 5.4 G/mm; however, a quantitative discussion is difficult due to the uncertainty of the discrimination of the multi-domain existence.

3.2.2. Comparison of Alternating Current Permeability and Magnetic Domain

Figure 20 indicates an AC permeability of the clustered element with the magnetization easy axis of almost θ = 90°. It indicates a measured complex permeability as a parameter of the intensity of the distributed field; the real part is in Figure 20a, and the imaginary part is in Figure 20b. Both of them are shown as a function of the frequency of the AC magnetic field.

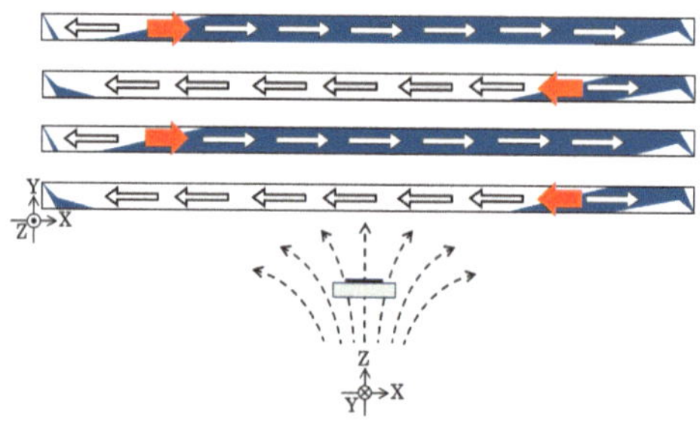

Figure 19. Schematic expression of the effect of the intensity of distributed magnetic field in the case of θ = 90°. The red arrow indicates the variation direction of the transition point.

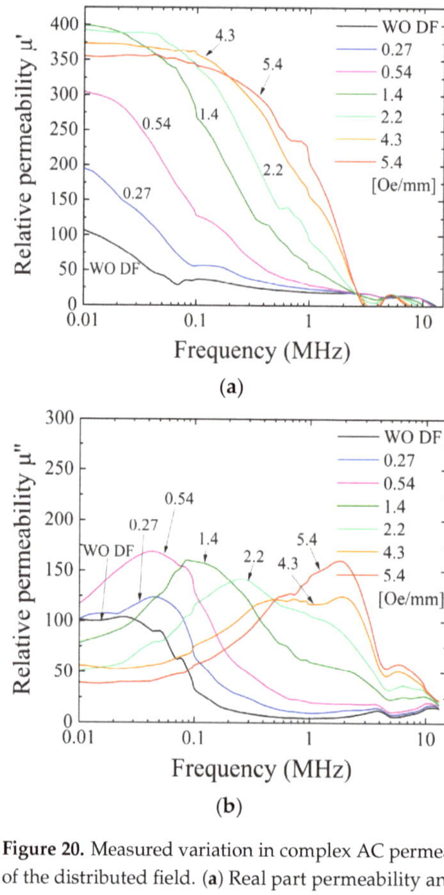

Figure 20. Measured variation in complex AC permeability for θ = 90° as a parameter of the intensity of the distributed field. (**a**) Real part permeability and (**b**) imaginary part permeability.

It is mentioned that the variation tendency is the same as that of the θ = 71° element, although it has a different domain transition property, especially for the structure of the multi-domain. The point of difference between them is the existence of the LLD stripe domain for θ = 71°, but the point of these phenomena is assumed to be the appearance of the multi-domain area and the enhancement of the effect of the domain wall movement. A comprehensive and detailed discussion is provided in the following discussion section.

4. Discussion

In this study, the effect of the distributed field on the magnetic domain variation for the clustered soft magnetic strips with a certain inclined magnetization easy axis was investigated. The observed variation in the magnetic domain was compared with the permeability variation, in order to reveal the effect of the distributed field on the AC permeability. The element, which has the in-plane inclined magnetization easy axis of θ = 71° and θ = 90°, and also has the width of the single strip of the clustered element as 20 μm, was experimentally confirmed as having low AC permeability in the frequency range from 10 kHz to 10 MHz, when it has a single domain structure in each strip. In the case of the distributed magnetic field when a suitable value is applied to the element, the original single domain with the magnetic momentum along the length direction changes to the multi-domain state. The sectional ratio of the multi-domain area changes as a function of the intensity of the distributed field, which was clearly observed in the case of θ = 71°. For the different easy axis cases, which were between θ = 71° and θ = 90°, the experimentally observed domain structures of the multi-domain were quite different, whereas the tendency of the permeability variation was almost the same under these two conditions. There is a possibility of the existence of an effect of the enhancement of the domain wall movement caused by the distributed field and the inclined easy axis in a certain domain structure.

Additionally, It should be mentioned that a typical domain transition was experimentally confirmed for the θ = 71° clustered element, which would seem to be the effect of the existence of the hidden stable multi-domain state. It was recently reported [44] that a small soft magnetic particle placed on the element was able to induce a domain transition from a single domain to a multi-domain. It was explained that a distributed magnetic field generated by the dipole magnetization of the particle enhanced the domain transition.

In this discussion, a comprehensive overview of the phenomenon concerning the element with the hidden multi-domain state and its neighboring condition is summarized. The future investigation of these phenomena is also discussed.

Firstly, the phenomenon of low AC permeability is discussed for the clustered element with the inclined easy axis. As explained in references [42,43], the low permeability typically appeared for the narrow strips, which have a width of 20 μm and the easy axis direction of more than θ = 70°. In the previous study [42], the parameters of the experiment ranged from strip widths of 20 μm to 100 μm and from θ = 61° to θ = 90° in the easy axis direction. The single domain in the strips, which causes the low AC permeability, was obtained when the easy axis was larger than θ = 71° and also when the width was 20 μm. The results are summarized in Figure 21. The Figure shows a matrix expression for the element width W and the easy axis direction θ. The "*M*" indicates a multi-domain state, and the "*S*" indicates a single-domain state. The hatched area indicates the low AC permeability conditions. Despite the single domain, the θ = 90° and the W = 50 μm element has a high-permeability property. The triangular-shaped residual magnetic domain in the edge area of the strip appeared to be the same as that of the 50 μm width and the 20 μm width; therefore, the element width would be expected to be the key parameter for the low AC permeability. The low permeability was not restricted by the existence of the hidden multi-domain state. The element with the hidden state is included in the condition range of the single domain because the magnetic domain of the condition apparently forms the single domain, and the strip width is narrow enough.

In the case of the narrow single domain, with a low AC permeability, it is possible to control the permeability by controlling the intensity of the distributed magnetic field within the surface normal magnetic field. It would be possible to control the permeability of the easy axis ranging from θ = 71° to θ = 90° without depending on the re-constructed multi-domain structure, which was shown in our study, including that of this paper. As a result of this study, the enhancement of permeability with application of the distributed magnetic field is caused by the formation of a multi-domain structure; this is assumed to be the effect of the enhancement of the domain wall movement of these domains.

	W= 20 μm	50 μm	100 μm
$\theta = 61°$	*M*	*M*	*M*
71°	***S***	*M*	*M*
90°	***S***	***S***	*M*

Figure 21. Matrix expression of magnetic domain state of the clustered element depending on element width and easy-axis direction θ. (M: multi-domain (LLD), S: single domain). The shaded part indicates the low AC permeability conditions.

It is well known that the magnetization process of soft magnetic materials consists of both the domain wall movement and the magnetization rotation. The high-permeability property of soft magnetic material is mainly caused by the domain wall movement. When the frequency of the external magnetic field increases, the domain wall movement is restricted by the generation of a microscopic eddy current, which appears in the vicinity of the domain wall. In this case, the permeability is dominated by the magnetization rotation [11]. The study in this paper probably includes a transition frequency range of these phenomena. At 10 kHz, which is the lowest frequency of this study, the domain wall movement would be dominant, due to the observation of the sensitivity of the permeability to the domain existence in the element. At around 10 MHz, which is the highest frequency of this study, a typical resonant profile of the complex permeability was observed [42], and the permeability decreased suddenly in the frequency occurring here. In the case of several hundred MHz, the domain walls are stuck in a certain position, which is the position formed initially in the stable energy state. In this case, the domain structure is observable even in the case that a high-frequency current is induced in the magnetic element [47]. In this study, the magnetization process is dominated by the domain wall movement; therefore the single-domain structure has low permeability and the multi-domain structure has high permeability; this comes from the existence of movable domain walls in the element.

The difference in the effect of the easy axis direction for the permeability variation generated by the intensity of the distributed field mainly appeared in an intensity value which increased the real permeability from a low to a high value. In the case of 71°, this is the increase from 0.14 to 0.27 Oe/mm given in Figure 15a. In the case of 90°, this is the increase from 0.27 to 0.54 Oe/mm given in Figure 20a. The inclined easy axis direction which would have hidden the stable state, has a lower value, almost half of the intensity of the distributed field for making the transient phenomenon of permeability. It is assumed that it comes from the easiness of generating the multi-domain structure for the 71° easy axis rather than the 90° one. It is experimentally observed in the domain photo as the portion of the multi-domain area, between Figures 11b and 16b, which is the condition of the intensity just after the permeability increase.

Concerning the formation and transition of the magnetic domain that appeared in the soft magnetic narrow strip and the magnetic energy estimation of these different states, the consideration of the threshold barrier that needs to be overcome to make the change in the domain state is important. It was discussed in my previous paper [48] with regard to the case of a single narrow strip. Based on the previous knowledge, the effect of the distributed field should be discussed. In this paper, the formation of the multi-domain for a 90° easy axis indicates that there is another multi-domain stable state existing in the distributed field, other than the ordinally longitudinal strip domain, such as that in Figure 17. Because of this, the energy difference between the initial domain state and the final state would be changed by applying the distributed field. It would be the reason for the different rising point of the permeability for the different easy axis directions. In order to make a strict discussion, a magnetic energy simulation is needed; this would be the subject of a future investigation.

For a brief comment about the effect of distance between the strips, a consideration was made in my previous paper based on magnetic field simulation [41]. It was mentioned that "There are two types of mutual interaction. One is a strong short-range effect of magnetic field generated at the edge of a single narrow strip, and the other is a cumulative effect of the widely spread weak field which come from the other adjacent many-body elements". In order to investigate the distance effect, we have to consider both the short-range field and the cumulative widely spread field, based on the previous consideration.

It is noticed that a newly reported typical phenomenon, which appears in the element with the hidden multi-domain state of $\theta = 71^\circ$ has been recently reported [44]. It is an appearance of the LLD multi-domain state, which happens when a small magnetic particle is placed just above the strip and controls the applied surface normal magnetic field. The parallel surface normal field makes the particle magnetize, and it generates a magnetic dipole field. The magnetic dipole of the magnetized particle generates a distributed magnetic field on the magnetic strip of the element; then, it has a possibility to enhance the re-construction of the LLD stripe domain. This phenomenon is assumed to be a typical one for the element with the hidden stable state. The appearance of the transited domain of a certain strip would possibly be detected by an electrical parameter such as impedance or inductance. The detection and discrimination of the appearance of the domain transition are two of the objectives of this paper for the sensor application of this phenomenon.

For further understanding of this physical phenomenon, a more detailed study should be conducted based on the knowledge of this paper and our related papers.

5. Conclusions

The correspondence of the AC permeability and the structure of magnetic domain as a function of the intensity of the distributed field was investigated for the element with an in-plane inclined magnetization easy axis. As a result, the extension of the domain area of the Landau–Lifshitz-like multi-domain (LLD) on the clustered narrow strips was observed for the element with $\theta = 71^\circ$, as a function of the intensity of the distributed magnetic field, which ranges from 0.89 G/mm to 5.39 G/mm and has a certain linear relation between the permeability variations. In the case of the magnetization easy axis of $\theta = 90^\circ$, the multi-domain was not the LLG stripe domain, but some typical structures of multi-domain appeared. This multi-domain extension was compared with the permeability variation, and it was revealed that they have a certain linear relation, within the intensity of the distributed field from 0.89 G/mm to 5.39 G/mm, which is the same property as $\theta = 71^\circ$. Based on the investigation, the comprehensive tendency of the domain transition and permeability variation was discussed, including the neighboring fabrication conditions for different strip widths and also for different easy axis angles, which surround the target element with the hidden stable multi-domain state of $\theta = 71^\circ$ and $W = 20$ μm. The clustered narrow strip element with the single domain, which includes the element with the hidden state, has a possibility to control the AC permeability by controlling the intensity of the distributed magnetic field. Fundamentally, the result leads to the application of this phenomenon to a tunable inductor, electromagnetic shielding, or a sensor for detecting and memorizing the existence of a distributed magnetic field generated in the vicinity of a magnetic nanoparticle.

Funding: This research received no external funding.

Informed Consent Statement: Not applicable.

Data Availability Statement: Data are contained within the article.

Acknowledgments: The author gratefully acknowledges the staff of ITIM (Industrial Technology Institute, Miyagi Prefectural Government) for their kind support of this research activity.

Conflicts of Interest: The author declares no conflicts of interest.

References

1. Kittel, C. Theory of the Structure of Ferromagnetic Domains in Films and Small Particles. *Phys. Rev. B* **1946**, *70*, 965–971. [CrossRef]
2. Aharoni, A. *Introduction to the Theory of Ferromagnetism*, 2nd ed.; Oxford University Press: Oxford, UK, 1996; ISBN 0 19 850809.
3. Venkat, G.; A Allwood, D.; Hayward, T.J. Magnetic domain walls: Types, processes and applications. *J. Phys. D Appl. Phys.* **2024**, *57*, 063001. [CrossRef]
4. Atkinson, D.; Allwood, D.; Xiong, G.; Cooke, M.; Faulkner, C.; Cowburn, R. Magnetic domain-wall dynamics in a submicrometre ferromagnetic structure. *Nat. Mater.* **2003**, *2*, 85–87. [CrossRef] [PubMed]
5. Tsoi, M.; Fontana, R.E.; Parkin, S.S.P. Magnetic domain wall motion triggered by an electric current. *Appl. Phys. Lett.* **2003**, *83*, 2617–2619. [CrossRef]
6. Grollier, J.; Boulenc, P.; Cros, V.; Hamzić, A.; Vaurès, A.; Fert, A.; Faini, G. Switching a spin valve back and forth by current-induced domain wall motion. *Appl. Phys. Lett.* **2003**, *83*, 509–511. [CrossRef]
7. Kläui, M.; Vaz, C.A.F.; Bland, J.A.C.; Wernsdorfer, W.; Faini, G.; Cambril, E.; Heyderman, L.J.; Nolting, F.; Rüdiger, U. Controlled and Reproducible Domain Wall Displacement by Current Pulses Injected into Ferromagnetic Ring Structures. *Phys. Rev. Lett.* **2005**, *94*, 106601. [CrossRef]
8. Hayashi, M.; Thomas, L.; Moriya, R.; Rettner, C.; Parkin, S.S.P. Current-Controlled Magnetic Domain-Wall Nanowire Shift Register. *Science* **2008**, *320*, 209–211. [CrossRef]
9. Allwood, D.A.; Xiong, G.; Cooke, M.D.; Faulkner, C.C.; Atkinson, D.; Vernier, N.; Cowburn, R.P. Submicrometer Ferromagnetic NOT Gate and Shift Register. *Science* **2002**, *296*, 2003–2006. [CrossRef]
10. Yamaguchi, M.; Suezawa, K.; Arai, K.I.; Takahashi, Y.; Kikuchi, S.; Shimada, Y.; Li, W.D.; Tanabe, S.; Ito, K. Microfabrication and characteristics of magnetic thin-film inductors in the ultrahigh frequency region. *J. Appl. Phys.* **1999**, *85*, 7919–7922. [CrossRef]
11. Chikazumi, S. *Physics of Ferromagnetism*, 2nd ed.; Graham, C.D., Jr., Ed.; Oxford University Press: Oxford, UK, 2009; ISBN 9780199564811.
12. Williams, H.J.; Bozorth, R.M.; Shockley, W. Magnetic Domain Patterns on Single Crystals of Silicon Iron. *Phys. Rev. B* **1949**, *75*, 155–178. [CrossRef]
13. Shilling, J.; Houze, G. Magnetic properties and domain structure in grain-oriented 3% Si-Fe. *IEEE Trans. Magn.* **1974**, *10*, 195–223. [CrossRef]
14. Zaykova, V.A.; Dragoshanskiy, Y.N.; Zhakov, S.V.; Filipov, B.N. On the role of domain structure in the formation of electromagnetic losses in Fe-3% Si crystals. *Fiz. Met. Metalloved.* **1977**, *43*, 979.
15. Maricic, A.; Spasojevic, M.; Arnaut, S.; Minic, D.; Ristic, M. The effect of structural changes on magnetic permeability of amorphous powder Ni80Co20. *Sci. Sinter.* **2008**, *40*, 303–309. [CrossRef]
16. Viau, G.; Fiévet-Vincent, F.; Fiévet, F.; Toneguzzo, P.; Ravel, F.; Acher, O. Size dependence of microwave permeability of spherical ferromagnetic particles. *J. Appl. Phys.* **1997**, *81*, 2749–2754. [CrossRef]
17. Kuo, Y.-M.; Duh, J.-G. Effects of the stacking levels on permeability characteristics and soft magnetic properties in FeCo-HfAlO/AlOx multilayered films. *Mater. Lett.* **2010**, *64*, 2167–2169. [CrossRef]
18. Yang, J.; Zhang, X.; Shi, M.; Yuan, S.; Wang, L.; Han, B. Modeling and Application of Magnetic Shaking for Improving Permeability and Performance of Magnetic Shields. *IEEE Trans. Ind. Electron.* **2023**, *early access*, 1–11. [CrossRef]
19. Williams, H.J.; Shockley, W.; Kittel, C. Studies of the Propagation Velocity of a Ferromagnetic Domain Boundary. *Phys. Rev. B* **1950**, *80*, 1090–1094. [CrossRef]
20. Berger, L. Motion of a magnetic domain wall traversed by fast-rising current pulses. *J. Appl. Phys.* **1992**, *71*, 2721–2726. [CrossRef]
21. Makar, J.; Tanner, B. The in situ measurement of the effect of plastic deformation on the magnetic properties of steel: Part II—Permeability curves. *J. Magn. Magn. Mater.* **1998**, *187*, 353–365. [CrossRef]
22. Takezawa, J. Yamasaki, Dynamic domain observation in narrow thin films. *IEEE Trans. Magn.* **2001**, *37*, 2034–2037. [CrossRef]
23. Sun, Z.; Kuramochi, H.; Mizuguchi, M.; Takano, F.; Semba, Y.; Akinaga, H. Magnetic properties and domain structures of FeSiB thin films. *Surf. Sci.* **2004**, *556*, 33–38. [CrossRef]
24. Shin, J.; Kim, S.; Hashi, S.; Ishiyama, K. Analysis of thin-film magneto-impedance sensor using the variations in impedance and the magnetic domain structure. *J. Appl. Phys.* **2014**, *115*, E507. [CrossRef]
25. Morikawa, T.; Nishibe, Y.; Yamadera, H.; Nonomura, Y.; Takeuchi, M.; Taga, Y. Giant magneto-impedance effect in layered thin films. *IEEE Trans. Magn.* **1997**, *33*, 4367–4372. [CrossRef]
26. Takezawa, M.; Kikuchi, H.; Yamaguchi, M.; Arai, K. Miniaturization of high-frequency carrier-type thin-film magnetic field sensor using laminated film. *IEEE Trans. Magn.* **2000**, *36*, 3664–3666. [CrossRef]
27. Zhong, Z.; Zhang, H.; Jing, Y.; Tang, X.; Liu, S. Magnetic microstructure and magnetoimpedance effect in NiFe/FeAlN multilayer films. *Sens. Actuators A Phys.* **2008**, *141*, 29–33. [CrossRef]
28. Pasynkova, A.A.; Lepalovskii, V.N.; Svalov, A.V.; Volochaev, M.N.; Kurlyandskaya, G.V. Magnetic Properties and Peculiarities of the Domain Structure of Multilayer Permalloy-Based GMI Elements. *Phys. Met. Met.* **2022**, *123*, 715–720. [CrossRef]
29. Kikuchi, H.; Ajiro, N.; Yamaguchi, M.; Arai, K.; Takezawa, M. Miniaturized high-frequency carrier-type thin-film magnetic field sensor with high sensitivity. *IEEE Trans. Magn.* **2001**, *37*, 2042–2044. [CrossRef]
30. Shin, J.; Miwa, Y.; Kim, S.H.; Hashi, S.; Ishiyama, K. Observation of the Magnetic Properties According to Changes in the Shape of Thin-Film Giant Magnetoimpedance Sensor. *IEEE Trans. Magn.* **2014**, *50*, 4005603. [CrossRef]

31. Wang, M.; Cao, Q.; Liu, S.; Wang, X.; Zhang, D.; Fang, Y.; He, X.; Chang, C.; Tao, Q.; Jiang, J. The effect of thickness and annealing treatment on microstructure and magnetic properties of amorphous Fe-Si-B-P-C thin films. *J. Non-Crystalline Solids* **2019**, *505*, 52–61. [CrossRef]
32. Svalov, A.V.; Aseguinolaza, I.R.; Garcia-Arribas, A.; Orue, I.; Barandiaran, J.M.; Alonso, J.; Fernández-Gubieda, M.L.; Kurlyandskaya, G.V. Structure and Magnetic Properties of Thin Permalloy Films Near the "Transcritical" State. *IEEE Trans. Magn.* **2010**, *46*, 333–336. [CrossRef]
33. Kikuchi, H.; Sumida, C. Observation of Static Domain Structures of Thin-Film Magnetoimpedance Elements with DC Bias Current. *IEEE Trans. Magn.* **2019**, *55*, 4001405. [CrossRef]
34. Kilic, U.; Ross, C.A.; Garcia, C. Tailoring the Asymmetric Magnetoimpedance Response in Exchange-Biased Ni-Fe Multilayers. *Phys. Rev. Appl.* **2018**, *10*, 034043. [CrossRef]
35. Dong, C.; Chen, S.; Zuyao, T.H. A simple model of giant magneto-impedance effect in amorphous thin films. *J. Magn. Magn. Mater.* **2002**, *250*, 288–294. [CrossRef]
36. Matsuo, T. Magnetization process analysis using a simplified domain structure model. *J. Appl. Phys.* **2011**, *109*, 07D332. [CrossRef]
37. Nakai, T. Magnetic Domain Observation of Stepped Magneto-Impedance Sensor with Subjecting to Normal Magnetic Field. *Proc. IEEE Sens.* **2015**, *2015*, 1461–1464.
38. Nakai, T. Magnetic Domain of Stepped Magneto-Impedance Sensor Controlled by a Normal Magnetic Field. *IEEJ Trans. Sens. Micromachines* **2018**, *138*, 23–29. [CrossRef]
39. Nakai, T. Magnetic domain transition controlled by distributed normal magnetic field for stepped magneto-impedance sensor. *Int. J. Appl. Electromagn. Mech.* **2019**, *59*, 105–114. [CrossRef]
40. Tejima, S.; Ito, S.; Mifune, T.; Matsuo, T.; Nakai, T. Magnetization analysis of stepped giant magneto impedance sensor using assembled domain structure model. *Int. J. Appl. Electromagn. Mech.* **2016**, *52*, 541–546. [CrossRef]
41. Nakai, T. Magnetic Domain Transition of Adjacent Narrow Thin Film Strips with Inclined Uniaxial Magnetic Anisotropy. *Micromachines* **2020**, *11*, 279. [CrossRef]
42. Nakai, T. High-Frequency Permeability of Adjacent Narrow Strips Having an Inclined Magnetic Anisotropy. *SPIN* **2023**, *13*, 2340003. [CrossRef]
43. Nakai, T. Permeability of Clustered Soft Magnetic Narrow Strips Controlled by a Surface Normal Magnetic Field. *IEEE Trans. Magn.* **2023**, *59*, 2001105. [CrossRef]
44. Nakai, T. Magnetic domain transition of clustered soft magnetic narrow strips caused by a magnetic small particle. *Proc. IEEE Sens.* **2023**, *2023*, 1934.
45. Landau, L.; Lifshitz, E. On the theory of the dispersion of magnetic permeability in ferromagnetic bodies. *Physik. Z. Sowjetunion* **1935**, *8*, 153–169.
46. Hubert, A.; Schafer, R. *Magnetic Domains: The Analysis of Magnetic Microstructures*; Springer: Berlin/Heidelberg, Germany; New York, NY, USA, 2009.
47. Nakai, T. Sensitivity of Thin Film Magnetoimpedance Sensor in 0.3 T Surface Normal Magnetic Field. *IEEJ Trans. Electr. Electron. Eng.* **2020**, *15*, 1230–1235. [CrossRef]
48. Nakai, T.; Ishiyama, K.; Yamasaki, J. Study of hysteresis for steplike giant magnetoimpedance sensor based on magnetic energy. *J. Magn. Magn. Mater.* **2008**, *320*, e958–e962. [CrossRef]

sensors

Article

A Spin Valve-Based Rhombus-Shaped Micro-Object Implementing a Full Wheatstone Bridge

Mikhail Milyaev *, Larisa Naumova, Anastasiya Germizina, Tatyana Chernyshova, Anastasia Pavlova, Tatiana Krinitsina, Vyacheslav Proglyado and Vladimir Ustinov

M.N. Mikheev Institute of Metal Physics, Ural Branch, Russian Academy of Sciences, S. Kovalevskoi Street 18, Ekaterinburg 620990, Russia; naumova@imp.uran.ru (L.N.); germizina@imp.uran.ru (A.G.); chernyshova@imp.uran.ru (T.C.); anastasia.pavlova.1988@gmail.com (A.P.); krinitsina@imp.uran.ru (T.K.); proglyado@imp.uran.ru (V.P.); ustinov@imp.uran.ru (V.U.)
* Correspondence: milyaev@imp.uran.ru; Tel.: +7-(343)-374-3881

Abstract: Spin valves with a synthetic antiferromagnet were fabricated via magnetron sputtering. It was shown that the fabricated spin valve layers had a perfect microstructure and smooth interfaces and therefore, an RKKY interaction dominated in the coupling of the ferromagnetic layers separated by a copper spacer. Rhombus-shaped micro-objects were fabricated from a single spin valve film. The thermomagnetic treatment procedure was found to form unidirectional anisotropy in the micro-object such that the values of the exchange bias fields in the rhombus' nonparallel sides were opposite in sign. For the CoFeNi/Ru/CoFeNi synthetic antiferromagnet, we determined the differences between the ferromagnetic layer thicknesses at which the thermomagnetic treatment formed the same exchange bias all over each rhombus' side. We also fabricated a sensor element in which each side of the rhombus was the shoulder of a Wheatstone bridge. After the thermomagnetic treatment procedure, each shoulder worked as an active magnetosensitive element, enabling the device to operate as a full Wheatstone bridge. The sensor output exhibited a step shape, high sensitivity to field changes, and significant magnetic hysteresis. Such characteristics are suitable for switching devices.

Keywords: spin valve; Wheatstone bridge; magnetic anisotropy; shape anisotropy; exchange bias field

Citation: Milyaev, M.; Naumova, L.; Germizina, A.; Chernyshova, T.; Pavlova, A.; Krinitsina, T.; Proglyado, V.; Ustinov, V. A Spin Valve-Based Rhombus-Shaped Micro-Object Implementing a Full Wheatstone Bridge. *Sensors* **2024**, *24*, 625. https://doi.org/10.3390/s24020625

Academic Editor: Galina V. Kurlyandskaya

Received: 29 November 2023
Revised: 25 December 2023
Accepted: 3 January 2024
Published: 18 January 2024

1. Introduction

Bridge circuit measurements have important advantages such as low measurement errors, high sensitivity, reductions in noise, and temperature drift [1–4]. A Wheatstone bridge electrical circuit consists of two parallel branches, each consisting of two arms containing a resistor element. If a Wheatstone bridge is used in magnetic field sensors then elements, the electrical resistivity of which depend on the applied magnetic field, are used as resistors in each arm. The supply voltage U_{in} is applied to the parallel branches and the potential difference U_{out} between the middle points of the branches is measured. The maximum value of U_{out}/U_{in} can be achieved if the resistance increases in two bridge arms and decreases in two other arms with a change in the magnetic field. In this case, all four sensitive elements contribute to the output signal, and the circuit is a full bridge. In microelectronics, a spin valve (SV) nanostructure with giant magnetoresistance (GMR) is often used as a magnetosensitive material [5–7].

An SV consists of two ferromagnetic (FM) layers, the free layer (FL) and the pinned layer (PL), which are separated by a nonmagnetic spacer, usually made from Cu. The magnetic moment ($\mathbf{M}_p$) of the pinned layer is fixed by the exchange interaction with the adjacent antiferromagnetic (AF) layer [8]. In this way, the magnetic reversal hysteresis loop for the PL is shifted. The high-field loop shift (H_{ex}) depends on the interaction between the AF and PL. For the FL, the field shift of the hysteresis loop from $H = 0$ (H_j) is small. In the fields $|H_j| < |H| < |H_{ex}|$, the magnetic moments of the FL and PL are opposite, and there is a plateau of the magnetoresistive curve corresponding to the maximum magnetoresistance (MR_{max}). H_j depends on the FM layers' interaction through the nonmagnetic spacer. These interactions are the result of competition between the dipolar ferromagnetic and oscillating RKKY exchange interactions [9]. If dipolar ferromagnetic interactions are dominant, then H_j and MR_{max} decrease with an increase in spacer thickness. If the standard deviation of the interlayer roughness is no more than 0.3–0.5 nm [10,11], then the RKKY interactions can be dominant. In this case, Hj changes periodically with an increase in the spacer thickness, enabling an SV with $Hj \approx 0$ and a large MR_{max} to be obtained [12].

The pinned layer in an SV can be replaced by a synthetic antiferromagnet (SAF) to reduce the magnetostatic interactions between the FM layers and increase the operating temperature range. An SAF consists of two FM layers coupled through a layer of Ru [13–15]. The thickness of the Ru layer (0.7–0.9 nm) corresponds to the maximum antiferromagnetic exchange interaction [16]. In this case, the FM layer adjacent to the AF layer is called the pinned layer (PL), and the second FM layer in the SAF is called the reference layer (RL).

The magnetic and magnetoresistive properties of an SV have several types of anisotropy: (1) uniaxial easy axis (EA) anisotropy, induced during deposition in a magnetic field; (2) unidirectional anisotropy with the pinning direction (PD), arising due to the interaction between the pinned and AF layers; and (3) shape anisotropy in micro-objects. The PD can be changed via thermomagnetic treatment (TMT) [1].

It is difficult to implement a full Wheatstone bridge during magnetic field microsensor fabrication from a single SV film because the resistance (R_n, n = 1, 2, 3, 4) of all the bridge arms changes equally with a change in the magnetic field, and all the dR_n/dH values are the same in sign. $dR_{1,3}/dH$ and $dR_{2,4}/dH$ should be opposite in sign to contribute to the output voltage U_{out} from all the bridge elements. This result is possible for the SVs in the Wheatstone bridge if their PDs are mutually opposite. The following methods are used to obtain this result: the two-stage deposition of SVs with different PDs [1] and EAs [17] or with different compositions [18,19] onto the corresponding parts of the substrate; TMT in the field corresponding to the spin-flop state in the SAF [16,17]; and the use of a permanent magnet to create additional oppositely directed field components in the active elements of the bridge [20].

In this work, spin valves with predominant RKKY interactions between the free and reference layers are studied. The formation of an exchange shift in the pinned layer under the combination of shape anisotropy and uniaxial anisotropy is also investigated. A thermomagnetic treatment mode to form an opposite exchange shift in the nonparallel sides of a rhombus-shaped micro-object was found, and the full Wheatstone bridge circuit was realized.

2. Materials and Methods

SVs with the composition $Ta(5)/[Ni_{80}Fe_{20}]_{60}Cr_{40}(5)/Co_{70}Fe_{20}Ni_{10}(t_{FL})/Cu(t_{Cu})/Co_{70}Fe_{20}Ni_{10}(t_{RL})/Ru(0.8)/Co_{70}Fe_{20}Ni_{10}(t_{PL})/Fe_{50}Mn_{50}(10)/Ta(5)$ were deposited on glass substrates via dc magnetron sputtering in a magnetic field of 80 Oe and applied in a film plane. Layer thicknesses are given in nm. A computer program controlled the process of deposition for the multilayer nanostructure. This program set the sequence of deposition of the layers, the time of deposition for each layer, the time of launching argon gas into the chamber, the power of the magnetrons, argon pressure, substrate temperature, and substrate rotation speed. The base pressure of the residual gasses in the sputtering chamber was less than 5×10^{-7} Pa. Sputtering was carried out at room temperature under an argon pressure of 0.1 Pa and magnetron power of 100 W. The nanolayer thickness was nominal and calculated from the measured deposition rate with a pre-assigned sputtering time. The applied experimental procedure allowed us to change the layer deposition time with a step of 0.1 s (corresponding to a step of 0.01 nm) at a copper deposition rate of 7 nm/min and a deposition rate of 2.8–3 nm/min for the rest of the layers. The materials' deposition rates were determined by making calibration films and measuring their thickness values using a Zygo NewView 7300 (Zygo, Middlefield, OH, USA) white light interferometer with an accuracy of ±0.02 nm at full film thickness (40–60) nm. The thicknesses of the Cu and FM layers were varied, and $Co_{70}Fe_{20}Ni_{10}$ was used as the FM layer. For this alloy, saturation magnetization was 13% higher and coercivity was 4.5 times lower than the values for $Co_{90}Fe_{10}$ commonly used in SVs [21]. The buffer layer $Ta/[Ni_{80}Fe_{20}]_{60}Cr_{40}$ promoted the formation of the <111> texture [22,23] and a decrease in crystallite size, as well as the interlayer roughness of the nanostructured film [24,25].

Analyses of the materials' microstructure were carried out via X-ray diffractometry on a DRON-3M (Bourevestnik, St. Petersburg, Russia) diffractometer under CoKα_1 radiation (λ = 1.7889 Å) and via transmission electron microscopy on a Tecnai G-30 (FEI, Hillsboro, OR, USA). A special technique was used to produce TEM samples (foil). The sample on a glass substrate was thinned using sandpaper, and then ion etching was carried out using a PIPS II 695 (GATAN, Pleasanton, CA, USA).

The magnetic microstructures of the micro-objects were investigated by using a scanning probe microscope Solver Next (NT-MDT, Zelenograd, Moscow, Russia) in the magnetic force microscopy (MFM) mode.

Magnetoresistive measurements and TMT were carried out in the setup based on a Bruker electromagnet and a LakeShore 336 temperature controller (Lake Shore Cryotronics, Westerville, OH, USA). TMT involved heating in a helium atmosphere to a temperature T_{TMT} = 448 K, which exceeded the blocking temperature T_b = 433 K for SVs based on an antiferromagnetic FeMn alloy [26]. The magnetoresistance was determined as $MR = (R(H) - R_s)/R_s$, where $R(H)$—resistance of the sample in a magnetic field, and R_s—resistance of the sample in a saturation field. The error of the relative resistance value was 0.05%. These experimental results were all obtained for the "Current In the Plane of the layers" (CIP) geometry.

Micro-objects were fabricated via laser lithography on DWL 66+ (Heidelberg Instruments Mikrotechnik GmbH, Heidelberg, Germany) and reactive ion etching on a PlasmaPro 80 RIE (Oxford Instruments, Abingdon-on-Thames, UK). The contact pads were fabricated using the lift-off procedure. Two types of micro-objects (Figure 1) were prepared. (1) V-shaped micro-objects were made from two microstrips forming a corner with Cu contact pads at the apex of the corner and ends of the microstrips. The corner angle (α) was 20 or 40°, and its bisector coincided with the EA. (2) Rhombus-shaped micro-objects included 315 µm long and 2 µm wide microstrips and Cu pads at the rhombus vertices. The acute angles were α = 20 or 40°, and the long rhombus diagonal coincided with the EA.

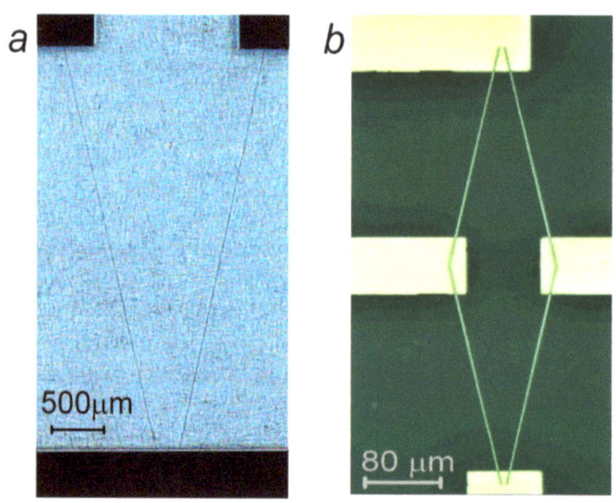

Figure 1. Images of V-shaped micro-object made from 2 microstrips (**a**) and rhombus-shaped micro-object made from 4 microstrips (**b**) with contact pads.

3. Results

3.1. Microstructural Studies

The $Co_{70}Fe_{20}Ni_{10}$ alloy, $[Ni_{80}Fe_{20}]_{60}Cr_{40}$ alloy, and Cu and $Fe_{50}Mn_{50}$ antiferromagnetic alloy have the same cubic face-centered (fcc) crystal structure and similar lattice parameter values. Figure 2a shows the coincidence of the angular position of the (111) peak for the bulk $Co_{70}Fe_{20}Ni_{10}$ alloy, the thin film of the $[Ni_{80}Fe_{20}]_{60}Cr_{40}$ alloy, and the SV containing the layers of these alloys. A <111> texture can be seen in the film and SV, with no other peaks in the fcc structure. The full width at the half-maximum of the rocking curve (ω-scan) for the (111) peak of SV is 4.4 degrees.

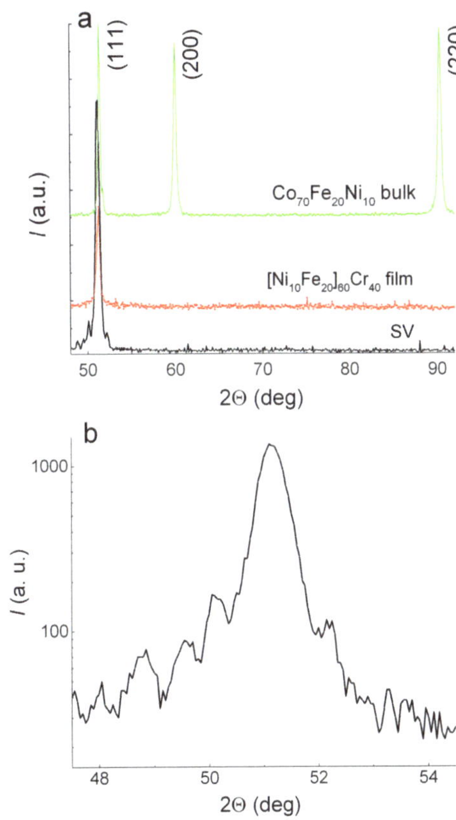

Figure 2. (**a**) X-ray diffraction patterns for the SV of the Ta(5)/NiFeCr(5)/CoFeNi(3.5)/Cu(3.3)/CoFeNi(3.5)/Ru(0.8)/CoFeNi(3)/FeMn(10)/Ta(5) composition (black line), NiFeCr film with a thickness of 60 nm (red line), and bulk sample of the CoFeNi alloy (green line). (**b**) Thickness oscillations (satellites) around the (111) peak of the SV diffraction pattern.

Thickness oscillations (satellites) are clearly visible around the (111) peak (Figure 2b) in the SV X-ray diffraction patterns. The presence of satellites shows that the SV films are of high quality and have smooth interfaces. Such satellites appear when X-ray radiation interferes on a layer with a uniform microstructure and smooth boundaries. The thickness (d) of such a layer can be estimated from the period of the oscillations with Selyakov's formula [27,28]:

$$d = \frac{\lambda}{\Delta(2\Theta)cos\Theta_B} \quad (1)$$

where λ is the X-ray wavelength, Θ_B is the Bragg angle for the (111) peak, and $\Delta(2\Theta)$ is the period of the oscillations. The value of d determined from formula (1) coincides with an accuracy of 0.1 nm for the total nominal thickness of the NiFeCr(5)/CoFeNi(3.5)/Cu(3.3)/CoFeNi(3.5) layers in the spin valve. Thus, in this part of the SV, the microstructure is uniform, and the microstructure of the layers become aligned.

Figure 3 shows the results of the TEM investigation. In the electron diffraction pattern Debye rings for NiFeCr, CoFeNi, Cu, and FeMn are common because they have the same fcc structure and similar lattice parameters. Note that the {220} ring is the brightest, while the {111} ring is weak. It is typical for the <111> texture and consistent with the results of the X-ray diffraction study. High-resolution images show thin parallel bands (Figure 3a). These are a direct resolution of the atomic plane projections on the image plane. Figure 3b shows the moiré lines. These lines are a result of the interference of electron beams diffracted by crystal lattices in adjacent layers. In multilayer structures parallel to each other, moiré lines appear if the lattice mismatch of the layers is low, and the number of dislocations and packaging defects is small.

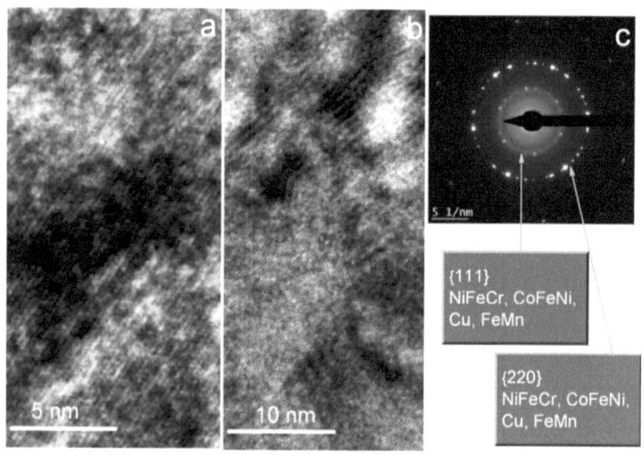

Figure 3. TEM high-resolution images (**a**,**b**) and electron diffraction pattern (**c**) for the SV of the Ta(5)/NiFeCr(5)/CoFeNi(3.5)/Cu(3.3)/CoFeNi(3.5)/Ru(0.8)/CoFeNi(3)/FeMn(10)/Ta(5) composition.

The studied SVs featured layers with a perfect microstructure and low lattice mismatch along with an excellent <111> texture. Such properties of the microstructure were obtained by using the layer materials with similar crystal structures and the buffer layer of the Ta(5)/NiFeCr(5) composition. In [12,24], a similar high-quality microstructure was achieved in the superlattices and spin valves sputtered onto the Ta/NiFeCr buffer layer. In the present investigation, such microstructural properties are necessary to obtain smooth interfaces and a high prevalence of RKKY interlayer coupling.

3.2. Exchange Coupling of Free and Reference Layers

Magnetoresistive curves were measured for SVs with $t_{Cu} = 1.8 \div 2.4$ nm in the field applied parallel to PD | | EA. Figure 4 shows the $MR(H)$ curve for the SV with $t_{Cu} = 2.2$ nm. Figure 4 shows how the values of MR_{max} and the shift of low field hysteresis loop (H_j) were estimated. For spin valves with weak interactions between the free and pinning layers the low field loop retains a rectangular shape in the case of small values for the angle of deviation PD from the EA [9,29].

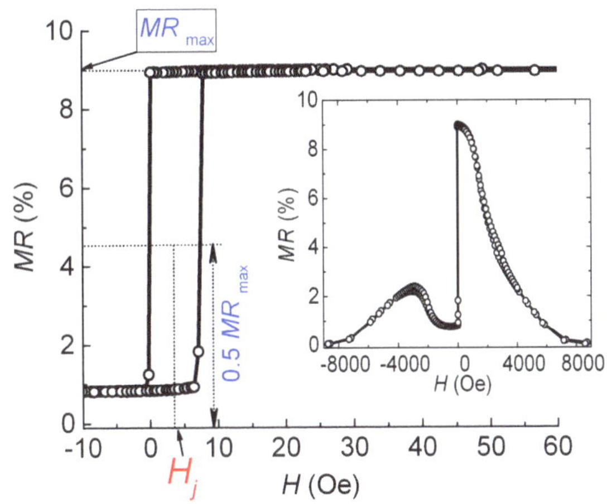

Figure 4. Low field part of the magnetoresistive curve for the SV of the Ta(5)/NiFeCr(5)/CoFeNi(5.5)/Cu(2.2)/CoFeNi(2.3)/Ru(0.8)/CoFeNi(2)/FeMn(10)/Ta(5) composition. The inset shows the $MR(H)$ curve in a wide range of fields.

The H_j value depends on the interaction between the free and reference FM layers. Ordinarily, if ferromagnetic dipolar coupling dominates, $H_j(t_{Cu})$ dependence decreases monotonously. In our case, $H_j(t_{Cu})$ dependence has an oscillating behavior. Thus, we may conclude that RKKY interlayer coupling dominates (Figure 5).

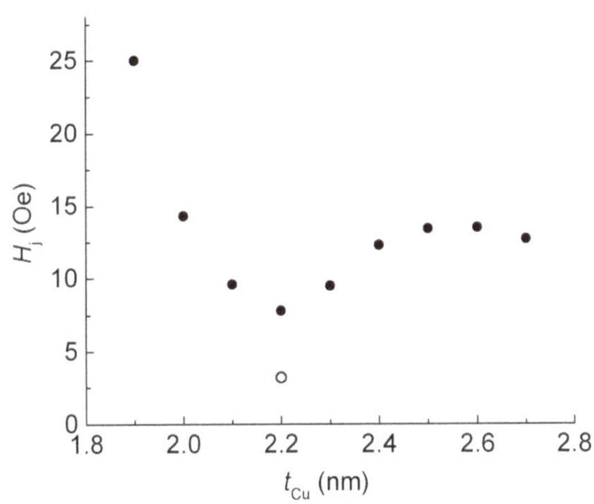

Figure 5. The dependence of the low field hysteresis loop shift on the thickness of the Cu layer for SVs of the Ta(5)/NiFeCr(5)/CoFeNi(4)/Cu(t_{Cu})/CoFeNi(4)/Ru(0.8)/CoFeNi(3.5)/FeMn(10)/Ta(5) composition (solid symbols) and the shift in the low field hysteresis loop for the SV of the Ta(5)/NiFeCr(5)/CoFeNi(5.5)/Cu(2.2)/CoFeNi(2.3)/Ru(0.8)/CoFeNi(2)/FeMn(10)/Ta(5) composition (open symbol).

The minimum value of H_j was obtained at t_{Cu} = 2.2 nm. Changing the thickness of the ferromagnetic layers in the SV is also an effective way to influence the amount of interlayer interaction [30]. Thus, we changed the thickness of the FM layers to increase MR_{max} and further reduce H_j. We achieved an H_j value close to $H = 0$ (open symbol in Figure 5) and MR_{max} = 9% at a t_{FL} of 5.5 nm and t_{PL} of 2 nm. In the SV of the obtained composition, we changed the reference layer thickness (t_{RL}). Thus, the difference ($t_{RL} - t_{PL}$) between the FM layers in the SAF and the total magnetic moment of the SAF changed. Figures 6 and 7 show the changes in the magnetoresistive curves and the values of H_j and MR_{max} that occurred due to increasing the value of t_{RL}.

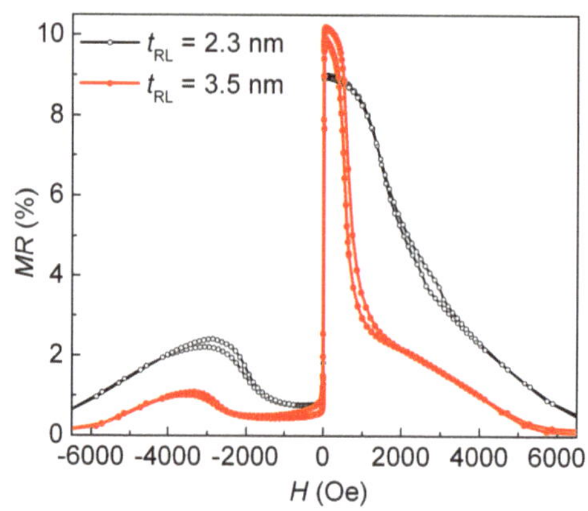

Figure 6. Magnetoresistive curves for SVs of the Ta(5)/NiFeCr(5)/CoFeNi(5.5)/Cu(2.2)/CoFeNi(t_{RL})/Ru(0.8)/CoFeNi(2)/FeMn(10)/Ta(5) composition for t_{RL} = 2.3 and 3.5 nm.

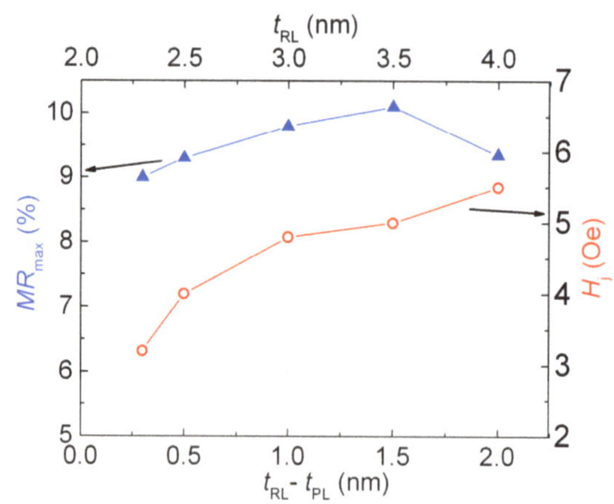

Figure 7. Dependencies of the low field hysteresis loop shift (red line and circles, shift values are given on the right axis) and the maximal magnetoresistance (blue line and triangles, magnetoresistance values are given on the left axis) on the reference layer thickness for the SVs of the Ta(5)/NiFeCr(5)/CoFeNi(5.5)/Cu(2.2)/CoFeNi(t_{RL})/Ru(0.8)/CoFeNi(2)/FeMn(10)/Ta(5) composition.

When the thickness and, accordingly, the magnetic moment of the reference layer increase, the interlayer coupling and H_j value naturally increase. The value of MR_{max} also varies slightly.

Further analyses will be carried out on this series of SVs with a small shift in the low field hysteresis loop and different values of the reference layer magnetic moment.

3.3. Change in the Pinning Direction in Spin Valve Films and Micro-Objects during Thermomagnetic Treatment

The thermomagnetic treatment procedure includes annealing at T_{TMT} and subsequent cooling in the applied magnetic field H_{TMT}. If $T_{TMT} > T_b$, the exchange interaction at the boundary of CoFeNi/FeMn is destroyed. Then, the initial unidirectional anisotropy and initial PD in the SV disappear. Cooling in the applied magnetic field then forms new unidirectional anisotropy and new PD1. The direction of PD1 coincides with the direction of the magnetic moment of the adjacent FM layer [31,32]. We performed two consecutive TMT procedures for the following objects: the V-shaped SV micro-object and the SV film. We performed the first TMT(1) with H_{TMT} = 9 kOe exceeding the magnetic saturation field and the second TMT(2) with $H_{TMT} \approx 0$. H_{TMT} was always directed perpendicular to the EA. After TMT(2), we investigated the magnetic structure of the V-shaped micro-object and SV film using magnetic force microscopy. Figure 8 shows the topography and corresponding MFM images obtained for the film and the V-shaped micro-object.

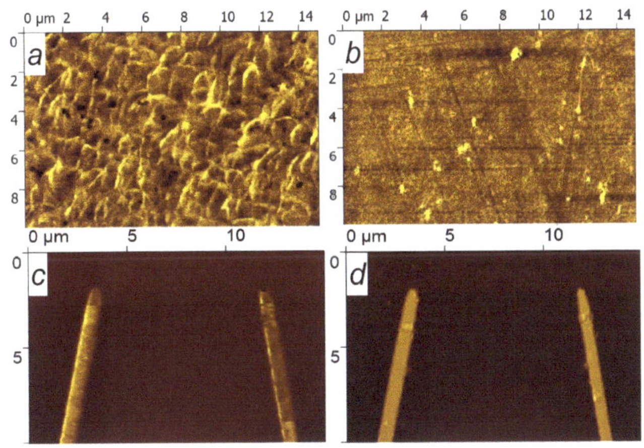

Figure 8. Topography (**b**,**d**) and MFM (**a**,**c**) images obtained for the film (**a**,**b**) and the V-shaped micro-object (**c**,**d**).

The MFM image of the film shows an irregular magnetic structure. The majority of each strip has one magnetic order with small areas of a different magnetic order. Magnetic contrast shows that the predominant magnetic order differs between the strips. In the MFM image (Figure 8c), the left strip is lighter than the right strip, while there is no such difference in the topographic image. Thus, after the same TMT, different magnetic structures were formed in the film and the V-shaped micro-object. The presumed reason for the observed differences is as follows.

Here, the magnetic moments of the pinned and reference layer ($\mathbf{M}_{PL}$ and $\mathbf{M}_{RL}$) are antiferromagnetically coupled and opposite to each other in $H_{TMT} \approx 0$. In the SV film during TMT(2) at $T = T_{TMT,}$ the uniaxial anisotropy controls the turn of $\mathbf{M}_{PL}$ and $\mathbf{M}_{RL}$; thus, clockwise and counterclockwise turns are equally probable (Figure 9). During subsequent cooling, the exchange interaction in the CoFeNi/FeMn boundary fixes this different magnetic ordering and forms the new pinning directions PD1 and PD2 in different regions of the film. PD1 and PD2 are antiparallel and collinear with EA. In the micro-objects, the shape anisotropy competes with the uniaxial anisotropy. If the shape anisotropy dominates, then it controls the turn of $\mathbf{M}_{PL}$ and $\mathbf{M}_{RL}$ during TMT(2) at $T = T_{TMT}$. With our experimental geometry (Figure 9), in one strip of the V-shaped micro-object, $\mathbf{M}_{PL}$ and $\mathbf{M}_{RL}$ turn clockwise, and in the other strip, they turn counterclockwise. Thus, after cooling, the new pinning directions PD1and PD2 are collinear with the strips. If the angle between the strips is 20 or 40°, the angle between PD1 and PD2 will be 160 or 140°, respectively.

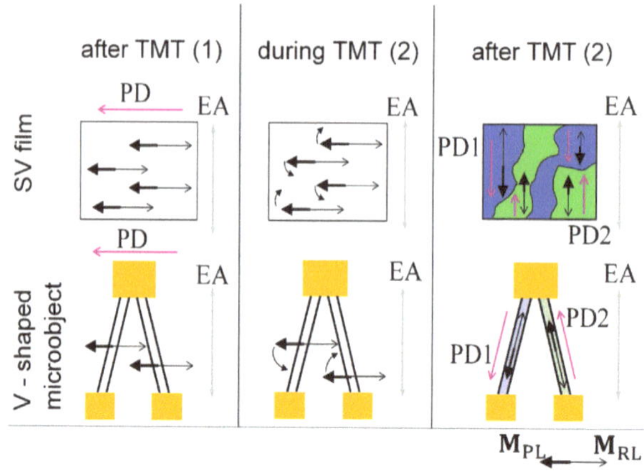

Figure 9. Schematic representation of the turn of reference (thin black arrow) and pinned (thick black arrow) layer magnetic moments in the SV film and SV micro-object during TMT. PD (pink arrow) changes and EA direction (grey arrow) is unchanged during TMT. Green and blue colors correspond to areas with different PDs.

Magnetoresistive curves measured for each strip of the V-shaped micro-object in the magnetic field applied parallel to EA are shown in Figure 10. The values of H_{ex} for two different strips are opposite in sign. This result is in good agreement with the fact that in one strip, the projection of PD onto the applied field is positive, and in the other, the projection of PD onto the applied field is negative.

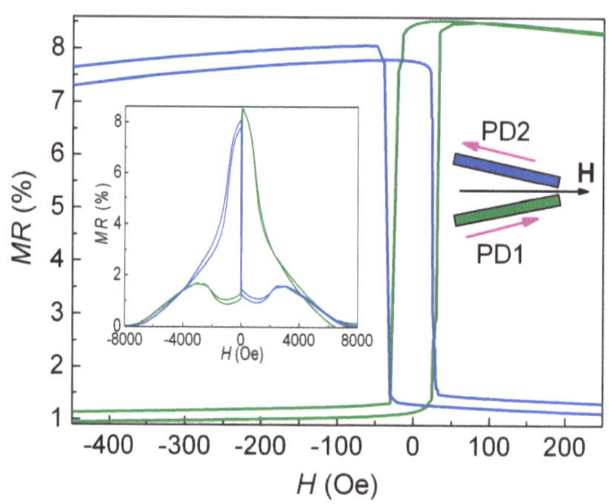

Figure 10. Low field parts of the magnetoresistive curves measured for each strip of the V-shaped SV micro-object. The applied magnetic field direction with respect to PD1 and PD2 is shown in the insert on the right. The inset on the left shows the *MR*(*H*) curves in a wide range of fields. The color of the curve corresponds to the color of strip.

In the low fields, $dMR/dH > 0$ for the green magnetoresistive curve, and $dMR/dH < 0$ for the blue magnetoresistive curve. Note that this PD arrangement was obtained in different strips of a single micro-object under two TMTs in a direction-fixed magnetic field.

3.4. Full Wheatstone Bridge Based on Rhombus-Shaped Spin Valve Micro-Object

TMT(1) and TMT(2) procedures were performed for the rhombus-shaped spin valve micro-object. Figure 11 shows the PD arrangement with rhombus sides, which can be expected if one considers the rhombus as two V-shaped micro-objects. This mutual PD arrangement is required to implement the Wheatstone bridge, in which $dR_{1,3}/dH$ and $dR_{2,4}/dH$ are opposite in sign. Thus, each SV element makes an active contribution to the output signal.

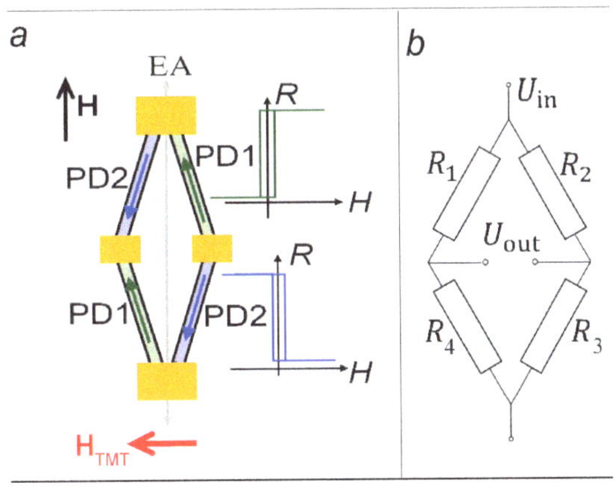

Figure 11. Scheme of PD arrangement in the rhombus-shaped micro-object sides after TMT(1) and TMT(2) (**a**) and the electrical circuit of the Wheatstone bridge (**b**). The inset in (**a**) shows $R(H)$ dependencies for R_1, R_3 (blue) and R_2, R_4 (green).

We applied the supply voltage (U_{in}) to the long diagonal of the rhombus and measured output voltage (U_{out}) in the short diagonal of the rhombus under the magnetic field swept along the EA (Figure 12).

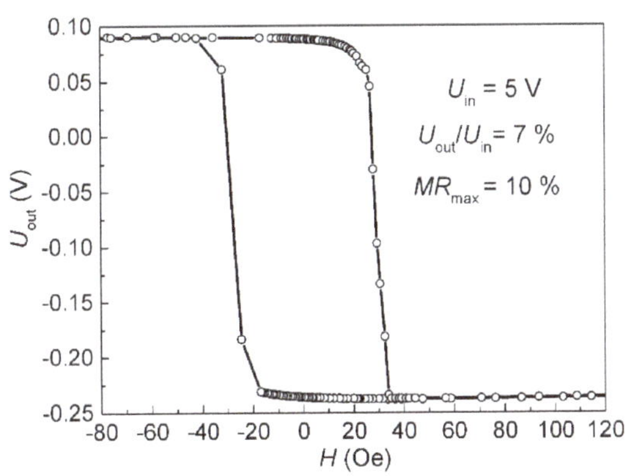

Figure 12. Wheatstone bridge output under the magnetic field sweep along EA for the rhombus-shaped micro-object based on the SV of the Ta(5)/NiFeCr(5)/CoFeNi(5.5)/Cu(2.2)/CoFeNi(3.5)/Ru(0.8)/CoFeNi(2)/FeMn(10)/Ta(5) composition.

The maximal output of the full Wheatstone bridge was estimated from $U_{out}/U_{in} = \Delta R/R$, where R is the resistance of each of the four elements, and ΔR is the resistance change in the applied magnetic field. For the half Wheatstone bridge with two active sensing elements, the maximal output was $U_{out}/U_{in} = 0.5\Delta R/R$ [33]. Thus, for the full Wheatstone bridge with four active SV elements, we obtained $U_{out}/U_{in} = MR_{max}$, where MR_{max} is the maximal magnetoresistance of each SV element. In our case, U_{out}/U_{in} = 7%, while MR_{max} = 10% (Figures 6 and 12, respectively). Thus, $0.5MR_{max} < U_{out}/U_{in} < MR_{max}$. A possible reason for this result could be the presence of small areas of undesirable PD orientation in the rhombus forming strips. These areas are visible in the image of the magnetic structure (Figure 8c) in the upper part of the left strip and in the lower part of the right strip.

3.5. Formation of Opposite Pinning Directions in Rhombus-Shaped Micro-Objects Based on Spin Valves with Different Thicknesses of the Reference Layer

An increase in t_{RL} led to an increase in the total effective magnetic moment ($\mathbf{M}_{eff}$) of antiferromagnetically coupled $\mathbf{M}_{PL}$ and $\mathbf{M}_{RL}$, with $M_{eff} = M_{RL} - M_{PL}$. During TMT(2) at $T > T_b$, the magnetic moments were arranged in a way that minimized the anisotropy anisotropic energy. We estimated the shape anisotropy field using the research in [34], in which the demagnetizing factors of the general ellipsoid were reported. For the long strip the demagnetizing factor was approximately t/w, where t is the thickness of the FM layer and w is the width of the strip. Hence, the anisotropy field was calculated as the sum of two terms: $H_a = 2K_u/M_{eff} + 4\,M_{eff}\,t/w$, where K_u is the uniaxial anisotropy constant, and $t = t_{RL} + t_{PL}$. We changed the relationship between the two terms of the anisotropy field by changing t_{RL}. In this way, we changed the magnetic moments' arrangement after TMT(2).

Rhombus-shaped micro-objects with an angle α = 20 and 40° were fabricated from films of SVs with t_{RL} = 2.3, 2.5, 3.0, 3.5, and 4 nm. TMT(1) and TMT(2) were performed consistently. Then, the Wheatstone bridge output voltage was measured in the short diagonal of the rhombus under a magnetic field swept along the EA. Figure 13 shows bar charts to compare the Wheatstone bridge U_{out}/U_{in} ratio with the MR_{max} value of the SV strips. For convenience of comparison, we also plotted the relation between U_{out}/U_{in} and MR_{max} as a function of the reference layer thickness (Figure 14).

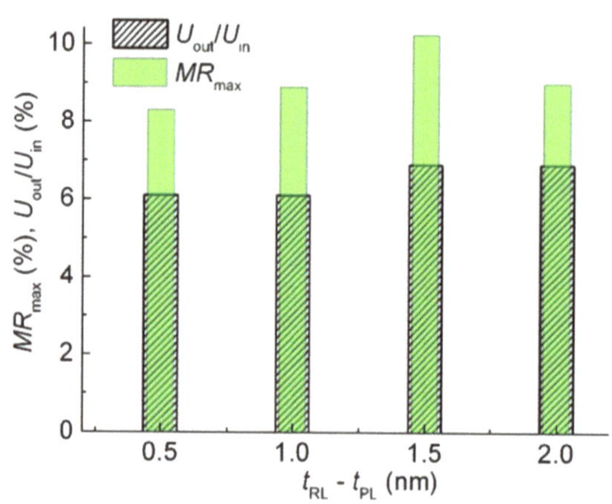

Figure 13. The dependencies of the Wheatstone bridge U_{out}/U_{in} ratio and SV strip MR_{max} value on the thickness of the reference layer for rhombus-shaped micro-objects with α = 40°.

For rhombus-shaped micro-objects with α = 20°, the minimal difference between U_{out}/U_{in} and MR_{max} was found for samples with the maximum t_{RL} and maximum difference in FM layer thicknesses in the SAF. The effect of shape anisotropy on the alignment of the magnetic moments under TMT likely increases with an increase in the total magnetic moment of the antiferromagnetically coupled M_{PL} and M_{RL}. For rhombus-shaped micro-objects with α = 40°, the difference between U_{out}/U_{in} and MR_{max} is practically the same under maximal and minimal t_{RL}. This result could be caused by different effects of uniaxial anisotropy due to the different micro-object geometries. When the angle α changes from 20 to 40°, the deviation of the strip from EA increases. As M_{eff} increases, coupled M_{PL} and M_{RL} can be set not only along the strip but also along EA in some areas of the strip.

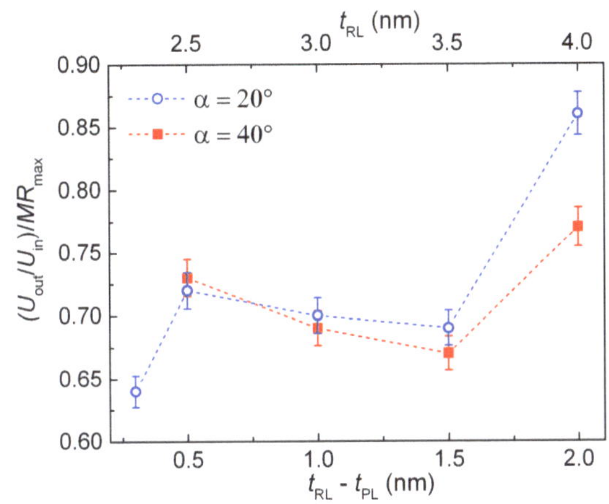

Figure 14. The dependencies of the relation between U_{out}/U_{in} and MR_{max} on the thickness of the reference layer for rhombus-shaped micro-objects with α = 20 and 40°.

4. Discussion

The proposed method for implementing a full Wheatstone bridge scheme with SVs as magnetically sensitive materials is simple. This method does not require the separate sputtering of the SV elements with an exchange bias that is opposite in sign or a separate TMT for each SV element. Nevertheless, there are a number of requirements for the successful application of this method. To prevent accidental splitting into magnetic domains, the microstructure of the layers should contain minimal defects, and the interlayer boundaries should be smooth. The SV elements with opposite exchange bias should change their magnetoresistance under the same low fields. Therefore, the shift of the low field hysteresis loop should be close to $H = 0$. To increase the ratio between U_{out}/U_{in} and MR_{max} for each SV element, it is necessary to increase the difference in the thicknesses of the reference and pinned layers and, accordingly, the total magnetic moment of the SAF. The proposed TMT procedure is also effective for SVs based on antiferromagnetic alloys with higher Neel temperatures and, accordingly, higher T_b values [35]. In this case, it is necessary to use a higher temperature $T_{TMT} > T_b$.

5. Conclusions

The results showed that shape anisotropy controls the turn of magnetic moments in spin valve micro-objects under the low magnetic fields.

We used the TMT procedure, which makes it possible to obtain exchange bias fields that are opposite in sign and a nearly collinear arrangement of the axes of uniaxial and unidirectional anisotropy in separate elements of the Wheatstone bridge circuit. The procedure includes two subsequent TMTs in the applied magnetic field perpendicular to the easy axis. During the first TMT, the field exceeded the saturation field, and during the second TMT, the field was close to $H = 0$. The formation of the opposite exchange bias in non-parallel sides of the rhombus-shaped micro-object was caused by the predominance of shape anisotropy in magnetic reversal and during TMT. The output signal of the complete Wheatstone bridge sensor had a step-like shape. Such characteristics are in demand for switching devices.

Author Contributions: V.U., supervision and project administration; V.P., magnetron sputtering SV films; L.N., design of experiments and analysis of the data, writing of the manuscript, and X-ray diffractometry; A.G., micro-object fabrication and magnetoresistance measurements; T.C., investigation, thermomagnetic treatment, figure representations, and analysis of the data; A.P., magnetic microscopic investigation via atomic force microscopy; T.K., microstructure studies via transmission electron microscopy; M.M., supervision, review, and editing. All authors have read and agreed to the published version of the manuscript.

Funding: This study was supported by the state assignment of the Ministry of Science and Higher Education of the Russian Federation (theme Spin, no. 122021000036-3).

Institutional Review Board Statement: Not applicable.

Informed Consent Statement: Not applicable.

Data Availability Statement: The original contributions presented in the study are included in the article, further inquiries can be directed to the corresponding author.

Conflicts of Interest: The authors declare no conflicts of interest.

References

1. Freitas, P.P.; Ferreira, R.; Cardoso, S. Spintronic sensors. *Proc. IEEE* **2016**, *104*, 1894–1918. [CrossRef]
2. Mabarroh, N.; Alfansuri, T.; Wibowo, N.A.; Istiqomah, N.I.; Tumbelaka, R.M.; Suharyadi, E. Detection of green-synthesized magnetite nanoparticles using spin-valve GMR-based sensor and their potential as magnetic labels. *J. Magn. Magn. Mater.* **2022**, *560*, 169645. [CrossRef]
3. Sánchez, J.; Ramírez, D.; Amaral, J.; Cardoso, S.; Freitas, P.P. Electrical ammeter based on spin-valve sensor. *Rev. Sci. Instrum.* **2012**, *83*, 105113. [CrossRef] [PubMed]
4. Prieto, J.L.; Evetts, J.E.; Blamire, M.G.; Rouse, N. Development of an integrated magnetic sensor with linear output based on spin valves. *J. Appl. Phys.* **2002**, *91*, 8578–8580. [CrossRef]
5. Carvalho, M.; Ribeiro, P.; Romão, V.; Cardoso, S. Smart fingertip sensor for food quality control: Fruit maturity assessment with a magnetic device. *J. Magn. Magn. Mater.* **2021**, *536*, 168116. [CrossRef]
6. Yan, S.; Cao, Z.; Guo, Z.; Zheng, Z.; Cao, A.; Qi, Y.; Leng, Q.; Zhao, W. Design and Fabrication of Full Wheatstone-Bridge-Based Angular GMR Sensors. *Sensors* **2018**, *18*, 1832. [CrossRef]
7. Suharyadi, E.; Alfansuri, T.; Handriani, L.S.; Wibowo, N.A.; Sabarman, H. Detection of Fe_3O_4/PEG nanoparticles using one and two spin-valve GMR sensing elements in Wheatstone bridge circuit. *J. Mater. Sci. Mater. Electron.* **2021**, *32*, 23958–23967. [CrossRef]
8. Dieny, B.; Speriosu, V.S.; Parkin, S.S.P.; Gurney, B.A.; Wilhout, D.R.; Maui, D. Giant magnetoresistance in soft ferromagnetic multilayers. *Phys. Rev. B* **1991**, *43*, 1297–1300. [CrossRef]
9. Kools, J.C.S. Exchange-Biased Spin-Valves for Magnetic Storage. *IEEE Trans. Magn.* **1996**, *32*, 3165–3184. [CrossRef]
10. Svalov, A.V.; Sorokin, A.N.; Savin, P.A.; García-Arribas, A.; Fernández, A.; Vas'kovskiy, V.O.; Kurlyandskaya, G.V. Co/Cu/Co Pseudo Spin-Valve System Prepared by Magnetron Sputtering with Different Argon Pressure. *Key Eng. Mater.* **2015**, *644*, 211–214. [CrossRef]
11. Öksüzoglu, R.M.; Sarac, U.; Yıldırım, M.; Çınar, H. Characterization of Microstructural and Morphological Properties in As-deposited Ta/NiFe/IrMn/CoFe/Ta Multilayer System. *J. Mater. Sci. Technol.* **2014**, *30*, 359–364. [CrossRef]
12. Naumova, L.I.; Milyaev, M.A.; Zavornitsin, R.S.; Pavlova, A.Y.; Maksimova, I.K.; Krinitsina, T.P.; Chernyshova, T.A.; Proglyado, V.V.; Ustinov, V.V. High-Sensitive Sensing Elements Based on Spin Valves with Antiferromagnetic Interlayer. Coupling. *Phys. Met. Metallogr.* **2019**, *120*, 653–659. [CrossRef]
13. Meguro, K.; Hoshiya, H.; Watanabe, K.; Hamakawa, Y.; Fuyama, M. Spin-Valve Films Using Synthetic Ferrimagnets for Pinned Layer. *IEEE Trans. Magn.* **1999**, *35*, 2925–2927. [CrossRef]
14. Leala, J.L.; Kryder, M.H. Spin valves exchange biased by Co/Ru/Co synthetic antiferromagnets. *J. Appl. Phys.* **1998**, *83*, 3720–3723. [CrossRef]
15. Negulescu, B.; Lacour, D.; Hehn, M.; Gerken, A.; Paul, J.; Duret, C. On the control of spin flop in synthetic antiferromagnetic films. *J. Appl. Phys.* **2011**, *109*, 103911. [CrossRef]
16. Huai, Y.; Zhang, J.; Anderson, G.W.; Rana, P.; Funada, S.; Hung, C.Y.; Zhao, M.; Tran, S. Spin-valve heads with synthetic antiferromagnet CoFe/Ru/CoFe/IrMn. *J. Appl. Phys.* **1999**, *85*, 5528–5530. [CrossRef]
17. Milyaev, M.A.; Naumova, L.I.; Zavornitsyn, R.S.; Maksimova, I.K.; Pavlova, A.Y.; Proglyado, V.V.; Ustinov, V.V. Use of a Spin-Flop State for the Creation of Spin-Valve Elements for a Full Wheatstone Bridge. *Phys. Met. Metallogr.* **2020**, *121*, 721–728. [CrossRef]
18. Ferreira, R.; Paz, E.; Freitas, P.; Ribeiro, J.; Germano, J.; Sousa, L. 2-Axis Magnetometers Based on Full Wheatstone Bridges Incorporating. Magnetic Tunnel Junctions Connected in Series. *IEEE Trans. Magn.* **2012**, *48*, 4107–4110. [CrossRef]
19. Luong, V.S.; Nguyen, A.T.; Tran, T.H.D. Antiparallel-Pinned Spin Valves with Modified Artificial Antiferromagnetic Layer for Full-Bridge Magnetic Sensors. *IEEE Trans. Magn.* **2018**, *54*, 4001705. [CrossRef]
20. Borole, U.P.; Khan, J.; Barshilia, H.C.; Chowdhury, P. Design, fabrication, and characterization of giant magnetoresistance (GMR) based open-loop current sensor with U-shaped current carrying conductor. *Sens. Actuators A* **2021**, *332*, 112103. [CrossRef]
21. Fukuzawa, H.; Iwasaki, H.; Koi, K.; Sahashi, M. Soft magnetic characteristics of an ultrathin CoFeNi free layer in spin-valve films. *J. Magn. Magn. Mater* **2006**, *298*, 65–71. [CrossRef]

22. Sugita, Y.; Kawawake, Y.; Satomi, M.; Sakakima, H. Thermal stability of PtMn based synthetic spin valves using thin oxide layer. *J. Appl. Phys.* **2001**, *89*, 6919–6921. [CrossRef]
23. Bannikova, N.S.; Milyaev, M.A.; Naumova, L.I.; Patrakov, E.I.; Proglyado, V.V.; Kamenskii, I.Y.; Ryabukhina, M.V.; Ustinov, V.V. Giant Magnetoresistance and Hysteresis Phenomena in CoFe/Cu Superlattices with Highly Perfect Crystallographic Texture. *Phys. Met. Metallogr.* **2018**, *119*, 1073–1078. [CrossRef]
24. Bannikova, N.S.; Milyaev, M.A.; Naumova, L.I.; Krinitsina, T.P.; Patrakov, E.I.; Proglyado, V.V.; Chernyshova, T.A.; Ustinov, V.V. NiFeCo/Cu superlattices with high magnetoresistive sensitivity and weak hysteresis. *Phys. Solid State* **2016**, *58*, 2011–2017. [CrossRef]
25. Diken, S.; Ozkaya, O.; Cakmaktepe, S. Deposition of $(Ni_{80}Fe_{20})_{100-x}Cr_x$ Alloy Thin Films for Potential Applications in Magnetic Recording Media. *IEEE Trans. Magn.* **2016**, *52*, 2003404. [CrossRef]
26. Berkowitz, A.E.; Takano, K. Exchange anisotropy—A review. *J. Magn. Magn. Mat.* **1999**, *200*, 552–570. [CrossRef]
27. Iveronova, V.I.; Revkevich, G.P. *Teoriya Rasseyaniya Rentgenovskikh Luchei [X-ray Scattering Theory]*, 2nd ed.; Revised and Expanded; Izdatel'stvo Moskovskogo Universiteta: Moscow, Russian, 1978; 278p.
28. Michaelsen, C. On the structure and homogeneity of solid solutions the limits of conventional X-ray diffraction. *Philos. Mag. A* **1995**, *7*, 813–828. [CrossRef]
29. Labrune, M.; Kools, J.C.S.; Thiaville, A. Magnetization rotation in spin-valve multilayers. *J. Magn. Magn. Mater.* **1997**, *171*, 1–15. [CrossRef]
30. Chernyshova, T.A.; Milyaev, M.A.; Naumova, L.I.; Proglyado, V.V.; Maksimova, I.K.; Pavlova, A.Y.; Blagodatkov, D.V.; Ustinov, V.V. Control of Low-Field Hysteresis Loop Shift of Spin Valves. *Phys. Met. Metallogr.* **2017**, *118*, 1209–1214. [CrossRef]
31. Tong, H.C.; Qian, C.; Miloslavsky, L.; Funada, S.; Shi, X.; Liu, F.; Dey, S. The spin flop of synthetic antiferromagnetic films. *J. Appl. Phys.* **2000**, *87*, 5055–5067. [CrossRef]
32. Beach, R.S.; McCord, J.; Webb, P.; Mauri, D. Orthogonal pinning of two ferromagnetic layers in a synthetic spin valve. *Appl. Phys. Lett.* **2002**, *80*, 4576–4578. [CrossRef]
33. Reig, C.; Cubells-Beltran, M.-D.; Munoz, D.R. Magnetic Field Sensors Based on Giant Magnetoresistance (GMR) Technology: Applications in Electrical Current Sensing. *Sensors* **2009**, *9*, 7919–7942. [CrossRef] [PubMed]
34. Osborn, J.A. Demagnetizing Factors of the General Ellipsoid. *Phys. Rev.* **1945**, *67*, 351–357. [CrossRef]
35. Stobiecki, F.; Stobiecki, T. Multilayer Structures with Giant Magnetoresistance. *Acta Phys. Pol. A* **2002**, *102*, 95–108.

sensors

MDPI

Article

Magnetoimpedance Effect in Cobalt-Based Amorphous Ribbons with an Inhomogeneous Magnetic Structure

Dmitry A. Bukreev *, Michael S. Derevyanko and Alexander V. Semirov

Department of Physics, Pedagogical Institute, Irkutsk State University, 664003 Irkutsk, Russia; mr.derevyanko@gmail.com (M.S.D.); semirov@mail.ru (A.V.S.)
* Correspondence: da.bukreev@gmail.com; Tel.: +7-9025688044

Abstract: The results of a computer simulation and experimental study of the magnetoimpedance effect (MI) in amorphous $Co_{68.5}Fe_{4.0}Si_{15.0}B_{12.5}$ and $Co_{68.6}Fe_{3.9}Mo_{3.0}Si_{12.0}B_{12.5}$ ribbons in the ac frequency range from 0.01 to 100 MHz are presented. It was found that the maximum MI value exceeds 200%, which may be of interest in the development of magnetic field sensors. It is also shown that practically significant characteristics of the MI response strongly depend on the ac frequency which is due to the inhomogeneous distribution of magnetic properties over the ribbon cross section. This distribution was studied using magnetoimpedance tomography based on the analysis of the experimental dependences of the reduced impedance on the ac frequency.

Keywords: magnetoimpedance tomography; magnetoimpedance effect; amorphous magnetically soft alloys; computer simulation; finite element method

Citation: Bukreev, D.A.; Derevyanko, M.S.; Semirov, A.V. Magnetoimpedance Effect in Cobalt-Based Amorphous Ribbons with an Inhomogeneous Magnetic Structure. *Sensors* **2023**, *23*, 8283. https://doi.org/10.3390/s23198283

Academic Editor: Daniel Ramos

Received: 1 September 2023
Revised: 29 September 2023
Accepted: 4 October 2023
Published: 7 October 2023

1. Introduction

The magnetoimpedance effect (MI) consists of changing the electrical impedance of a ferromagnetic conductor under the action of an external magnetic field [1]. If we confine ourselves to conductors of planar geometry, then MI is most intensively studied in amorphous magnetically soft ribbons based on cobalt and iron [2,3], including those after nanocrystallization [4,5], as well as in thin magnetic films [6,7] and multilayer films [8–10]. In this case, the MI is most often studied in the so-called longitudinal configuration, when the alternating current and the external magnetic field are oriented along the same axis [11].

The relative change in the impedance in the mentioned objects can reach hundreds of percents when the magnetic field intensity changes by units of A/m. For this reason, the development of highly sensitive magnetic field sensors based on MI is being actively discussed because it can be in demand both in technical [12–14] and biomedical applications [15–17].

The MI can be clearly explained in terms of the skin effect. The distribution of an alternating electric current (ac) over the cross section of a conductor with magnetic permeability μ and specific conductivity σ depends on the ac frequency f and can be characterized by the thickness of the skin layer [18]:

$$\delta = \sqrt{\frac{1}{\pi f \sigma \mu_0 \mu}}, \quad (1)$$

where μ_0 is the magnetic constant.

Simplistically, it can be assumed that the alternating current is mainly concentrated in the surface layer of the conductor with a thickness equal to the thickness of the skin layer, i.e., the effective cross section of the conductor is determined by δ. Thus, it follows from expression (1) that if the magnetic permeability of the conductor changes, then its effective cross section changes. Consequently, the impedance also changes, i.e., MI is observed.

With a uniform distribution of the magnetic permeability over the cross section of a planar conductor with a thickness of $2b$, its impedance, taking into account the external inductance L_{ext}, can be represented as follows [18,19]:

$$\dot{Z} = R_{DC} kb \cot kb + j \cdot 2\pi f L_{ext}, \quad (2)$$

where R_{DC} is the ribbon resistance to direct current and $k = (1 - j)/\delta$, j is an imaginary unit. External inductance is due to the magnetic flux created by the electric current outside the conductor. In the case of a rectangular conductor with length l, width a and thickness $2b$, the value of L_{ext} can be written as follows [20]:

$$L_{ext} = \frac{\mu_0 l}{2\pi}\left(\ln\frac{2l}{2b+a} + \frac{1}{2}\right), \quad (3)$$

Equation (2) shows that the contribution of an external inductance increases with an ac frequency increase. With the usual parameters for samples of cobalt-based ribbons (a = 1 mm, $2b$ = 25 μm and σ = 600 kS/m), the contribution of the external inductance at an ac frequency of 1 MHz is 10 times less than R_{DC}, that at the frequency of 10 MHz becomes comparable and that at the frequency of 100 MHz exceeds R_{DC} by 10 times. The contribution of the external inductance is often not considered when modeling MI at low ac frequencies [19]. However, at high frequencies, this contribution must be considered.

To date, theoretical concepts have been developed on how to determine the orientation of an easy magnetization axis and the magnitude of magnetic anisotropy on MI of planar magnetically soft conductors [19,21]. The frequency dispersion of the magnetic permeability and different orientations of the magnetization in neighboring magnetic domains were considered in [22]. The MI was also considered in the presence of an inhomogeneous external magnetic field [23]. The theoretical models of MI are of interest in multilayer films, the electrical and magnetic properties of which vary over the cross section [10,24]. A computer simulation using the finite element method also recommended itself in the study of MI response of such objects [8,25,26]. The effect of various coatings on the MI of amorphous magnetically soft ribbons was also studied using a computer simulation [27]. However, the influence of the nonuniform distribution of magnetic properties over the cross section of the ribbons on the MI has not been discussed either theoretically or with the help of computer simulations, although, this issue is relevant, as shown below.

Quenching stresses are irregularly distributed over the thickness of the amorphous ribbon. According to [28,29], the quenching stresses reach their maximum modulus near the surface of the ribbon, while they are minimal in its central part. Since there is no magnetocrystalline anisotropy in amorphous ribbons, the magnetoelastic anisotropy caused by quenching stresses leads to the formation of an inhomogeneous magnetic structure. It was shown when studying amorphous FePC ribbons that the distribution of the magnetic anisotropy over the thickness of the ribbon is inhomogeneous and asymmetric [30]. Its value is minimal near the middle of the ribbon and increases as it approaches the ribbon's surfaces. This distribution of the magnetic anisotropy correlates with the quenching stress distribution over the ribbons' cross sections described above.

Another factor affecting the distribution of magnetic properties over the cross section of the ribbon is surface irregularities [29,31,32]. The effect of irregularities on the magnetic anisotropy parameters of an amorphous ribbon is especially strong in the case of zero magnetostriction [32]. Surface irregularities can also cause an increase in the magnetic anisotropy dispersion [31,33], which can strongly affect the MI, especially at high frequencies [34]. Polishing and other modifications of the ribbon surface lead to a significant change in the MI response [35,36].

If the magnetic parameters are not uniformly distributed over the ribbon cross section, then different regions of the ribbon, which may have different magnetic anisotropy parameters, are involved in the formation of the MI response at different ac frequencies due to different skin layer thicknesses. In this case, it is fair to expect that the nature of the MI response depends on the ac frequency. This is also supported by numerous experimenta results, for example [37]. Thus, the inhomogeneity of the ribbon magnetic structure mus be considered when developing MI sensors. Knowing the distribution of the magneti properties over the ribbon cross section and how it changes depending on the choice o manufacturing conditions and further heat treatment, it is possible to purposefully tun the characteristics of MI sensors based on amorphous ribbons for a specific task.

To establish the distribution of the magnetic properties over the cross section of amor phous and electrodeposited wires, a method called magnetic impedance tomography (MIT was proposed [38]. This method is based on the analysis of the impedance dependence on the ac frequency. In this paper, we propose an implementation of this method fo amorphous ribbons.

2. Samples, Experimental Methods and Computer Simulation

2.1. Description of the Samples

The ribbons $Co_{68.5}Fe_{4.0}Si_{15.0}B_{12.5}$ (S0) and $Co_{68.6}Fe_{3.9}Mo_{3.0}Si_{12.0}B_{12.5}$ (S1) were pre pared by rapid quenching using the Cu weal technique. The nominal widths of the S0 and S1 ribbons were 0.71 mm and 0.78 mm, respectively. The thicknesses of the S0 and S1 ribbons were 24 and 26 µm, respectively. Samples 30 mm long were cut from the original rib bon. Designations of the samples, their geometrical parameters, values of magnetostriction and specific conductivity are given in Table 1.

Table 1. Sample parameters for experimental study.

Sample	Composition	Length l_0, mm	Width a, mm	Thickness $2b$, µm	Magnetostriction Sign	Specific Conductivity, kS/m
S0	$Co_{68.5}Fe_{4.0}Si_{15.0}B_{12.5}$	30	0.71	24	–	615
S1	$Co_{68.6}Fe_{3.9}Mo_{3.0}Si_{12.0}B_{12.5}$	30	0.78	26	+	618

2.2. Experimental Methods

The modulus of the electrical impedance Z was measured using the measuring com plex of magnetoimpedance spectroscopy. A measuring setup was developed by the authors of this article. A photograph of the setup is shown in Figure 1; its detailed description is given in [11]. The distance between the contacts of the measuring cell was l = 24 mm The effective value of the alternating current was equal to 1 mA. The alternating current frequency, f, varied in the range 0.01–80 MHz. An external magnetic field, H, was gen erated by a pair of Helmholtz coils. Its maximum strength was $H_{max} = \pm 12.3$ kA/m The alternating current and the external magnetic field were oriented along the length of the sample.

MI was calculated using the following formula:

$$\Delta Z/Z(H) = \frac{Z(H) - Z(H_{max})}{Z(H_{max})} \times 100\%. \quad (4)$$

The magnetic hysteresis loops were obtained using the induction method. The remag netizing magnetic field oriented along the length of the sample varied with a frequency of 1 kHz, and its amplitude was 1.2 kA/m.

Saturation magnetostriction, λ_S, was determined using the approach of changing the peak field in the $\Delta Z/Z(H)$ dependence under the action of tensile mechanical stresses γ This approach was proposed in [39]. The maximum values of the mechanical stresses, γ_{max}, were 575 MPa and 480 MPa for the S0 and S1 ribbons, respectively.

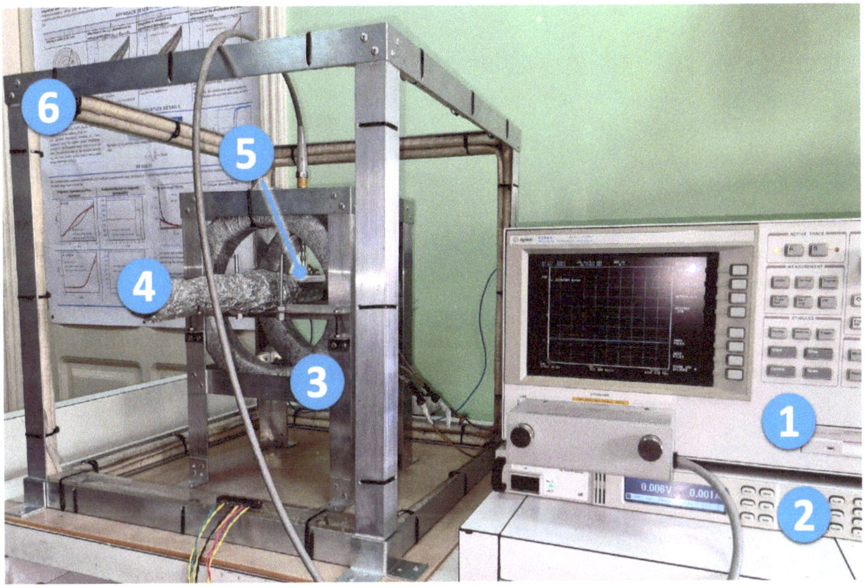

Figure 1. Photograph of the measuring setup. 1—Agilent 4294A impedance analyzer (Keysight Technologies, Santa Rosa, CA, USA); 2—Agilent N6700B (Keysight Technologies, Santa Rosa, CA, USA) power supply for Helmholtz coils; 3—Helmholtz coils; 4—pipe-holder of the measuring cell, which can also be used as part of a heating system (more details in [11]); 5—measuring cell; 6—three pairs of Helmholtz coils to compensate for geomagnetic and effective laboratory fields (coils' power supplies are not shown).

2.3. Computer Simulation of the MI

A computer simulation of the MI was performed using the finite element method in Comsol Multiphysics software in the ac frequency range from 0.01 to 80 MHz (license no. 9602434). The ribbon model had a rectangular section $2b \times a$ and length l. The section of the model was divided into a stack of $2n - 1$ layers of a rectangular section, arranged symmetrically with respect to its middle (Figure 2b). The width and length of each layer were a and l, respectively. For each layer, the value of the transverse magnetic permeability μ_i was set according to the following system of equations:

$$\begin{cases} \mu_1, \ if \ |y| \leq |y_1|; \\ \mu_2, \ if \ |y_1| < |y| \leq |y_2|; \\ \quad \cdots \\ \mu_i, \ if \ |y_{i-1}| < |y| \leq |y_i|; \\ \quad \cdots \\ \mu_n, \ if \ |y_{n-1}| < |y| \leq |y_n|, \end{cases} \tag{5}$$

where y_i is the coordinate of the outer boundary of the i-th layer. Obviously, $|y_n| = b$. The specific conductivities of the layers were assumed to be the same and equal to σ (see Table 1). Models with n from 1 to 6 were considered.

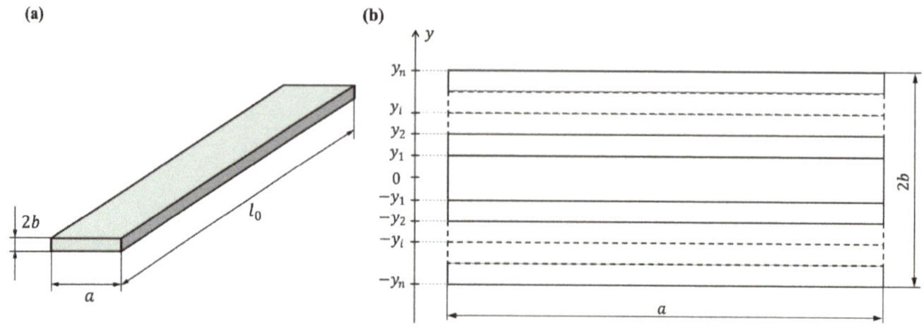

Figure 2. Schematic representation of the ribbon (**a**) and splitting the ribbon model into layers (**b**).

To simulate the magnetic flux outside the ribbon in Comsol Multiphysics, the circular section Air with radius 2*a* was used (Figure 3). The magnetic permeability of this region is 1, the permittivity is 1 and the specific conductivity is 0. The closing of the magnetic field lines at infinity was considered using the Infinite Element Domain tool (outer cylindrical layers of the model in Figure 3). It should be noted that, we made sure that the size of finite element mesh elements was less than the thickness of the skin layer. Figure 3c also shows an example of calculating the distribution of magnetic induction in and around the ribbon.

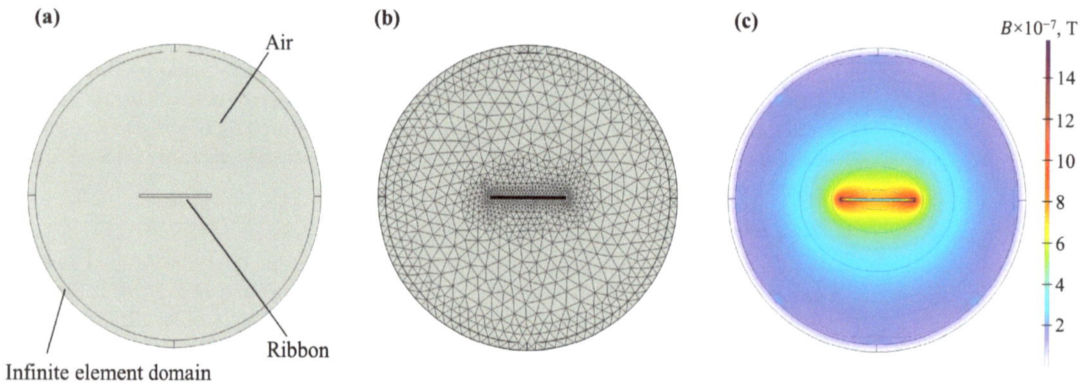

Figure 3. A model for calculating the MI response of the ribbon with the finite element method using Comsol Multiphysics (**a**), fragmentation into a finite element mesh (**b**) and the result of calculating the magnetic induction in the ribbon and the surrounding space (**c**).

2.4. Implementation of Magnetic Impedance Tomography of Amorphous Ribbons

MIT was used in this work to estimate the distribution of magnetic permeability over the amorphous ribbons' cross sections. This method was implemented as described below.

1. The distribution of the magnetic permeability was set according to the system of Equation (5) and Table 2. In this case, μ_i varied from 1 to 25,000.
2. Using solutions for the electric and magnetic field obtained with Comsol Multiphysics for various combinations of μ_i values, the dependences of the reduced impedance on the ac frequency $Z(f)/R_{DC}$ were calculated.
3. In the array of the simulated $Z(f)/R_{DC}$ dependences, we found the one that had the smallest absolute deviation from the $Z(f)/R_{DC}$ dependence obtained experimentally.
4. The combination of μ_i values at which the simulated dependence $Z(f)/R_{DC}$ has the smallest deviation from the experimental one is, presumably, an approximation of the actual distribution of the magnetic permeability over the ribbon cross section.

Table 2. Coordinates of the layers of ribbon models S0 and S1 when performing MIT.

Sample	Layer Boundary Coordinates, μm					
	y_1	y_2	y_3	y_4	y_5	y_6
S0	4	6	8	10	11	12
S1	4	6	8	10	12	13

The dependence of the magnetic permeability on the ac frequency can be described using the Landau–Lifshitz–Gilbert equation (see for example [40]). The greater the Gilbert damping parameter k_G included in this equation, the more the magnetic permeability changes with frequency. For amorphous CoFeSiB alloys, the typical k_G value is around 0.03 [41]. With this parameter value, the magnetic permeability module at the frequency of 80 MHz differs from the magnetic permeability module at the frequency of 0.01 MHz by several percent. Therefore, to simplify the modeling in this work, we neglect the frequency dependence of the magnetic permeability modulus. However, it should be noted that the real and especially imaginary component of the magnetic permeability changes with the ac frequency much more significantly than the modulus. Therefore, taking into account magnetic permeability dependence on the ac frequency is strictly necessary when determining its components using MIT. Obviously, for this, it will be necessary to analyze not the impedance modulus, but the resistance and reactance. The determination of the magnetic permeability components using MIT will be discussed in further articles devoted to the development of this method.

3. Results and Discussion

3.1. Experimental Results

The saturation magnetization of both ribbons is about $M_S \approx 560$ kA/m (Figure 4). The coercive force in both cases is about 50 A/m. Thus, according to the magnetic hysteresis loops, the ribbons have pronounced soft magnetic properties.

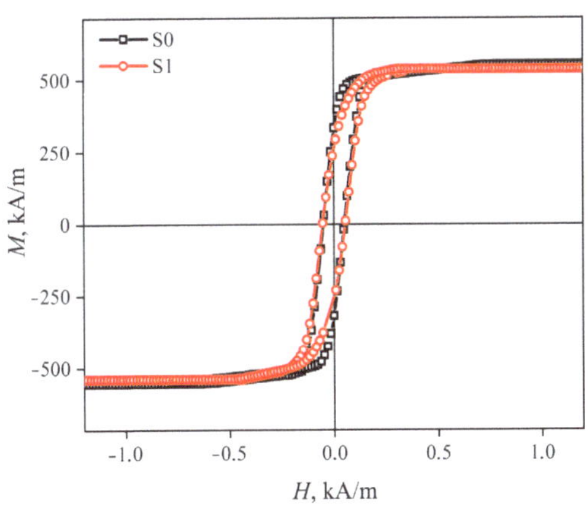

Figure 4. Magnetic hysteresis loops of S0 and S1 ribbons.

The saturation magnetostriction, determined as described in Section 2.2, is $-0.18 \cdots 10^{-7}$ and $+0.59 \cdots 10^{-7}$ for samples S0 and S1, respectively. It should be noted that the obtained magnetostriction values are in good agreement with the results obtained for similar ribbons based on the analysis of changes in hysteresis loops under the action of tensile mechanical stresses [42].

The value of $(\Delta Z/Z)_{max}$ corresponds to the peak in the magnetic field dependence of the MI and reaches its maximum value at an ac frequency of about 8 MHz in the case of S1 ribbons (Figure 5). In this case, $(\Delta Z/Z)_{max}$ exceeds 200%, which may be of practical interest.

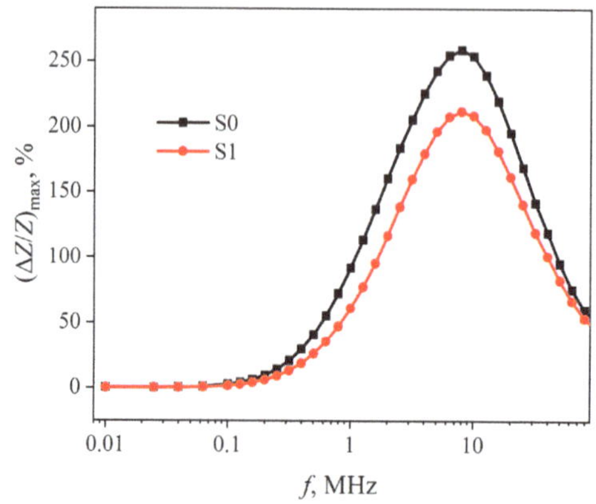

Figure 5. Frequency dependences of the highest MI value $(\Delta Z/Z)_{max}(f)$ in S0 and S1 ribbons.

In the case of the S0 ribbons, the dependence $(\Delta Z/Z)_{max}(f)$ has the same characteristics as in the case of S1 ribbons (Figure 5). However, the highest MI value in these ribbons exceeds 250%, which is observed at an ac frequency of about 8 MHz.

Both in the case of the S1 ribbons and of the S0 ribbons, the MI dependences on the external magnetic field strength $\Delta Z/Z(H)$ in the entire ac frequency range have two peaks. This is expressed as the stronger the $\Delta Z/Z(H)$, the higher the f. In addition, an increase in the magnetic field strength H_p is observed with an increase in the frequency of the alternating current. For S1 ribbons, the minimum value of the peak field is $H_{p1} \approx 60$ A/m and the maximum value is $H_{p2} \approx 320$ A/m (Figure 6b). In turn, for S0 ribbons, these values are approximately 20 and 260 A/m, respectively (Figure 6a). The described changes in H_p with ac frequency increasing, as well as an increase in the increasing portion of the $\Delta Z/Z(H)$ dependence may be associated with the inhomogeneous distribution of the magnetic permeability over the cross section of the S0 and S1 ribbons [38].

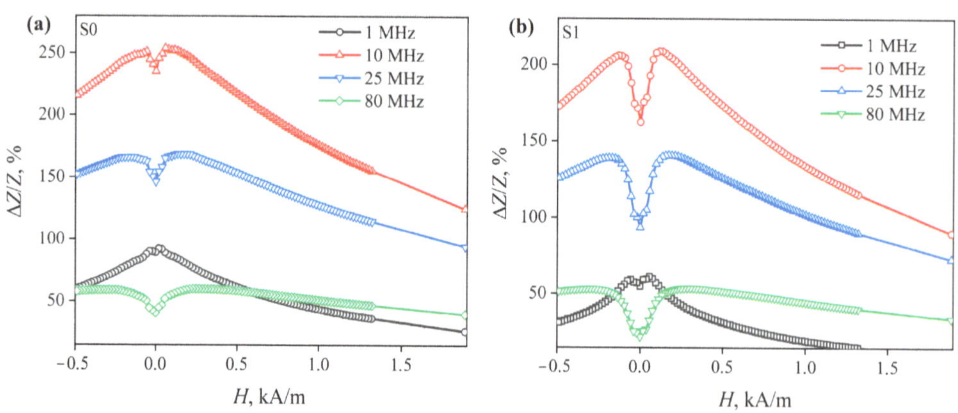

Figure 6. Dependences of the magnetoimpedance effect in the S0 (**a**) and S1 (**b**) ribbons on the external magnetic field strength, obtained at various ac frequencies.

3.2. MIT Results

Details of the magnetic permeability distribution over the ribbon cross section were restored using MIT (see Section 2.4) based on the analysis of the experimental $Z(f)/R_{DC}$

dependences (Figures 7 and 8, solid lines). These dependencies consist of two sections: an almost horizontal section $Z/R_{DC} \approx 1$, which at a certain alternating current frequency f_0, smoothly transforms into an increasing section. It should be noted that there are features such as kinks and changes in the angle of inclination in low magnetic fields in the increasing section of the $Z(f)/R_{DC}$ dependence. It may indicate a non-uniform distribution of the magnetic permeability over the cross section of the ribbon [38]. In magnetic fields close to H_{max}, the increasing section is almost a straight line, indicating a distribution of magnetic permeability that is close to uniform.

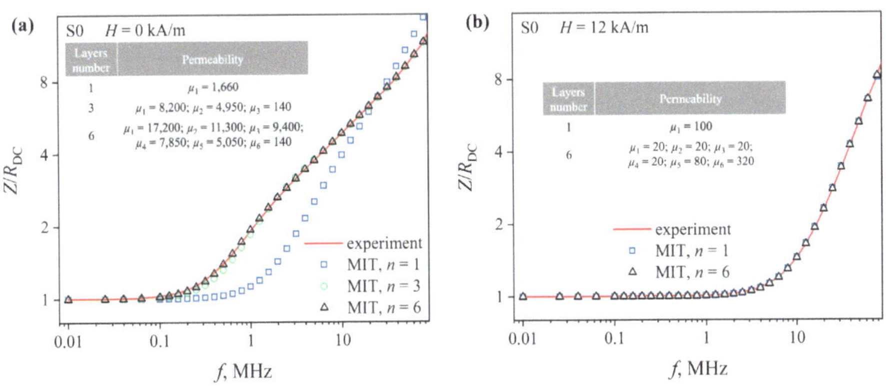

Figure 7. Frequency dependences of the reduced impedance $Z(f)/R_{DC}$ of the S0 ribbons. The dependences were obtained in magnetic fields of 0 (**a**) and 12 kA/m (**b**). Lines are experimental dependences; markers—dependencies restored using MIT. The tables show the magnetic permeability values reconstructed using MIT.

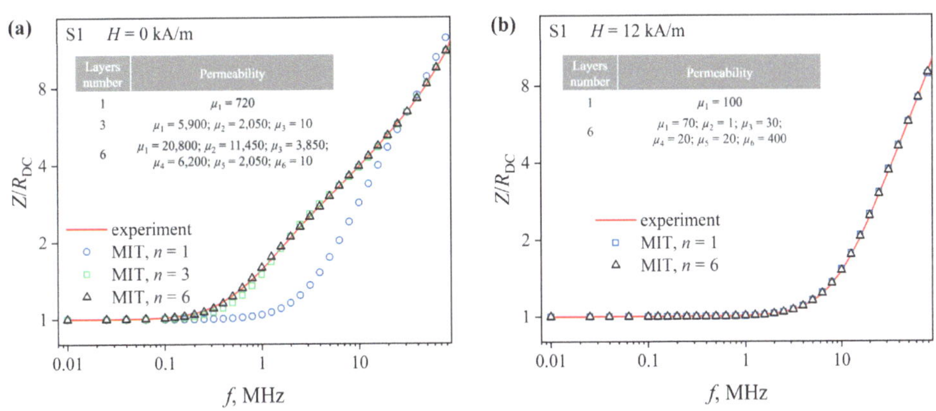

Figure 8. Frequency dependences of the reduced impedance $Z(f)/R_{DC}$ of the S1 ribbons. The dependences were obtained in magnetic fields of 0 (**a**) and 12 kA/m (**b**). Lines are experimental dependences; markers—dependencies restored using MIT. The tables show the magnetic permeability values reconstructed using MIT.

The $Z(f)/R_{DC}$ dependences reconstructed from the MIT results at n = 1 (uniform distribution of the magnetic permeability over the ribbon cross section) differ significantly from the experimental ones obtained in low magnetic fields (Figures 7a and 8a, markers). The relative deviation for some ac frequencies exceeds 80% and confirms the conclusion about the inhomogeneous magnetic structure of the ribbons. An increase in n in the model leads to a decrease in the deviation of the reconstructed $Z(f)/R_{DC}$ dependences from the experimental ones. Thus, at n = 3, the $Z(f)/R_{DC}$ dependences reconstructed using MIT deviate from the experimental ones by no more than 6%. Increasing the number of layers to six made it possible to ensure that the deviation did not exceed 3% over the entire studied range of the magnetic fields.

At the same time, at H_{max}, the $Z(f)/R_{DC}$ dependences, constructed from the MIT results, even at $n = 1$, deviate from the experimental ones by no more than 3% (Figures 7b and 8b markers). This confirms the conclusion that with external magnetic field strength increasing, the distribution of the magnetic permeability over the ribbons cross section becomes more uniform. When using the model with $n = 6$, the deviation of the reconstructed and experimental dependencies does not exceed 1%.

It should also be noted that the deviation of the simulated and experimental dependences may be due to the fact that the frequency dispersion of the magnetic permeability, including the dispersion of the magnetic permeability associated with the motion of domain walls, is not taken into account. Moreover, symmetrical models were used for MIT while the distribution of magnetic permeability can be asymmetric due to the asymmetric distribution of hardening stresses and the different state of the contact and free surfaces of the ribbon [28–33].

Magnetic permeability distributions reconstructed using MIT for $n = 6$ are shown in Figure 9 for the S0 ribbon and in Figure 10 for the S1 ribbon.

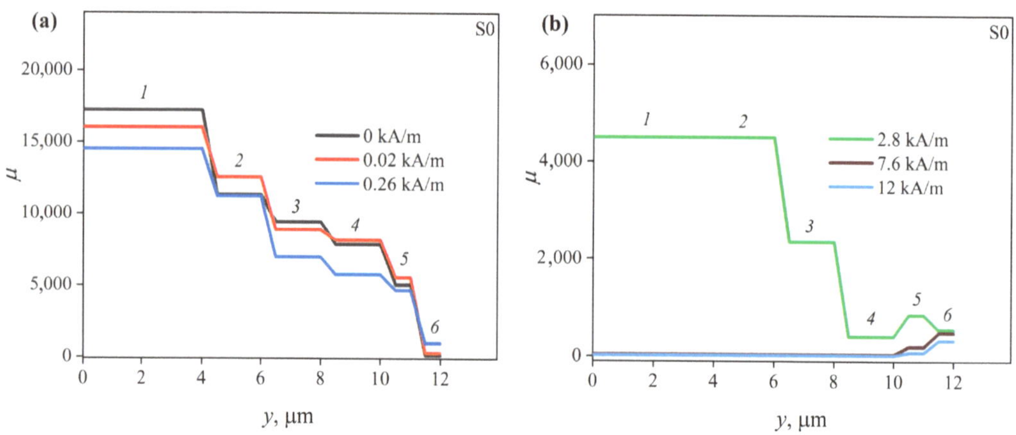

Figure 9. Distributions of the magnetic permeability over the S0 ribbon cross section in the magnetic fields with strengths (**a**) 0, 0.02 and 0.26 kA/m and (**b**) 2.80, 7.6 and 12.4 kA/m.

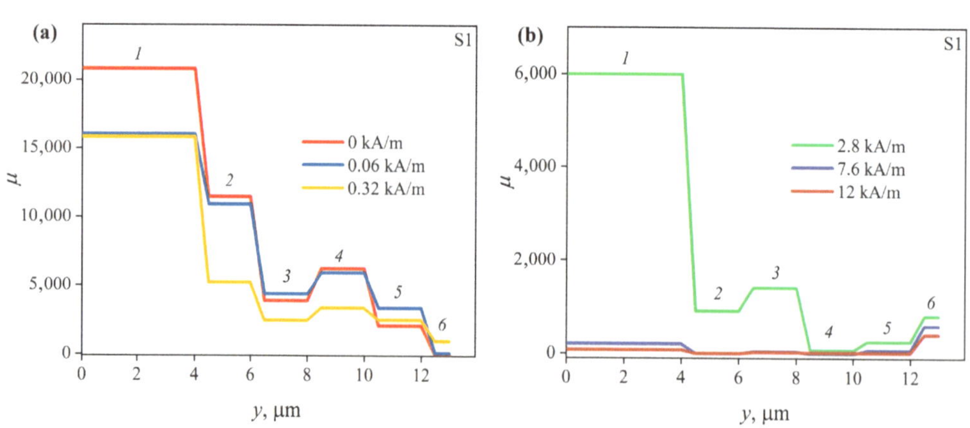

Figure 10. Distributions of magnetic permeability over the S1 ribbon cross section in magnetic fields with strengths (**a**) 0, 0.06 and 0.32 kA/m and (**b**) 2.80, 7.6 and 12.4 kA/m.

According to the MIT results, the permeability of the S0 ribbon surface layer that is 1 μm thick is about 140 in the absence of the external magnetic field (Figure 9). The permeability of the next layer is much higher—more than 4000. With further advancement in depth, the magnetic permeability gradually increases, reaching 17,000 in the central part of the ribbon.

The permeability of the surface layer of the S1 ribbon is also low—about 10 (Figure 10). However, the details of the magnetic permeability distribution in the inner regions of the S1 ribbon are somewhat different from those of the S0 ribbon. Thus, at $H = 0$, the magnetic permeability first increases from 2000 to 6200, then decreases to 3800 and then increases again when moving deep into the ribbon from layer 5 to layer 2 (see Table 2). The magnetic permeability of the central layer exceeds 20,000. Note that internal quenching stresses at some distance from the ribbon surface can change their sign, passing through 0 [28,29]. In this region, the magnetoelastic energy $3\lambda_s\gamma_{in}/2$ (γ_{in} is the internal quenching stress) can also be near zero. This, apparently, determines the higher magnetic permeability in layer 4 compared to that in neighboring layers.

Next, consider how the magnetic permeability changes with a change in the external magnetic field.

As H increases, the magnetic permeability of the outer layer of the S0 ribbon first increases, reaching a maximum value of about 950 at $H = H_{p2}$, and then decreases. It reaches 300 at H_{max} (Figure 9). The magnetic permeabilities of layers 4–5, as well as the permeability of layer 2, reach their maximum values in lower fields, at $H = H_{p1}$. The magnetic permeabilities of the remaining layers decrease with increasing external magnetic field strength. Note that the inner layers, in contrast to the surface ones, are close to saturation at H_{max}. Their magnetic permeabilities are about 1.

In the case of ribbon S1, an increase in the magnetic field strength from 0 to H_{p1} leads to an increase in the magnetic permeability of layers 3, 5 and 6. The permeability of layer 4 practically does not change in this case. The magnetic permeability of layer 5 increases most noticeably. The magnetic permeability, as in the case of S0 ribbons, decreases in the entire ribbon with the subsequent increase in H, except for the surface layer. Its permeability continues to increase, reaching a value of about 950 and then decreasing to about 400 at H_{max}.

Probably, the increase in the MI at low frequencies is mainly due to an increase in the magnetic permeability of the inner regions of the ribbon both in the case of the S1 ribbon and in the case of the S0 ribbon. The nature of the $\Delta Z/Z(H)$ dependence at high frequencies is mainly determined by the magnetic field dependence of the magnetic permeability surface layer.

The peculiarities of the change in the magnetic permeability of the surface layers of both ribbons indicates a predominantly transverse orientation of the EMA in them. At the same time, the maximum values of these permeabilities are small. This is probably due to the significant anisotropy dispersion caused by non-uniform relief of the ribbon. Also, the magnetic permeabilities of the surface layers of the ribbons are far from saturation, even at H_{max}, due to the anisotropy dispersion. This, along with the high contribution of the external inductance $2\pi f L_{ext}$ (see expression (2)) limits the maximum MI at high ac frequencies (Figure 5), which must be taken into account when developing MI sensors.

The magnetic anisotropy of the ribbon inner regions is predominantly longitudinal. This is indicated by high values of the magnetic permeability and its decrease with increasing H. An increase in the permeability of some inner layers with a change in the magnetic field from 0 to H_{p1} is presumably associated with a decrease in the magnetic anisotropy dispersion. Based on this, it can be assumed that quenching stresses do not have a decisive effect on the orientation of the EMA. Otherwise, the orientation of the EMA in the inner and outer regions of the S0 and S1 ribbons would be different since the signs of their magnetostriction constants are different. It is likely that the EMA orientation is strongly affected by the anisotropy of the ribbons' shape. However, the energy of magnetoelastic anisotropy can affect the values of magnetic permeability due to the quenching stresses.

The magnetoelastic energy $3\lambda_s\gamma_{in}/2$ of the S0 ribbon is smaller, since its magnetostriction is smaller in modulus. This leads to the fact that the magnetic permeability of the inner region of the S0 ribbon is higher than that of the S1 ribbon. According to the literature [28,29], the quenching stresses in the central part are very low, which seems to be responsible for the high magnetic permeability of both ribbons.

The practical significance of MIT deserves special attention. As is known, amorphous soft magnetic ribbons obtained by rapid quenching from a melt are characterized by a noticeable scatter of functional characteristics even within the same batch. Despite this, a minimum scatter of metrological characteristics must be ensured when manufacturing devices using these materials, especially MI magnetic field sensors. This can be achieved by selecting pieces of ribbons with similar distributions of magnetic characteristics using MIT. In addition, MIT may be in demand when monitoring changes in the magnetic structure during heat treatment, which is often used to improve the functional properties of amorphous ribbons [43,44].

4. Conclusions

Thus, based on the analysis of the experimental $\Delta Z/Z(H)$ and $Z(f)/R_{DC}$ dependences it was shown that $Co_{68.5}Fe_{4.0}Si_{15.0}B_{12.5}$ (S0) amorphous ribbons with magnetostriction $\lambda_s \approx -0.18 \cdot 10^{-7}$ and $Co_{68.6}Fe_{3.9}Mo_{3.0}Si_{12.0}B_{12.5}$ (S1) with magnetostriction $\lambda_s \approx +0.59 \cdot 10^{-7}$ have an inhomogeneous magnetic structure. Details of the distribution of magnetic permeability over the ribbons' cross section were reconstructed using MIT based on an analysis of the experimental $Z(f)/R_{DC}$ dependences obtained in the ac frequency range from 0.01 to 80 MHz in magnetic fields of various strengths.

According to the MIT, the magnetic permeability distribution over the cross section of both ribbons is nonuniform. The surface layers of both ribbons have predominantly transverse magnetic anisotropy. In this case, the low values of their magnetic permeability indicate a significant dispersion of the magnetic anisotropy associated with the surface topography. The magnetic permeability of the inner layers of both ribbons is much higher than that of the surface ones, and the nature of its change in the magnetic field indicates that the magnetic anisotropy of the inner layers is predominantly longitudinal.

The influence of magnetostriction is manifested in the fact that the magnetic permeability of the inner layers of the $Co_{68.5}Fe_4Si_{15.0}B_{12.5}$ ribbon is generally higher than that of the $Co_{68.6}Fe_{3.9}Mo_{3.0}Si_{12.0}B_{12.5}$ ribbon, since its magnetostriction is lower in the modulus respectively, and the magnetoelastic energy associated with quenching stresses is less.

The distribution of the magnetic properties over the cross section of amorphous magnetically soft ribbons, reconstructed using MIT, must be taken into account when choosing the optimal operating modes for MI sensors. It should also be noted that the MIT can be in demand in the manufacture of devices based on amorphous ribbons to ensure the minimum spread of their characteristics.

Author Contributions: Conceptualization, D.A.B.; Investigation, M.S.D.; Methodology, D.A.B., M.S.D. and A.V.S.; Project administration, D.A.B.; Software, M.S.D.; Supervision, D.A.B.; Validation, D.A.B., M.S.D. and A.V.S.; Visualization, M.S.D.; Writing—original draft, D.A.B. and M.S.D.; Writing—review and editing, A.V.S. All authors have read and agreed to the published version of the manuscript.

Funding: The modeling and measurements of the impedance were carried out at Irkutsk State University and supported by the Russian Science Foundation (grant no. 22-22-00709, https://rscf.ru/project/22-22-00709/ (accessed on 5 October 2023).

Institutional Review Board Statement: Not applicable.

Informed Consent Statement: Not applicable.

Data Availability Statement: The data are available from the corresponding author upon reasonable request.

Conflicts of Interest: The authors declare no conflict of interest.

References

1. Beach, R.S.; Berkowitz, A.E. Sensitive field- and frequency-dependent impedance spectra of amorphous FeCoSiB wire and ribbon (invited). *J. Appl. Phys.* **1994**, *76*, 6209–6213. [CrossRef]
2. Kurlyandskaya, G.V.; Lezama, L.; Pasynkova, A.A.; Volchkov, S.O.; Lukshina, V.A.; Larrañaga, A.; Dmitrieva, N.V.; Timofeeva, A.V.; Orue, I. Amorphous FeCoCrSiB Ribbons with Tailored Anisotropy for the Development of Magnetic Elements for High Frequency Applications. *Materials* **2022**, *15*, 4160. [CrossRef] [PubMed]
3. Arzuza, L.C.C.; Béron, F.; Pirota, K.R. High-frequency GMI hysteresis effect analysis by first-order reversal curve (FORC) method. *J. Magn. Magn. Mater.* **2021**, *534*, 168008. [CrossRef]
4. Tejedor, M.; Hernando, B.; Sánchez, M.; Prida, V.; Garcia-Beneytez, J.; Vázquez, M.; Herzer, G. Magnetoimpedance effect in zero magnetostriction nanocrystalline $Fe_{73.5}Cu_1Nb_3Si_{16.5}B_6$ ribbons. *J. Magn. Magn. Mater.* **1998**, *185*, 61–65. [CrossRef]
5. Dwevedi, S.; Markandeyulu, G.; Ohodnicki, P.R.; Leary, A.; McHenry, M.E. Stress-MI and domain studies in Co-based nanocrystalline ribbons. *J. Magn. Magn. Mater.* **2011**, *323*, 1929–1933. [CrossRef]
6. Kikuchi, H.; Kamata, S.; Takahashi, Y.; Nakai, T.; Hashi, S.; Ishiyama, K. Investigation of driving power dependence on magnetoimpedance properties of thin-film elements with uniaxial anisotropy. *IEEE Trans. Magn.* **2015**, *51*. [CrossRef]
7. García, D.; Muñoz, J.; Kurlyandskaya, G.; Vázquez, M.; Ali, M.; Gibbs, M.R. Induced anisotropy, magnetic domain structure and magnetoimpedance effect in CoFeB amorphous thin films. *J. Magn. Magn. Mater.* **1999**, *191*, 339–344. [CrossRef]
8. García-Arribas, A.; Fernández, E.; Svalov, A.; Kurlyandskaya, G.V.; Barandiaran, J.M. Thin-film magneto-impedance structures with very large sensitivity. *J. Magn. Magn. Mater.* **2016**, *400*, 321–326. [CrossRef]
9. Chen, J.-A.; Ding, W.; Zhou, Y.; Cao, Y.; Zhou, Z.-M.; Zhang, Y.-M. Stress-impedance effects in sandwiched FeCuNbCrSiB/Cu/FeCuNbCrSiB films. *Mater. Lett.* **2006**, *60*, 2554–2557. [CrossRef]
10. Antonov, A.S.; Gadetskii, S.N.; Granovskii, A.B.; D'yachkov, A.L.; Paramonov, V.P.; Perov, N.S.; Prokoshin, A.F.; Usov, N.A.; Lagar'kov, A.N. Giant magnetoimpedance in amorphous and nanocrystalline multilayers. *Phys. Met. Metallogr.* **1997**, *83*, 612–618.
11. Bukreev, D.A.; Derevyanko, M.S.; Moiseev, A.A.; Semirov, A.V.; Savin, P.A.; Kurlyandskaya, G.V. Magnetoimpedance and Stress-Impedance Effects in Amorphous CoFeSiB Ribbons at Elevated Temperatures. *Materials* **2020**, *13*, 3216. [CrossRef] [PubMed]
12. Zhan, Z.; Yaoming, L.; Jin, C.; Yunfeng, X. Current sensor utilizing giant magneto-impedance effect in amorphous ribbon toroidal core and CMOS inverter multivibrator. *Sens. Actuators A Phys.* **2007**, *137*, 64–67. [CrossRef]
13. Gazda, P.; Szewczyk, R. Novel Giant Magnetoimpedance Magnetic Field Sensor. *Sensors* **2020**, *20*, 691. [CrossRef] [PubMed]
14. Feng, Z.; Zhi, S.; Wei, M.; Zhou, Y.; Liu, C.; Lei, C. An integrated three-dimensional micro-solenoid giant magnetoimpedance sensing system based on MEMS technology. *Sens. Actuators A Phys.* **2019**, *299*, 111640. [CrossRef]
15. Kurlyandskaya, G.V.; Sánchez, M.L.; Hernando, B.; Prida, V.M.; Gorria, P.; Tejedor, M. Giant-magnetoimpedance-based sensitive element as a model for biosensors. *Appl. Phys. Lett.* **2003**, *82*, 3053–3055. [CrossRef]
16. Yang, H.; Chen, L.; Lei, C.; Zhang, J.; Li, D.; Zhou, Z.-M.; Bao, C.-C.; Hu, H.-Y.; Chen, X.; Cui, F.; et al. Giant magnetoimpedance-based microchannel system for quick and parallel genotyping of human papilloma virus type 16/18. *Appl. Phys. Lett.* **2010**, *97*, 043702. [CrossRef]
17. Yang, Z.; Wang, H.; Dong, X.; Yan, H.; Lei, C.; Luo, Y. Giant magnetoimpedance based immunoassay for cardiac biomarker myoglobin. *Anal. Methods* **2017**, *9*, 3636–3642. [CrossRef]
18. Landau, L.D.; Lifshitz, E.M. *Electrodynamics of Continuous Media*; Pergamon Press: Oxford, UK, 1960.
19. Kraus, L. Theory of giant magneto-impedance in the planar conductor with uniaxial magnetic anisotropy. *J. Magn. Magn. Mater.* **1999**, *195*, 764–778. [CrossRef]
20. Kalantarov, P.L.; Tseytlin, L.A. *Raschet Induktivnostey [Calculation of inductances]*, 3rd ed.; Energoatomizdat, Leningrad; 1986; 488 p., (Rus).
21. Pirota, K.R.; Kraus, L.; Knobel, M.; Pagliuso, P.G.; Rettori, C. Angular dependence of giant magnetoimpedance in an amorphous Co-Fe-Si-B ribbon. *Phys. Rev. B.* **1999**, *60*, 6685–6691. [CrossRef]
22. Dong, C.; Chen, S.; Hsu, T.Y. A modified model of GMI effect in amorphous films with transverse magnetic anisotropy. *J. Magn. Magn. Mater.* **2003**, *263*, 78–82. [CrossRef]
23. Rouhani, A.A.; Matin, L.F.; Mohseni, S.M.; Zoriasatain, S. A Domain Dynamic Model Study of Magneto-impedance Sensor in the Presence of Inhomogeneous Magnetic Fields. *J. Supercond. Nov. Magn.* **2021**, *34*, 571–580. [CrossRef]
24. Buznikov, N.A.; Kurlyandskaya, G.V. Magnetoimpedance in Symmetric and Non-Symmetric Nanostructured Multilayers: A Theoretical Study. *Sensors* **2019**, *19*, 1761. [CrossRef] [PubMed]
25. García-Arribas, A.; Barandiarán, J.M.; de Cos, D. Finite element method calculations of GMI in thin films and sandwiched structures: Size and edge effects. *J. Magn. Magn. Mater.* **2008**, *320*, e4–e7. [CrossRef]
26. Li, B.; Kosel, J. Three dimensional simulation of giant magneto-impedance effect in thin film structures. *J. Appl. Phys.* **2011**, *109*, 07E519. [CrossRef]
27. Volchkov, S.O.; Pasynkova, A.A.; Derevyanko, M.S.; Bukreev, D.A.; Kozlov, N.V.; Svalov, A.V.; Semirov, A.V. Magnetoimpedance of CoFeCrSiB Ribbon-Based Sensitive Element with FeNi Covering: Experiment and Modeling. *Sensors* **2021**, *21*, 6728. [CrossRef]
28. Bengus, V.Z.; Duhaj, P.; Korolkova, E.B.; Ocelík, V. Internal Stress Contribution to the Yield Stress Anisotropy of Amorphous Alloy Ribbons. *Solid State Phenom.* **1993**, *35–36*, 575–580. [CrossRef]

29. Tejedor, M.; Garcı, J.A.; Carrizo, J.; Elbaile, L.; Santos, J.D. Effect of residual stresses and surface roughness on coercive force in amorphous alloys. *J. Appl. Phys.* **2002**, *91*, 8435. [CrossRef]
30. Takahashi, M.; Miyazaki, T. Magnetic Anisotropy in an Amorphous $Fe_{80}P_{13}C_7$ Alloy. *Jpn. J. Appl. Phys.* **1979**, *18*, 743–752. [CrossRef]
31. Tsukahara, S.; Satoh, T.; Tsushima, T. Magnetic anisotropy distribution near the surface of amorphous ribbons. *IEEE Trans. Magn.* **1978**, *14*, 1022–1024. [CrossRef]
32. Kraus, L.; Tomáš, I.; Keatociivílová, E.; Speingmann, B.; Müller, K. Magnetic anisotropy caused by oriented surface roughness of amorphous ribbons. *Phys. Status Solidi.* **1987**, *100*, 289–299. [CrossRef]
33. Vavassori, P.; Callegaro, L.; Puppin, E.; Malizia, F.; Ronconi, F. Surface magnetic characterization of FeB amorphous ribbons. *J. Magn. Magn. Mater.* **1996**, *157–158*, 171–172. [CrossRef]
34. Semirov, A.V.; Bukreev, D.A.; Moiseev, A.A.; Derevyanko, M.S.; Kudryavtsev, V.O. Relationship Between the Temperature Changes of the Magnetostriction Constant and the Impedance of Amorphous Elastically Deformed Soft Magnetic Cobalt-Based Ribbons. *Russ. Phys. J.* **2013**, *55*, 977–982. [CrossRef]
35. Amalou, F.; Gijs, M.A.M. Giant magnetoimpedance of chemically thinned and polished magnetic amorphous ribbons. *J. Appl. Phys.* **2001**, *90*, 3466–3470. [CrossRef]
36. Lotfollahi, Z.; Amirabadizadeh, A.; Safronov, A.P.; Beketov, I.V.; Kurlyandskaya, G.V. Magnetoimpedance Effect in CoFeMoSiB As-Quenched and Surface Modified Amorphous Ribbons in the Presence of Igon Oxide Nanoparticles of Water-Based Ferrofluid. *J. Sensors.* **2017**, *2017*, 1–9. [CrossRef]
37. Yang, Z.; Chlenova, A.A.; Golubeva, E.V.; Volchkov, S.O.; Guo, P.; Shcherbinin, S.V.; Kurlyandskaya, G.V. Magnetoimpedance Effect in the Ribbon-Based Patterned Soft Ferromagnetic Meander-Shaped Elements for Sensor Application. *Sensors* **2019**, *19*, 2468. [CrossRef]
38. Bukreev, D.A.; Derevyanko, M.S.; Moiseev, A.A.; Svalov, A.V.; Semirov, A.V. The Study of the Distribution of Electrical and Magnetic Properties over the Conductor Cross-Section Using Magnetoimpedance Tomography: Modeling and Experiment. *Sensors* **2022**, *22*, 9512. [CrossRef]
39. Knobel, M.; Gómez-Polo, C.; Vázquez, M. Evaluation of the linear magnetostriction in amorphous wires using the giant magneto-impedance effect. *J. Magn. Magn. Mater.* **1996**, *160*, 243–244. [CrossRef]
40. Usov, N.A.; Antonov, A.S.; Lagar'kov, A.N. Theory of giant magneto-impedance effect in amorphous wires with different types of magnetic anisotropy, *J. Magn. Magn. Mater.* **1998**, *185*, 159–173. [CrossRef]
41. Ménard, D.; Britel, M.; Ciureanu, P.; Yelon, A. Giant magnetoimpedance in a cylindrical magnetic conductor. *J. Appl. Phys.* **1998**, *84*, 2805–2814. [CrossRef]
42. Lotfollahi, Z.; García-Arribas, A.; Amirabadizadeh, A.; Orue, I.; Kurlyandskaya, G.V. Comparative study of magnetic and magnetoimpedance properties of CoFeSiB-based amorphous ribbons of the same geometry with Mo or W additions. *J. Alloys Compd.* **2017**, *693*, 767–776. [CrossRef]
43. Liu, J.; Du, Z.; Jiang, S.; Shen, H.; Li, Z.; Xing, D.; Ma, W.; Sun, J. Tailoring giant magnetoimpedance effect of Co-based microwires for optimum efficiency by self-designed square-wave pulse current annealing. *J. Magn. Magn. Mater.* **2015**, *385*, 145–150. [CrossRef]
44. Tehranchi, M.M.; Ghanaatshoar, M.; Mohseni, S.M.; Coisson, M.; Vázquez, M. Temperature dependence of magnetoimpedance in annealed Co-based ribbons. *J. Non. Cryst. Solids.* **2005**, *351*, 2983–2986. [CrossRef]

sensors

Article

Longitudinal Spin Seebeck Effect Thermopiles Based on Flexible Co-Rich Amorphous Ribbons/Pt Thin-Film Heterostructures

Marcio A. Correa [1,2], Andrey V. Svalov [3], Armando Ferreira [2,4], Matheus Gamino [1], Edimilson F. da Silva [1], Felipe Bohn [1], Filipe Vaz [2,4], Danniel F. de Oliveira [5] and Galina V. Kurlyandskaya [3,*]

1 Departamento de Física, Universidade Federal do Rio Grande do Norte, Natal 59078-900, RN, Brazil; marciocorrea@fisica.ufrn.br (M.A.C.); mgamino@fisica.ufrn.br (M.G.); efelix@fisica.ufrn.br (E.F.d.S.); felipebohn@fisica.ufrn.br (F.B.)
2 Centro de Física das Universidades do Minho e do Porto (CF-UM-UP), Universidade do Minho, 4710-057 Braga, Portugal; armando.f@fisica.uminho.pt (A.F.)
3 Institute of Natural Sciences and Mathematics, Ural Federal University, 620002 Ekaterinburg, Russia; andrey.svalov@urfu.ru
4 LaPMET—Laboratório de Física para Materiais e Tecnologias Emergentes, Universidade do Minho, 4710-057 Braga, Portugal
5 Departamento de Ciências dos Materiais, Universidade Federal da Paraíba, João Pessoa 58059-900, PB, Brazil; danniel.oliveira@acadmico.ufpb.br
* Correspondence: galinakurlyandskaya@urfu.ru

Abstract: Thermoelectric phenomena, such as the Anomalous Nernst and Longitudinal Spin Seebeck Effects, are promising for sensor applications in the area of renewable energy. In the case of flexible electronic materials, the request is even larger because they can be integrated into devices having complex shape surfaces. Here, we reveal that Pt promotes an enhancement of the thermoelectric response in Co-rich ribbon/Pt heterostructures due to the spin-to-charge conversion. Moreover, we demonstrated that the employment of the thermopiles configuration in this system increases the induced thermoelectric current, a fact related to the considerable decrease in the electric resistance of the system. By comparing present findings with the literature, we were able to design a flexible thermopile based on LSSE without the lithography process. Additionally, the thermoelectric voltage found in the studied flexible heterostructures is comparable to the ones verified for rigid systems.

Keywords: anomalous Nernst effect; spintronics; thermoelectric conversion; flexible magnetic materials; magnetic properties; magnetic sensors

Citation: Correa, M.A.; Svalov, A.V.; Ferreira, A.; Gamino, M.; Silva, E.F.d.; Bohn, F.; Vaz, F.; de Oliveira, D.F.; Kurlyandskaya, G.V. Longitudinal Spin Seebeck Effect Thermopiles Based on Flexible Co-Rich Amorphous Ribbons/Pt Thin-Film Heterostructures. *Sensors* **2023**, *23*, 7781. https://doi.org/10.3390/s23187781

Academic Editor: Evangelos Hristoforou

Received: 28 August 2023
Revised: 8 September 2023
Accepted: 8 September 2023
Published: 10 September 2023

1. Introduction

Co-rich amorphous ribbons have excellent ferromagnetic properties. They can be employed in a broad range of technological applications, including wound transformers, inductors, magnetic shielding [1,2], and high-frequency magnetic sensing devices [3–6]. In recent years, they were also extensively studied with focus on biomedical applications for sensitive elements of magnetoimpedance-based devices for both label-free and magnetic label detection [7]. The amorphous Co-rich alloys often present high saturation magnetization, high magnetic permeability, and low magnetic hysteresis losses [8–10]. Furthermore, the flexibility of these materials in combination with the corrosion stability of Cr-doped compositions makes them promising candidates for sensor applications, particularly for covering of the curved surfaces or surfaces having even more complex geometry. All these features are interesting for thermoelectric applications based on the Longitudinal Spin Seebeck Effect (LSSE) [11–13] and Anomalous Nernst Effect (ANE) [14–17].

While there is a limited number of experimental works examining thermoelectric effects in ferromagnetic ribbons [18,19], the existing literature presents promising findings.

Recently, we have explored the ANE effect in $Fe_3Co_{67}Cr_3Si_{15}B_{12}$ amorphous ribbons [18], disclosing the influence of the relaxation and annealing under stress of the amorphous samples on their thermoelectric response. In particular, for temperature differences of $\Delta T \sim 22$ K, thermoelectric signals of about 30 μV were achieved. However, it is worth mentioning that to reach such ΔT levels, the top of the ribbon may often be heated up to the high temperatures of about 80 °C. Considering that the Co-rich ribbons are not protected with a cap layer, surface oxidation can take place. Therefore, this issue must be considered for future sensor applications, especially if long-time stability is expected [20].

It is well known that an in-vacuum annealing process can be used for the improvement of the magnetic properties of the ribbons. This is not the case of the heating in thermoelectric experiments, such as LSSE and ANE measurements, that are carried out in air, hence inducing high levels of oxidation in the ribbons. A possible solution to overcome this issue consists of the use of the protecting covering layer. While the deposition of an insulating layer appears, in some aspects, as an interesting alternative for protection, it drastically modifies the electrical properties of the surface, making the thermoelectric measurements a hard task. In this sense, the usage of metallic protecting layers with high corrosion stability is preferred. Additionally, some specific metallic layers allow the amplification of the thermoelectric signal when a heterostructure ribbon/metallic layer is considered. This is the case for materials with high spin-orbit coupling and positive spin hall angle, which promote a raise in the spin-to-charge current conversion. Remarkably, Platinum (Pt) seems to be the best solution for both protecting against oxidation [21–26] and promoting spin-to-charge conversion [27–29].

From the theoretical point of view, both ANE and LSSE have similar descriptions, in which a magnetic material is submitted to a thermal gradient $\vec{\nabla} T$. As a consequence, an electric field $\vec{E}$ is induced and can be detected as a thermoelectric voltage V. Despite this similarity, the origin of these two effects is fundamentally different. The ANE electrical field comes from the relation between the magnetization of the material and the thermal gradient, and it is given by [29,30],

$$\vec{E}_{ANE} = -S_{ANE}\left(\vec{m} \times \vec{\nabla} T\right), \tag{1}$$

where $\vec{m}$ is the magnetization vector with saturation magnetization m_s, and S_{ANE} is the Anomalous Nernst Effect coefficient, which brings to light the effective efficiency of the thermoelectric conversion in the magnetic system. The LSSE is in turn connected to the generation of pure spin current, which is converted into charge current through the Inverse Spin Hall Effect (ISHE),

$$\vec{J}_c = \frac{2e}{\hbar}\theta_{SH}\left(\vec{J}_s \times \hat{\sigma}\right), \tag{2}$$

in which $\vec{J}_c$ is the charge current, e is the elementary charge, $\hbar$ corresponds to the Planck constant divided by 2π, θ_{SH} is the spin Hall angle, $\hat{\sigma}$ represents the magnetic polarization, and finally $\vec{J}_S$ is the spin current that flows to a metallic layer placed onto the magnetic one. Phenomenologically, the LSSE electrical field can be written as

$$\vec{E}_{LSSE} = -S_{LSSE}\left(\hat{\sigma} \times \vec{\nabla} T\right), \tag{3}$$

where S_{LSSE} is the LSSE coefficient.

Experimentally, $\hat{\sigma}$ can be changed by the application of an external magnetic field $\vec{H}$, which modifies the magnetization $\vec{m}$ of the material. Taking a closer look at Equations (1) and (3), one can notice a strict relation between both effects. The consequence is that, for a *metallic* ferromagnetic/Pt heterostructure, both effects are accessed during the thermoelectric experiment, bringing the contributions of ANE and LSSE simultaneously. While some studies found in the literature have focused on the separation of such contributions [31], for the case of metallic ribbon/Pt heterostructures, we concentrate on obtaining reliable estimates of an effective thermoelectric coefficient S_{eff}.

At the same time, for a practical application, it is necessary to improve the thermoelectric signal of the system. In this sense, design thermopile seems to be a promising path for the increase of the electrical signal. This geometry is explored in distinct studies in the literature [32–35], in which the magnetic system is deposited onto rigid substrate and/or produced using lithography processes. For instance, recently, Weng et al. [35] demonstrated the efficiency on the ANE thermopile composed by a single ferromagnetic element. In this study they showed a linear increase of the thermoelectric signal as the length of the thermopile increases. Kim et al. [33] in turn presented a spin thermopile system in which exchange bias Pt/CoFeB multilayers are explored. In the results, the authors demonstrated a linear increase of thermoelectric signal with the number of bilayers. In the same sense, Uchida et al. [32] used Pt/Nb thermopile based on $Y_3Fe_5O_{13}$ (YIG) to induce an improvement of the thermoelectric signal of the proposed system. From the results, a remarkable increase in the induced thermoelectric voltage of Pt/Nb thermopile in comparison with a single YIG/Pt system was observed.

In the present study, we designed, developed, and comparatively analyzed bare Co-rich ribbon and ribbon/Pt heterostructures and explored the Anomalous Nernst Effect and Longitudinal Spin Seebeck Effect showing that the Pt cap layer promotes an enhancement of the thermoelectric response. Further, we demonstrated the advantages of designing thermopiles consisting of Co-rich ribbon/Pt heterostructures in a parallel association, which can considerably improve the thermoelectric current response.

2. Materials and Methods

We investigated flexible $Fe_3Co_{67}Cr_3Si_{15}B_{12}$ (self-fabricated by the authors at UPV-EHU, Leioa, Spain) amorphous ribbons with a thickness of $t_F \approx 0.24$ μm and width of $w \approx 0.8$ mm (Figure 1a). According to the existing studies, the ribbons of this composition have a very small negative magnetostriction coefficient $\lambda_s = -1 \times 10^{-7}$ [36,37]. For this reason, we expect that the thermoelectric response does not show considerable change with stress applications.

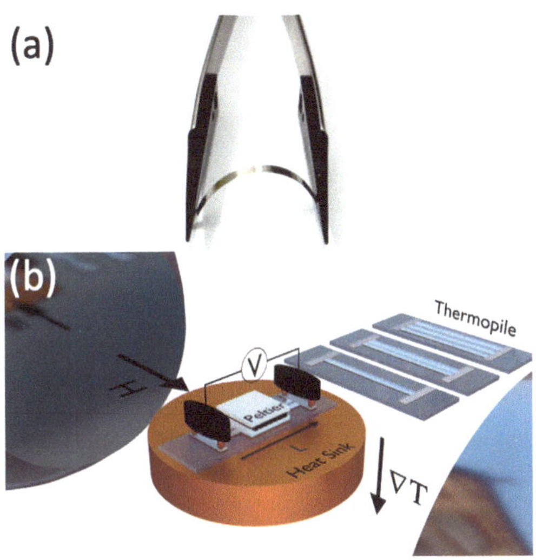

Figure 1. (**a**) Image of bare $Fe_3Co_{67}Cr_3Si_{15}B_{12}$ flexible ribbon. (**b**) Schematic representation of the thermoelectric experimental setup. In particular, the arrow characterizing the field corresponds to the direction labeled as $\varphi_H = 90°$. We show the sketch of the thermopiles explored in this study (Pt, Pt2, Pt3 for one, two, and three ribbons, respectively).

The samples were prepared by a rapid quenching technique, following a procedure similar to that described in Refs. [18,37,38], thus obtaining ribbons with a length of up to 10 m. Amorphous ribbons were prepared by the melt-spinning technique onto a rotating roller. The master alloy of desired composition was melted in a quartz crucible using an inductor coil and ejected by Ar pressure jump onto a copper roller rotating at 470 rad/s in a vacuum chamber. High-quality ribbons with shine-free surfaces were obtained.

Here, no annealing was performed to modify the structure and magnetic properties of the amorphous as-cast ribbons. To cap the ribbons, the Magnetron Sputtering technique was employed. Specifically, we considered a 6 nm thick Platinum layer as a cap layer deposited with the following experimental parameters: base pressure of 7×10^{-8} Torr deposition pressure of 3×10^{-3} Torr with 20 sccm Ar flow, and 50W set in the DC source resulting in a deposition rate of 0.98 Å/s. During the deposition, the ribbons were kept in a constant rotation to avoid any anisotropy induction due to the residual magnetic field of the magnetron sputtering guns. In particular, the Pt thickness was carefully chosen allowing the achievement of good efficiency in the spin-charge current during the LSSE measurements [39].

Regarding the characterization of the samples, first magnetization curves at room temperature were obtained using a Vibrating Sample Magnetometer (VSM, Lake Shore 7404 Westerville, OH, USA). Amorphous ribbons of 13 mm lengths were employed. Experiments were performed with the magnetic field applied along the main axis to the ribbon ($\varphi_H = 0°$) and perpendicular to the long side in-plane of the ribbon ($\varphi_H = 90°$). The thermoelectric voltage measurements were obtained using a homemade system. In this case, the thermal gradient ∇T was applied by using a micro-Peltier module (see Figure 1b), which heats the top of the heterostructure. At the same time, the bottom of the glass substrate was put in thermal contact with a heat sink. In particular, a glass substrate was considered to avoid the electrical contact of the sample with the metallic heat sink. To improve thermal conductivity, thermal paste was used in the system. The voltage detection was done with silver conductive glue and using gold spring contacts distanced $L = 11$ mm from each other. From theory [30], the thermoelectric voltage is described as follows:

$$V = -\int_0^L \vec{E} \cdot d\vec{l}, \tag{4}$$

where $d\vec{l}$ is the length differential element integrated along the distance between electrical contacts L. The magnetic field is applied through an electromagnet controlled by using a Kepco bipolar source (BOP 20/20). The thermoelectric signal is measured by using a high-precision $6\frac{1}{2}$ digits multimeter (Keithley). The sample rotation was performed by a high-resolution step motor. All the experiments were controlled through a homemade LabView software (LabVIEW Free online IDE for Learner). During the thermoelectric measurements, each experimental point was an average of 10 measurements. A similar procedure was employed for the magnetic characterization, in which each experimental point was an average of 20 measurements. For this reason, we believe that the experimental results reflect the real efficiency of the proposed thermopile in these specific experimental conditions (room temperature, controlled humidity and pressure).

The experiments were performed using both the bare $Fe_3Co_{67}Cr_3Si_{15}B_{12}$ amorphous ribbon and the $Fe_3Co_{67}Cr_3Si_{15}B_{12}$/Pt heterostructure for comparative ANE and ANE + LSSE responses. Prototypes with one (Pt), two (Pt2), and three (Pt3) $Fe_3Co_{67}Cr_3Si_{15}B_{12}$/Pt heterostructures in a parallel configuration were tested. The $Fe_3Co_{67}Cr_3Si_{15}B_{12}$/Pt heterostructures connected by using conductive silver glue, onto a glass substrate, are depicted in Figure 1b.

3. Results and Discussion

Figure 2 shows the magnetization curves for the bare ribbon (without an additional cap layer) and the $Fe_3Co_{67}Cr_3Si_{15}B_{12}$/Pt heterostructure with the external magnetic field applied along the $\varphi_H = 90^\circ$ and 0° directions. Remarkably, for each magnetic field orientation, the curves measured for the bare ribbon and heterostructure are similar. It is an indicator that the presence of the Pt layer in the heterostructure does not affect the magnetic response of the system. Nevertheless, we find differences in the magnetization curves when the responses for distinct field orientations are compared. For both samples, the coercive field H_c was close to 1.2 Oe, while the saturation field H_s was 35 Oe and 125 Oe when the experiments were performed at $\varphi_H = 0^\circ$ and 90°, respectively. It is interesting to notice that the rotation of the field does not cause significant changes in the coercive field but leads to strong modifications in the saturation field of the samples. This observation suggests that the effective magnetic anisotropy is longitudinal and primarily influenced by the shape of the samples. These findings are in concordance with previous studies reported in the literature [18].

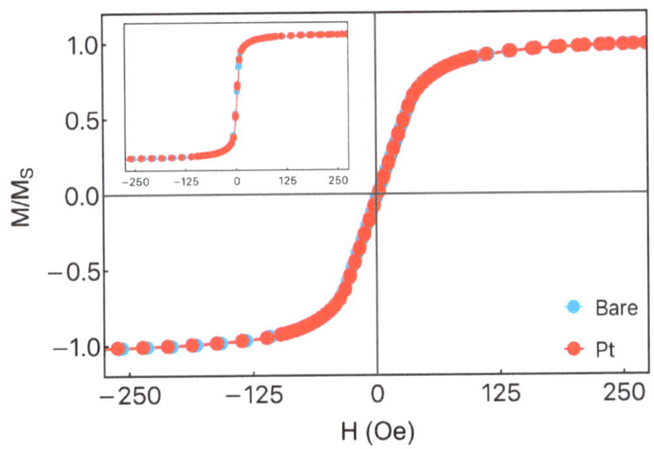

Figure 2. Normalized magnetization curve as a function of the external magnetic field taken at $\varphi_H = 90^\circ$ for the bare ribbon and $Fe_3Co_{67}Cr_3Si_{15}B_{12}$/Pt heterostructure. In the inset, we show the corresponding magnetization response for $\varphi_H = 0^\circ$.

Figure 3 shows the thermoelectric response for the bare $Fe_3Co_{67}Cr_3Si_{15}B_{12}$ ribbon. The thermoelectrical measurements were carried out varying the external magnetic field H and temperature difference ΔT between the top of the sample and the bottom of the glass substrate. For the bare ribbon, the sample structure allows us to assess uniquely the ANE. From Figure 3a, we disclose the thermoelectric voltage as a function of the external magnetic field for selected values of the temperature difference ΔT. The field alignment was set to $\varphi_H = 90^\circ$ to improve the thermoelectric signal. At $\varphi_H = 90^\circ$, the electrical field $\vec{E}$ is induced in the same direction of the electrical contacts $\vec{L}$ (see Figure 1b). For $\Delta T = 0$ K, we observe no thermoelectric signal, as expected. Nevertheless, thermoelectric voltage increases linearly as ΔT rises. Moreover, it is possible to observe the connection between the thermoelectric and magnetization curves, irrespective of the ΔT value. From Figure 3b, in turn, we show the thermoelectric voltage as a function of the magnetic field angle, at $\Delta T = 13$ K, and external magnetic field intensity of H = 400 Oe, where the sample is saturated magnetically. This behavior is expected, since for fixed ΔT and H values, Equations (1), (3) and (4) lead to a sine/cosine-shaped curve [29,30]. These initial results suggested the use of thermopiles configuration to investigate the response of $Fe_3Co_{67}Cr_3Si_{15}B_{12}$/Pt heterostructures and the role of the Pt cap layer in the thermoelectric effects, as well as exploring a thermopiles configuration.

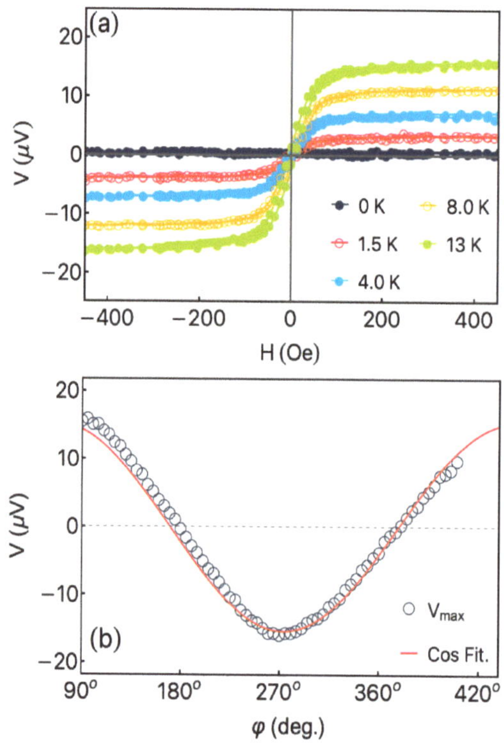

Figure 3. (**a**) Thermoelectric current as a function of the external magnetic field for selected values of the temperature difference ΔT for the bare $Fe_3Co_{67}Cr_3Si_{15}B_{12}$ ribbon. The measurements were obtained for $\varphi_H = 90^\circ$. This configuration maximizes the induced electric field and, consequently, the thermoelectric voltage/current, as depicted in Equation (1). (**b**) Angular dependence of the I_{max} for the bare ribbon. The solid line is the expected behavior predicted by Equations (1), (2) and (4) for a magnetically saturated system.

Regarding the sample structure, the deposition of the Pt onto the ribbon allows us to assess both LSSE and ANE phenomena. Here, it is worth remarking that Platinum has a large and positive spin Hall angle θ_{SHE}, so an increase in the thermoelectric response can be observed due to the spin-to-charge current conversion by ISHE. Moreover, the metallic Pt layer modifies the surface of the heterostructure, hence causing a decrease in the electric resistance and yielding an increase in the measured thermoelectric *current*. Therefore, in addition to protecting the ribbon surface against oxidation, the Pt layer can enhance the induced signal.

These initial results gave us the motivation to use thermopiles configuration to investigate the response of $Fe_3Co_{67}Cr_3Si_{15}B_{12}$/Pt heterostructures and the role of the Pt cap layer on the thermoelectric effects, as well as explore a thermopiles configuration.

Figure 4 shows the thermoelectric response for the $Fe_3Co_{67}Cr_3Si_{15}B_{12}$/Pt heterostructure. First of all, Figure 4a shows results for a single thermopile. The experimental setup with $\varphi_H = 90^\circ$ induces again the mirroring of the magnetic and thermoelectric responses. From a general point of view, the thermoelectric behavior is similar to the one verified for the bare ribbon, as expected. Nevertheless, it is possible to verify a small increase in the thermoelectric current when the heterostructure is considered. This feature is a signature of the thermoelectric contribution from the Longitudinal Spin Seebeck Effect associated with spin-to-charge current conversion [30].

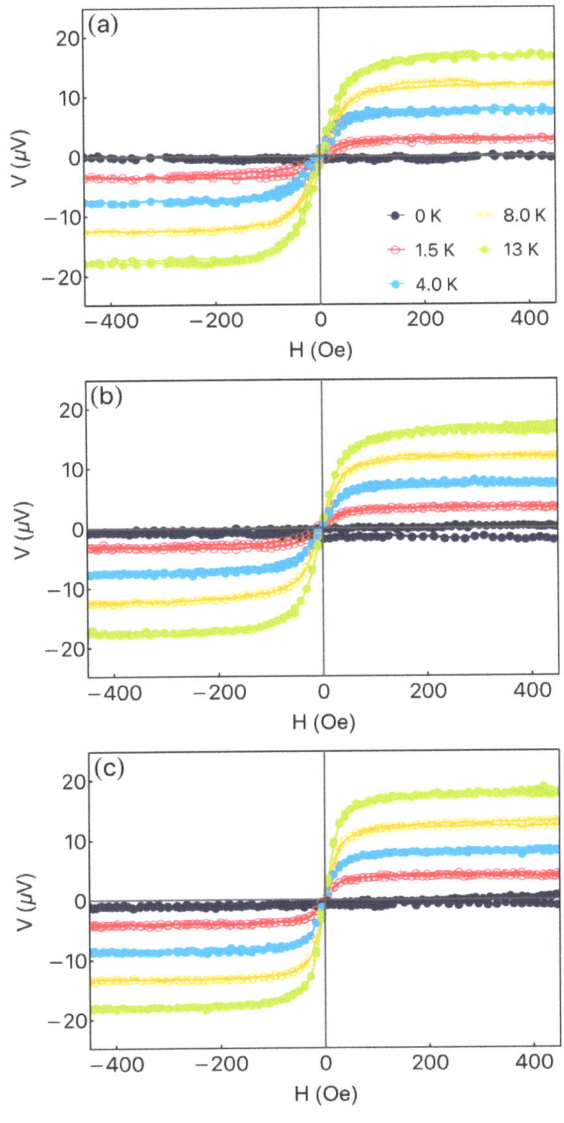

Figure 4. Thermoelectric response for the $Fe_3Co_{67}Cr_3Si_{15}B_{12}$/Pt heterostructure. Specifically, we address the thermoelectric voltage as a function of the external magnetic field, at $\varphi_H = 90°$, for distinct ΔT values for the $Fe_3Co_{67}Cr_3Si_{15}B_{12}$/Pt heterostructure, considering: (**a**) single thermopile (Pt); (**b**) double thermopile (Pt2); (**c**) triple thermopile (Pt3).

Looking at Figure 4a–c, one can see quite a similar evolution of the curves when different numbers of heterostructures composing the thermopiles are used. This behavior is expected once the parallel association of thermopiles does not modify the value of the induced voltage. On the other hand, from the point of view of the values of the thermoelectric current, the results are very promising, and the details of the discussion will be given below.

Figure 5 shows the maximum thermoelectric voltage V_{max} as a function of the temperature difference ΔT achieved for the bare amorphous ribbon and $Fe_3Co_{67}Cr_3Si_{15}B_{12}$/Pt heterostructure for a distinct number of thermopiles. The solid lines indicate the best linear fit for the experimental data. From the obtained results, it is possible to observe that the bare $Fe_3Co_{67}Cr_3Si_{15}B_{12}$ ribbon has a slope smaller than the ones found for the $Fe_3Co_{67}Cr_3Si_{15}B_{12}$/Pt heterostructures. Moreover, for all thermopiles, no significant modification of the slope is verified, which indicates that the effective thermoelectric coefficient S_{eff} for these systems is similar, as expected. In particular, the slope of the obtained curves is the signature of the energy conversion efficiency for the designed heterostructure.

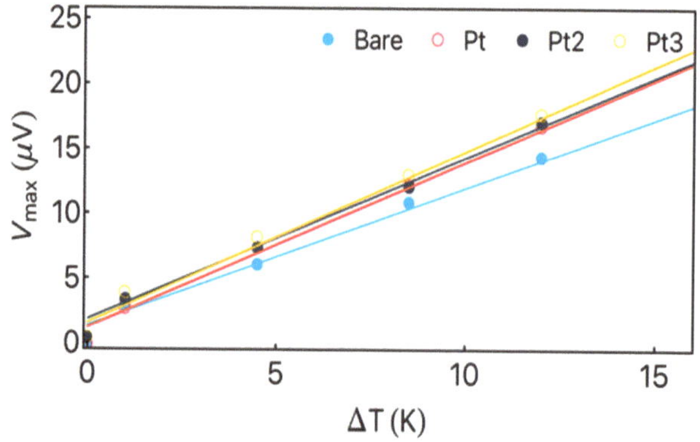

Figure 5. V_{max} as a function of temperature difference ΔT measured during the experiment for the bare ribbon and for the heterostructures in the configurations of thermopiles. The solid line indicates the best linear fit of the data.

From the relation between V_{max} and ΔT, we are able to calculate the effective thermoelectric coefficient S_{eff} for each system. To this end, the temperature difference ΔT_H on the heterostructure was estimated. In the case under consideration, the material was magnetic. Experimentally, it is given by following equation [15,18]:

$$\Delta T_H = \frac{t_H K_s}{t_s K_H} \Delta T, \tag{5}$$

where t_H and K_H are the thickness and thermal conductivity of the heterostructure, respectively, while t_s and K_s are corresponding quantities for the substrate. Then, from the dependence of V_{max} with ΔT_H, S_{eff} is obtained from the equation:

$$S_{eff} = \frac{t_R V_{max}}{L \Delta T_H}. \tag{6}$$

Then, assuming $t_s = 0.15$ mm, $K_s = 1.15$ W/mK [18], $K_H = 133.79$ W/mK, and $t_H = 0.24$ μm, we found the S_{eff} values of around 1.75 μV/mK for the bare ribbon, and an average value of 2.01 μV/mK for the $Fe_3Co_{67}Cr_3Si_{15}B_{12}$/Pt heterostructures. Notice the remarkable enhancement of S_{eff} for such heterostructures, which is a signature of the overlap of the ANE and LSSE effects contributing to the thermoelectric voltage.

Figure 6a brings to light the ratio of the thermoelectric *current* by the temperature difference as a function of the external magnetic field for the bare ribbon and thermopiles. For this purpose, we consider the remarkable decrease in the electrical resistance of the thermopile as the number of ribbon increases. In particular, here, we observed a decrease from $R = 1.76\ \Omega$ for Pt1 (single $Fe_3Co_{67}Cr_3Si_{15}B_{12}$/Pt heterostructure) to $R = 1.18\ \Omega$ for Pt2, and $R = 0.90\ \Omega$ for Pt3.

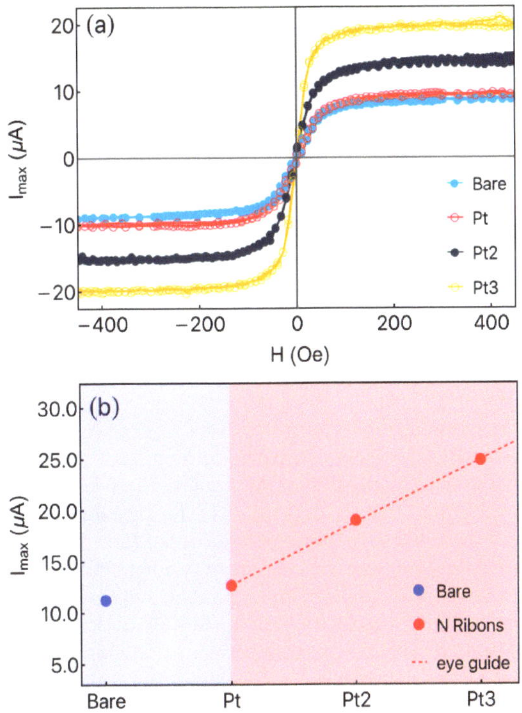

Figure 6. (**a**) The amplitude of the thermoelectric current as a function of the external magnetic field for the bare ribbon and thermopiles Pt, Pt2, and Pt3. (**b**) Maximum thermoelectric current for the bare ribbon (blue rectangle part) and thermopiles (red rectangle part). The dashed line is just a guide for the eyes.

Figure 6b shows the maximum I_{max} for the bare ribbon (blue region) and thermopiles (red region). The significant increase of this parameter shows a promising path for exploring these $Fe_3Co_{67}Cr_3Si_{15}B_{12}$/Pt heterostructures that are highly efficient for green energy devices. In the figure, the dashed line is an eye guide drawn using a linear function, as predicted by the increase of the current in an ohmic system. The observed behavior is mainly connected to the decrease in the electrical resistance in the thermopile when the parallel association is considered. It is important to point out that, if considered a parallel association of bare ribbons, a similar effect is reached.

Our findings bring to light the multifunctionality of the Pt cap layer in the system. Here, the Pt layer improves the protection of the ribbon against oxidation, leads to an increase in the thermoelectric voltage due to the spin-to-charge conversion, and decreases the electrical resistance of the heterostructure (1.90 Ω for the bare ribbon to 1.76 Ω for the heterostructure).

Obtained results are promising when compared with previous studies. The linear increase in the thermoelectric current associated with the number of ribbons in the thermopile is similar to the one found in the literature [32–35]. Moreover, we are able to reach values of around 20 μV for thermoelectric voltage, a value similar to those found in Ref. [35]. However, in our case, although the number of ribbons can limit the thermopile dimensions for a given application, no lithography process was required for the production process.

However, as aforementioned, flexible electronic materials become more requested in functional electronic units because they can be integrated into devices having complex shape surfaces. Flexibility of a magnetic field's sensitive elements is a very attractive property for different types of magnetic sensors [38,40]. Although we did not take advantage of flexible design for thermopiles based on Co-rich amorphous ribbons/Pt thin-film heterostructures in the present study, it corresponds to an issue to be highlighted. Th thermopiles exploring flexible ferromagnetic ribbons were studied in the present work. It i hard to make a quantitative comparison with other types of devices having thermoelectri responses. However, amorphous ribbons have much higher thermal conductivity in com parison with any flexible polymer substrate, and, therefore, the devices based on them ma have an advantage of symmetry of thermoelectric response. Certainly, it is a promising d rection for research and applications in the future. In addition, a large variety of functiona cover layers can be considered in a search for the particular performances requested b the applications. The present findings can motivate new groups to investigate this type o heterostructures for future sensor applications with mass production. In addition, there ar applications for which cost effectiveness might be less important if efficiency or flexibilit is guaranteed. One of the examples is wearable devices.

4. Conclusions

In summary, bare Co-rich ribbons without covering and ribbon/Pt heterostructure were considered to explore thermoelectric voltage conversion. In particular, we investigate the Anomalous Nernst Effect and Longitudinal Spin Seebeck Effect in such systems. First it was observed that the deposition of the very thin Pt layer onto the ribbon surface doe not influence the magnetic response of the ferromagnetic amorphous ribbon. Next, we revealed that the thermoelectric results mirror the magnetization curves for a particula magnetic field orientation. In a specific condition, we observed the maximum induce voltage in the thermoelectric experiments. Remarkably, we found that the presence of the P layer, beyond protecting against surface oxidation, facilitates the spin-to-charge conversion increasing the thermoelectric voltage induced in the system. Further, we demonstrate that the employment of thermopiles consisting of Co-rich ribbon/Pt heterostructures in a parallel association increases the thermoelectric current induced in the system, a fac related to the considerable decrease in the electric resistance in the system. Specifically, a linear increase in the thermoelectric current signal was observed with the increase in the number of ribbons in the thermopile, a fact that is relevant for sensor applications, althoug the number of ribbons can limit the thermopile dimensions. Our findings bring to light noticeable efficiency of the Co-rich ribbon/Pt heterostructure to be used for the conversion of thermoelectric energy through ANE or LSSE.

Author Contributions: Conceptualization, M.A.C., A.F. and G.V.K.; software, M.A.C., A.V.S., G.V.K. F.B., F.V. and A.F.; validation, M.A.C., G.V.K., F.B., F.V. and A.F.; formal analysis, M.A.C., G.V.K., F.B. F.V. and A.F.; investigation, M.A.C., D.F.d.O., G.V.K., F.B., A.V.S., F.V., M.G., E.F.d.S. and A.F.; resources M.A.C., G.V.K., F.B., F.V. and A.F.; data curation, M.A.C.; writing—original draft preparation, M.A.C. G.V.K., F.B. and A.F.; writing—review and editing, M.A.C., A.V.S. and G.V.K.; supervision, G.V.K. and M.A.C.; project administration, M.A.C., G.V.K., F.B., F.V. and A.F.; funding acquisition. All authors have read and agreed to the published version of the manuscript.

Funding: This research was funded by different sources. M.A.C. thanks CAPES (8887.573100/2020-00) and CNPq. A.F. thanks the FCT (CTTI-31/18-C.F. (2) junior researcher contract). A.V.S. and G.V.K were supported in the frame of the Priority-2030 Program of Ural Federal University.

Institutional Review Board Statement: Not applicable.

Informed Consent Statement: Not applicable.

Data Availability Statement: Data available from the corresponding author upon reasonable request

Conflicts of Interest: The authors declare no conflict of interest.

References

1. Nykyruy, Y.; Mudry, S.; Kulyk, Y.; Prunitsa, V.; Borysiuk, A. Magnetic properties and nanocrystallization behavior of Co-based amorphous alloy. *Phys. Chem. Solid State* **2023**, *24*, 106–113. [CrossRef]
2. Murugaiyan, P.; Mitra, A.; Panda, A.K.; Kumar, A.; Roy, R.K.; Manna, K.; Srivastava, S.K. Electromagnetic interference shielding effectiveness of amorphous and nanocomposite soft magnetic ribbons. *Phys. B Condens. Matter* **2019**, *568*, 13–17. [CrossRef]
3. Beach, R.; Berkowitz, A. Sensitive field-and frequency-dependent impedance spectra of amorphous FeCoSiB wire and ribbon. *J. Appl. Phys.* **1994**, *76*, 6209–6213. [CrossRef]
4. Volchkov, S.O.; Pasynkova, A.A.; Derevyanko, M.S.; Bukreev, D.A.; Kozlov, N.V.; Svalov, A.V.; Semirov, A.V. Magnetoimpedance of CoFeCrSiB Ribbon-Based Sensitive Element with FeNi Covering: Experiment and Modeling. *Sensors* **2021**, *21*, 6728. [CrossRef]
5. Makhotkin, V.E.; Shurukhin, B.P.; Lopatin, V.A.; Marchukov, P.Y.; Levin, Y.K. Magnetic field sensors based on amorphous ribbons. *Sens. Act. A Phys.* **1991**, *27*, 759–762. [CrossRef]
6. Gazda, P.; Nowicki, M. Giant Stress-Impedance Effect in CoFeNiMoBSi Alloy in Variation of Applied Magnetic Field. *Materials* **2021**, *14*, 1919. [CrossRef]
7. Semirov, A.V.; Bukreev, D.A.; Moiseev, A.A.; Derevyanko, M.S.; Kudryavtsev, V.O. Relationship Between the Temperature Changes of the Magnetostriction Constant and the Impedance of Amorphous Elastically Deformed Soft Magnetic Cobalt-Based Ribbons. *Russ. Phys. J.* **2013**, *55*, 977–982. [CrossRef]
8. Kurlyandskaya, G.V.; Sánchez, M.L.; Hernando, B.; Prida, V.M.; Gorria, P.; Tejedor, M. Giant-magnetoimpedance-based sensitive element as a model for biosensors. *Appl. Phys. Lett.* **2003**, *82*, 3053–3055. [CrossRef]
9. Yang, Z.; Lei, C.; Zhou, Y.; Cheng Sun, X. Study on the giant magnetoimpedance effect in micro-patterned Co-based amorphous ribbons with single strip structure and tortuous shape. *Microsyst. Technol.* **2015**, *21*, 1995–2001. [CrossRef]
10. Knobel, M.; Vazquez, M.; Kraus, L. Giant magnetoimpedance. In *Handbook of Magnetic Materials*; Buschow, K.H.J., Ed.; Elsevier: Amsterdam, The Netherlands, 2003; Volume 15, pp. 497–563.
11. Cox, C.D.W.; Caruana, A.J.; Cropper, M.D.; Morrison, K. Anomalous Nernst effect in Co_2MnSi thin films. *J. Phys. D Appl. Phys.* **2020**, *53*, 035005. [CrossRef]
12. Tian, D.; Li, Y.; Qu, D.; Jin, X.; Chien, C.L. Separation of spin Seebeck effect and anomalous Nernst effect in Co/Cu/YIG. *App. Phys. Lett.* **2015**, *106*, 212407. [CrossRef]
13. Uchida, K.; Xiao, J.; Adachi, H.; Ohe, J.; Takahashi, S.; Ieda, J.; Ota, T.; Kajiwara, Y.; Umezawa, H.; Kawai, H.; et al. Spin Seebeck insulator. *Nat. Mater.* **2010**, *9*, 894–897. [CrossRef]
14. Chen, T.; Minami, S.; Sakai, A.; Wang, Y.; Feng, Z.; Nomoto, T.; Hirayama, M.; Ishii, R.; Koretsune, T.; Arita, R.; et al. Large anomalous Nernst effect and nodal plane in an iron-based kagome ferromagnet. *Sci. Adv.* **2022**, *8*, eabk1480. [CrossRef] [PubMed]
15. Melo, A.S.; de Oliveira, A.B.; Chesman, C.; Pace, R.D.D.; Bohn, F.; Correa, M.A. Anomalous Nernst effect in stressed magnetostrictive film grown onto flexible substrate. *Sci. Rep.* **2019**, *9*, 15338. [CrossRef] [PubMed]
16. Asaba, T.; Ivanov, V.; Thomas, S.M.; Savrasov, S.Y.; Thompson, J.D.; Bauer, E.D.; Ronning, F. Colossal anomalous Nernst effect in a correlated noncentrosymmetric kagome ferromagnet. *Sci. Adv.* **2021**, *7*, eabf1467. [CrossRef]
17. Xing, Y.; Feng Sun, Q.; Wang, J. Nernst and Seebeck effects in a graphene nanoribbon. *Phys. Rev. B* **2009**, *80*, 235411. [CrossRef]
18. Correa, M.A.; Ferreira, A.; Souza, A.L.R.; Neto, J.M.D.; Bohn, F.; Vaz, F.; Kurlyandskaya, G.V. Anomalous Nernst Effect in Flexible Co-Based Amorphous Ribbons. *Sensors* **2023**, *23*, 1420. [CrossRef]
19. Geishendorf, K.; Vir, P.; Shekhar, C.; Felser, C.; Facio, J.I.; van den Brink, J.; Nielsch, K.; Thomas, A.; Goennenwein, S.T.B. Signatures of the Magnetic Entropy in the Thermopower Signals in Nanoribbons of the Magnetic Weyl Semimetal $Co_3Sn_2S_2$. *Nano Lett. J.* **2020**, *20*, 300–305. [CrossRef]
20. Kai, W.; Lin, P.; Chen, W.; Kao, P.; Huang, R.; Liaw, P. Air-oxidation of a Co-based amorphous ribbon at 400–600 °C. *J. Alloys Compd.* **2011**, *509*, S179–S183. [CrossRef]
21. Park, D.; Kim, C.; Kim, W.; Hong, J. Study of GMI-valve characteristics in the Co-based amorphous ribbon by ferromagnetic resonance. *J. Magn. Magn. Mater.* **2007**, *310*, 2295–2297. [CrossRef]
22. Egbu, J.; Ohodnicki, P.R., Jr.; Baltrus, J.P.; Talaat, A.; Wright, R.F.; McHenry, M.E. Analysis of surface roughness and oxidation of FeNi-based metal amorphous nanocomposite alloys. *J. Alloys Compd.* **2022**, *912*, 165155. [CrossRef]
23. Clark, A.; Zhu, A.; Sun, K.; Petty, H.R. Cerium oxide and platinum nanoparticles protect cells from oxidant-mediated apoptosis. *J. Nanoparticle Res.* **2011**, *13*, 5547–5555. [CrossRef]
24. Xu, M.; Chen, F.; Wang, T.; Yu, B.; Zhao, Z.; Zhou, L.; Hua, D. Platinum Nanoparticles Anchored on Covalent Triazine Frameworks Modified Cordierite for Efficient Oxidation of Hydrogen Isotopes. *ACS Appl. Nano Mater.* **2023**, *6*, 867–874. [CrossRef]
25. Pragnell, W.; Evans, H.; Williams, A. Oxidation protection of Sm_2Co_{17}-based alloys. *J. Alloys Compd.* **2012**, *517*, 92–97. [CrossRef]
26. Castel, V.; Vlietstra, N.; Youssef, J.B.; van Wees, B.J. Platinum thickness dependence of the inverse spin-Hall voltage from spin pumping in a hybrid yttrium iron garnet/platinum system. *Appl. Phys. Lett.* **2012**, *101*, 132414. [CrossRef]
27. Sun, Y.; Chang, H.; Kabatek, M.; Song, Y.Y.; Wang, Z.; Jantz, M.; Schneider, W.; Wu, M.; Montoya, E.; Kardasz, B.; et al. Damping in Yttrium Iron Garnet Nanoscale Films Capped by Platinum. *Phys. Rev. Lett.* **2013**, *111*, 106601. [CrossRef]
28. Adachi, H.; Ichi Uchida, K.; Saitoh, E.; Maekawa, S. Theory of the spin Seebeck effect. *Rep. Prog. Phys.* **2013**, *76*, 036501. [CrossRef]
29. Snyder, G.J.; Toberer, E.S. Complex thermoelectric materials. *Nat. Mater.* **2008**, *7*, 105–114. [CrossRef] [PubMed]
30. Uchida, K.; Takahashi, S.; Harii, K.; Ieda, J.; Koshibae, W.; Ando, K.; Maekawa, S.; Saitoh, E. Observation of the spin Seebeck effect. *Nature* **2008**, *455*, 778–781. [CrossRef]

31. Holanda, J.; Santos, O.A.; Cunha, R.O.; Mendes, J.B.; Rodríguez-Suárez, R.L.; Azevedo, A.; Rezende, S.M. Longitudinal spir Seebeck effect in permalloy separated from the anomalous Nernst effect: Theory and experiment. *Phys. Rev. B* **2017**, *95*, 21442 [CrossRef]
32. Uchida, K.; Nonaka, T.; Yoshino, T.; Kikkawa, T.; Kikuchi, D.; Saitoh, E. Enchancement of Spin-Seebeck Voltage by Spin-Ha Thermopile. *Appl. Phys. Exp.* **2012**, *5*, 093001. [CrossRef]
33. Kim, J.M.; Jeon, C.Y.; Kim, D.J.; Van, P.C.; Jeong, J.R.; Park, B.G. Amplification of Spin Thermoelectric Signal in Multilayer Spi Thermopiles. *ACS Appl. Electron. Mater.* **2020**, *2*, 2906–2912. [CrossRef]
34. Kim, D.J.; Lee, K.D.; Surabhi, S.; Yoon, S.G.; Jeong, J.R.; Park, B.G. Utilization of Antiferromagnetic IrMn electrode in Spir thermoelectric devices and their beneficial hybrid for thermopiles. *Adv. Func. Mater.* **2016**, *26*, 5507. [CrossRef]
35. Weng, T.W.; Chuang, T.C.; Qu, D.; Huang, S.Y. Anomalous Nernst thermopile made of single element iron. *J. Magn. Magn. Mater* **2022**, *563*, 169892. [CrossRef]
36. Fal Miyar, V.; Cerdeira, M.A.; García, J.A.; Potatov, A.P.; Pierna, A.R.; Marzo, F.F.; Barandiarán, J.M.; Kurlyandskaya, G.V. Gian magnetoimpedance of electrochemically surface modified Co-based amorphous ribbons. *IEEE Trans. Magn.* **2008**, *44*, 4476–4479 [CrossRef]
37. Kurlyandskaya, G.V.; Lezama, L.; Pasynkova, A.A.; Volchkov, S.O.; Lukshina, V.A.; Larrañaga, A.; Dmitrieva, N.V.; Timofeeva A.V.; Orue, I. Amorphous FeCoCrSiB Ribbons with Tailored Anisotropy for the Development of Magnetic Elements for High Frequency Applications. *Materials* **2022**, *15*, 4160. [CrossRef] [PubMed]
38. Kurlyandskaya, G.; Dmitrieva, N.; Zayarnaya, T.; Lukshina, V.; Potapov, A. The thermomechanical treatment of an amorphous Co-based alloy with a low Curie temperature. *J. Magn. Magn. Mater.* **1996**, *160*, 307–308. [CrossRef]
39. Althammer, M.; Meyer, S.; Nakayama, H.; Schreier, M.; Altmannshofer, S.; Weiler, M.; Huebl, H.; Geprägs, S.; Opel, M.; Gross, R. et al. Quantitative study of the spin Hall magnetoresistance in ferromagnetic insulator/normal metal hybrids. *Phys. Rev. B* **2013** *87*, 224401. [CrossRef]
40. Melzer, M.; Kaltenbrunner, M.; Makarov, D.; Karnaushenko, D.; Karnaushenko, D.; Sekitani, T.; Someya, T.; Schmidt, O.G Imperceptible magnetoelectronics. *Nat. Commun.* **2015**, *6*, 6080. [CrossRef]

MDPI AG
Grosspeteranlage 5
4052 Basel
Switzerland
Tel.: +41 61 683 77 34

Sensors Editorial Office
E-mail: sensors@mdpi.com
www.mdpi.com/journal/sensors

www.ingramcontent.com/pod-product-compliance
Lightning Source LLC
Chambersburg PA
CBHW041357210926
51850CB00013B/3105